HALOPHYTES FOR FOOD SECURITY IN DRY LANDS

HALOPHYTES FOR FOOD SECURITY IN DRY LANDS

Edited by

MUHAMMAD AJMAL KHAN
MUNIR OZTURK
BILQUEES GUL
MUHAMMAD ZAHEER AHMED

ELSEVIER

AMSTERDAM • BOSTON • HEIDELBERG • LONDON • NEW YORK • OXFORD
PARIS • SAN DIEGO • SAN FRANCISCO • SINGAPORE • SYDNEY • TOKYO
Academic Press is an imprint of Elsevier

Academic Press is an imprint of Elsevier
The Boulevard, Langford Lane, Kidlington, Oxford OX5 1GB
225 Wyman Street, Waltham MA 02451

Library of Congress Cataloging-in-Publication Data
A catalog record for this book is available from the Library of Congress

British Library Cataloguing-in-Publication Data
A catalogue record for this book is available from the British Library

ISBN: 978-0-12-801854-5

For information on all Academic Press publications
visit our website at http://store.elsevier.com.

Working together
to grow libraries in
developing countries

www.elsevier.com • www.bookaid.org

Publisher: Nikki Levy
Acquisition Editor: Nancy Maragioglio
Editorial Project Manager: Billie Jean Fernandez
Production Project Manager: Melissa Read
Designer: Maria Ines Cruz
Printed and bound in the United States of America

CONTENTS

Foreword by Sheikha Abdulla Al Misnad...xiii
Foreword by Eiman Al-Mustafawi xv
List of Contributors xvii
Introduction xxi

1 Characterization and Function of Sodium Exchanger Genes in *Aeluropus Lagopoides* Under NaCl Stress.........................1

Muhammad Zaheer Ahmed, Bilquees Gul, M. Ajmal Khan and Kazuo N. Watanabe

1.1 Introduction 1
1.2 Materials and Methods 3
1.3 Results... 5
1.4 Discussion... 12
References 14

2 Multi-Temporal Soil Salinity Assessment at a Detailed Scale for Discriminating Halophytes Distribution 17

Jorge Batlle-Sales, Juan Bautista Peris and María Ferrandis

2.1 Introduction 17
2.2 Objective 19
2.3 Methodology 19
2.4 Results and Discussion 24
2.5 Conclusions 33
References 34

3 Nutritional Value of *Chenopodium Quinoa* Seeds Obtained from an Open Field Culture Under Saline Conditions 37

Meryem Brakez, Salma Daoud, Moulay Chérif Harrouni, Naima Tachbibi and Zahra Brakez

3.1 Introduction 37
3.2 Materials and Methods 39
3.3 Results and Discussion 41
3.4 Conclusion 44
References 45

4 Halophytes and Saline Vegetation of Afghanistan, a Potential Rich Source for People 49
Siegmar-W. Breckle

4.1 Introduction 49
4.2 Methods 50
4.3 Results.. 50
4.4 Discussion... 57
4.5 Conclusions 63
Acknowledgments 63
References 63

5 Comparison of Seed Production and Agronomic Traits of 20 Wild Accessions of *Salicornia Bigelovii* Torr. Grown Under Greenhouse Conditions 67
Cylphine Bresdin, Edward P. Glenn and J. Jed Brown

5.1 Introduction 67
5.2 Materials and Methods 68
5.3 Results.. 72
5.4 Discussion... 77
5.5 Conclusion 80
5.6 Acknowledgment 81
References 81

6 Carbon Mitigation: A Salt Marsh Ecosystem Service in Times of Change 83
Isabel Caçador, Bernardo Duarte, João Carlos Marques and Noomene Sleimi

6.1 Salt Marshes: Key Ecosystems.. 83
6.2 Salt Marsh Sediments: Sinks or Sources? 84
6.3 Halophytes: An Efficient Carbon Pump 88
6.4 Out-Welling Carbon 90
6.5 Hydrological Control of Carbon Stocks 91
6.6 Global Warming and Carbon Stocks 93
6.7 CO_2 Rising in Salt Marshes: Improvement or Constraint?.... 97
6.8 Final Remarks 104
References 105

7 Food Security in the Face of Salinity, Drought, Climate Change, and Population Growth...**111**
John Cheeseman

7.1 Introduction 111
7.2 The Problem of Food Security 112
7.3 The Problem of Salinity in Agriculture 116
7.4 Fitting Crops to the Environment—A Place for Halophytes?....... 117
7.5 Concluding Remarks 120
References 121

8 The Importance of Mangrove Ecosystems for Nature Protection and Food Productivity: Actions of UNESCO's Man and the Biosphere Programme..**125**
Miguel Clüsener-Godt and María Rosa Cárdenas Tomažič

8.1 UNESCO Normative Tools to Ensure the Protection of the Environment and Its Wise Use... 125
8.2 The MAB and Its World Network of Biosphere Reserves 126
8.3 Distribution and Socio-Economic and Environmental Importance of Mangrove Ecosystems 129
8.4 Actions of UNESCO's MAB 133
8.5 Actions in Biosphere Reserves 135
8.6 Conclusion 139
References 139

9 The Potential Use of Halophytes for the Development of Marginal Dry Areas in Morocco...**141**
Salma Daoud, Khalid Elbrik, Naima Tachbibi, Laila Bouqbis, Meryem Brakez and Moulay Chérif Harrouni

9.1 Introduction 141
9.2 Bio-Climate in Morocco 142
9.3 Biodiversity in Morocco 143
9.4 Vulnerability of Morocco to Climate Variations.................... 143
9.5 Problems of Salinity in Morocco... 144
9.6 Agriculture in Massa and Drâa Valleys................................ 147
9.7 Potential Use of Halophytes in Areas Affected by Salinity 148
9.8 Youth Potential in Arid Areas in Morocco............................ 149
9.9 Conclusion 153
References 154

10 Halophyte Transcriptomics: Understanding Mechanisms of Salinity Tolerance 157

Joann Diray-Arce, Bilquees Gul, M. Ajmal Khan and Brent Nielsen

10.1 Introduction 157
10.2 Transcriptome Sequencing Overview 158
10.3 Applications of RNA Studies 158
10.4 NGS Approaches for Salt-Tolerance Studies 160
10.5 Genes Involved in General Metabolism 164
10.6 Regulatory Molecules 169
10.7 LEA Protein Coding Genes 169
10.8 Other Genomic Elements 170
10.9 Pathways 170
10.10 Conclusions and Future Directions 171
Acknowledgment 171
References 171

11 Sustainable Diversity of Salt-Tolerant Fodder Crop—Livestock Production System Through Utilization of Saline Natural Resources: Egypt Case Study 177

Hassan M. El Shaer and A.J. Al Dakheel

11.1 Introduction 177
11.2 Egypt's General Characteristics 178
11.3 General Characteristics of Project Location in Sinai Region 179
11.4 Main Activities and Results 181
11.5 Conclusions 192
Acknowledgments 193
References 193

12 Insights into the Ecology and the Salt Tolerance of the Halophyte *Cakile Maritima* Using Multidisciplinary Approaches 197

Karim Ben Hamed, Ibtissem Ben Hamad, François Bouteau and Chedly Abdelly

12.1 Introduction 197
12.2 Latitudinal Distribution and Taxonomic Diversity 198
12.3 Dispersal and Environmental Adaptation 199

12.4 Basis of the Tolerance to Salinity 202
12.5 *Cakile maritima*: Model Halophyte for Future Research
     in Salt-Stress Physiology...209
References 210

13 Exogenous Chemical Treatments Have Differential Effects in Improving Salinity Tolerance of Halophytes 213

Abdul Hameed, Bilquees Gul and M. Ajmal Khan

13.1 Introduction 213
13.2 Materials and Methods...214
13.3 Results..217
13.4 Discussion..221
13.5 Conclusions 225
Acknowledgments 225
References 225

14 Food and Water Security for Dry Regions: A New Paradigm............231

M. Ajmal Khan

14.1 Introduction 231
14.2 Water and Food Production 231
14.3 Conventional Solutions 232
14.4 Nonconventional Solutions.....................................232
14.5 Potential Uses of Halophytes 233
14.6 What We Have Done 236
14.7 Future Directions, Pitfalls, and Possibilities 238
References 240

15 Genetic and Environmental Management of Halophytes for Improved Livestock Production......................................243

David G. Masters and Hayley C. Norman

15.1 Introduction 243
15.2 Potential Forage and Crop Solutions........................244
15.3 Halophytes for Livestock..246
15.4 Current Limitations in the Use of Halophytes for Livestock
     Production..246

15.5 Genetic Improvement of Halophytes for Livestock 249
15.6 Environmental Manipulation...252
15.7 Conclusions 254
References.. 254

16 Drought and Salinity Differently Affect Growth and Secondary Metabolites of *"Chenopodium Quinoa* Willd" Seedlings **259**

Adele Muscolo, Maria Rosaria Panuccio, Angelo Maria Gioffrè
and Sven-Erik Jacobsen

16.1 Introduction 259
16.2 Materials and Methods 261
16.3 Results..264
16.4 Discussion...269
16.5 Conclusion 272
References.. 273

17 Germination Eco-Physiology and Plant Diversity in Halophytes of Sundarban Mangrove Forest in Bangladesh...............**277**

A.K.M. Nazrul Islam

17.1 Introduction 277
17.2 Materials and Methods 280
17.3 Results and Discussion 280
References.. 288

18 Halophytic Plant Diversity of Unique Habitats in Turkey: Salt Mine Caves of Çankırı and Iğdır **291**

Münir Öztürk, Volkan Altay, Ernaz Altundağ and Salih Gücel

18.1 Introduction 291
18.2 Study Areas 293
18.3 Halophyte Diversity 297
18.4 Economical Evaluations..297
18.5 Medicinal and Aromatic Halophytes in Çankırı
     and Iğdır Provinces 305
18.6 Conclusions 311
References.. 312

19 Halophytes as a Possible Alternative to Desalination Plants: Prospects of Recycling Saline Wastewater During Coal Seam Gas Operations **317**

Suresh Panta, Peter Lane, Richard Doyle, Marcus Hardie, Gabriel Haros and Sergey Shabala

19.1 Introduction 317
19.2 Materials and Methods 319
19.3 Results and Discussion 322
19.4 Conclusions 327
Acknowledgments 327
References 328

Index...331

FOREWORD BY
SHEIKHA ABDULLA AL MISNAD

Water scarcity is one of the defining issues for the future of Gulf Cooperation Council (GCC) countries. With the rapid pace of urban development and population growth in this region, the demand for water will only increase. Desalination of water for agricultural and domestic use is not without substantial financial cost and grave environmental implications. Both food and water security are key for Qatar's future and for its development plans. Innovative solutions are urgently needed.

Through academic programs and research initiatives, Qatar University has been contributing to the multi-faceted issue of sustainable development, with special emphasis on the roles of education, science, and technology. In November 2012, Qatar University (QU) and the Qatar National Food Security Program hosted the *International Conference on Food Security in Dry Lands*. Based on the conviction that high-quality scientific research is essential for finding sustainable development solutions in dry lands, QU created a Centre for Sustainable Development to address water and food security and wider environmental management issues and to link research with human, social, and economic developments in Qatari society. In May 2014, Qatar Shell Professorial Chair in Sustainable Development organized another conference on *Halophytes for Food Security in Dry Lands* with the participation of scientists from all over the world. This book was born out of the ideas and discussions at that conference and the pressing need for creative and context-appropriate solutions.

One such innovative idea is the use of vast resources of ground saline water or seawater for the production of economically important crops from the indigenous Qatari plants distributed in coastal and inland *sabkha* salt marshes and deserts. *Halophytes* are a group of plants that are naturally equipped with the mechanisms to survive under highly saline and arid conditions and produce high biomass. This high productivity could be used as fodder, forage, biofuel, turf, medicine, edible and essential oils, and biodiesel. The scientific community has made limited but steady progress in developing these salt-tolerant plant species as cash-crops, and attempts are ongoing to enhance research and implementation in farming and landscaping. Throughout the

Arabian Peninsula, promising results have been seen with certain halophytic species. This area therefore holds exciting potential explored throughout the conference papers.

Many of the participants of the May 2014 *Halophytes for Food Security in Dry Lands* conference have contributed to this volume and to enriching knowledge about halophyte productivity in the harsh Qatari environment. The editors have already produced four volumes on the *Sabkha* Ecosystem in regions of the world, including the Arabian Peninsula and adjacent countries. This volume is a continuation of those efforts. Importantly, the conference was followed-up with promising collaborations and research funding proposals around developing nonconventional crops that can alleviate some of the chronic food and water security issues in the region.

The professional contributions that have gone into the production of this volume are immense, and I encourage students and scientists to make use of this rich resource in the search for innovative and much-needed models to achieve food security in dry lands of this region and the rest of the world.

Sheikha Abdulla Al Misnad, Ph.D.
President, Qatar University,
Doha, Qatar

FOREWORD BY EIMAN AL-MUSTAFAWI

Since one of the tenets that Qatar's National Vision 2030 (QNV2030) resets on is advancing sustainable development, there has been an urgent need for new interdisciplinary approaches for food and water security enhancement.

To serve the needs of Qatar, the College of Arts and Sciences at Qatar University launched the Center for Sustainable Development to produce with our partners to make an interdisciplinary contribution towards promoting sustainable development in Qatar, and the Gulf region, with a focus on food security, given its importance both for current and future generations.

Qatar is a water-scarce country where per capita availability of water is amongst the lowest in the world. The population of Qatar has grown rapidly (as of 2015) to over 2 million, compared with a few hundred thousand over the last two decades. Most food is imported and the source of fresh water is through desalinating seawater into fresh water. This desalination process requires a huge amount of energy that substantially increases CO_2 emissions, which contribute to the challenge of global warming.

An innovative focus of our food security program has been to examine the possibility of developing coastal salt deserts into man-made ecosystems for agricultural productivity, with the food supply requirements of the growing human population in mind. It is encouraging that studies undertaken in this arid region have revealed that various medicinal/aromatic plants can be cultivated easily on slightly saline-alkaline soils using seawater irrigation. Many salt-tolerant plant taxa found in nature can be domesticated to provide better economic returns. Whilst initial results are encouraging, what is needed is vision, planning, and the involvement of scientific and agricultural authorities and politicians.

The Qatar Shell Professorial Chair in Sustainable Development, housed in the College of Arts and Sciences, organized an International Conference on *Halophytes for Food Security in Dry Lands* from May 12–13, 2014, Doha, at which distinguished scientists, participants, and contributors from all over the world were present. The theme of this conference was very timely: no longer do we merely try to understand the importance of halophytes for sustainable development, but we have also started to understand the tremendous importance of sabkha for the conservation of halophyte biodiversity. Halophytes hold significant potential to counteract adverse environmental impacts, such as climate change, marine discharge waters, ecosystem restoration, and the enhancement of primary productivity. It is for these reasons that this important volume includes all aspects of halophyte biology spanning from ecosystem to molecular levels. This information can be useful in making crop plants for food consumption salt-tolerant. This volume also contributes to our understanding of the economic significance of halophytes for food security in dry regions.

It is on this hopeful note that I offer my thanks to the editors and the authors for their contributions to the scientific community, given their recommendations and suggestions for future research. Overall, I am hopeful that if halophytes are properly utilized, it could be a blessing for dry lands and food security.

Dr. Eiman Al-Mustafawi
Dean, College of Arts and Science,
Qatar University, Doha, Qatar

LIST OF CONTRIBUTORS

Chedly Abdelly Laboratoire des Plantes Extrêmophiles, Centre de Biotechnologie de Borj Cédria, Hammam Lif, Tunisia

Muhammad Zaheer Ahmed Institute of Sustainable Halophyte Utilization, University of Karachi, Karachi, Pakistan; Gene Research Center, University of Tsukuba, Tsukuba City, Ibaraki, Japan

A.J. Al Dakheel International Center for Biosaline, Dubai, UAE

Volkan Altay Biology Department, Science and Arts Faculty, Mustafa Kemal University, Antakya-Hatay, Turkey

Ernaz Altundağ Biology Department, Science and Arts Faculty, Duzce University, Duzce, Turkey

Jorge Batlle-Sales Department of Vegetal Biology, University of Valencia, Valencia, Spain

Laila Bouqbis Polydisciplinary Faculty, Ibn Zohr University, Taroudant, Morocco

François Bouteau Institut des Energies de Demain, Université Paris Diderot, Sorbonne Paris Cité, Paris, France

Meryem Brakez Laboratory of Plant Biotechnologies, Faculty of Sciences, Ibn Zohr University, Agadir, Morocco

Zahra Brakez Laboratory of Cell Biology & Molecular Genetics, Faculty of Sciences, Ibn Zohr University, Agadir, Morocco

Siegmar-W. Breckle Department of Ecology, University of Bielefeld, Bielefeld, Germany

Cylphine Bresdin Environmental Research Laboratory of the University of Arizona, Tucson, AZ, USA

J. Jed Brown Center for Sustainable Development, College of Arts and Sciences, Qatar University, Doha, Qatar

Isabel Caçador Marine and Environmental Sciences Centre, Faculty of Sciences of the University of Lisbon, Lisbon, Portugal

John Cheeseman Department of Plant Biology, University of Illinois at Urbana-Champaign, Urbana, IL, USA

Miguel Clüsener-Godt UNESCO Man and the Biosphere Programme, Division of Ecological and Earth Sciences, Paris, France

Salma Daoud Laboratory of Plant Biotechnologies, Faculty of Sciences, Ibn Zohr University, Agadir, Morocco

Joann Diray-Arce Department of Microbiology and Molecular Biology, Brigham Young University, Provo, UT, USA

Richard Doyle School of Land and Food, University of Tasmania, Hobart, TAS, Australia

Bernardo Duarte Marine and Environmental Sciences Centre, Faculty of Sciences of the University of Lisbon, Lisbon, Portugal

Hassan M. El Shaer Desert Research Center, Mataria, Cairo, Egypt

Khalid Elbrik Faculty of Sciences, Ibn Zohr University, Agadir, Morocco

María Ferrandis Department of Vegetal Biology, University of Valencia, Valencia, Spain

Angelo Maria Gioffrè Department of Plant and Environmental Sciences, Faculty of Science, University of Copenhagen, Tåstrup, Denmark

Edward P. Glenn Environmental Research Laboratory of the University of Arizona, Tucson, AZ, USA

Salih Gücel Institute of Environmental Sciences, Near East University, Lefkoşa, Northern Cyprus

Bilquees Gul Institute of Sustainable Halophyte Utilization, University of Karachi, Karachi, Pakistan

Ibtissem Ben Hamad Laboratoire des Plantes Extrêmophiles, Centre de Biotechnologie de Borj Cédria, Hammam Lif, Tunisia; Institut des Energies de Demain, Université Paris Diderot, Sorbonne Paris Cité, Paris, France

Karim Ben Hamed Laboratoire des Plantes Extrêmophiles, Centre de Biotechnologie de Borj Cédria, Hammam Lif, Tunisia

Abdul Hameed Institute of Sustainable Halophyte Utilization, University of Karachi, Karachi, Pakistan

Marcus Hardie School of Land and Food, University of Tasmania, Hobart, TAS, Australia

Gabriel Haros The Punda Zoie Company Pty Ltd, Melbourne, VIC, Australia

Moulay Chérif Harrouni Hassan II Agronomic and Veterinary Institute, Agadir, Morocco

A.K.M. Nazrul Islam Ecology Laboratory, Department of Botany, University of Dhaka, Dhaka, Bangladesh

Sven-Erik Jacobsen Department of Plant and Environmental Sciences, Faculty of Science, University of Copenhagen, Tåstrup, Denmark

M. Ajmal Khan Institute of Sustainable Halophyte Utilization, University of Karachi, Karachi, Pakistan; Centre for Sustainable Development, College of Arts and Sciences, Qatar University, Doha, Qatar

Peter Lane School of Land and Food, University of Tasmania, Hobart, TAS, Australia

João Carlos Marques Marine and Environmental Sciences Centre, Faculty of Sciences and Technology, University of Coimbra, Coimbra, Portugal

David G. Masters School of Animal Biology, The University of Western Australia, Crawley, WA, Australia; CSIRO Agriculture, Wembley, WA, Australia

Adele Muscolo Department of Agriculture, Mediterranea University, Reggio Calabria, Italy

Brent Nielsen Department of Microbiology and Molecular Biology, Brigham Young University, Provo, UT, USA

Hayley C. Norman CSIRO Agriculture, Wembley, WA, Australia

Suresh Panta School of Land and Food, University of Tasmania, Hobart, TAS, Australia

Maria Rosaria Panuccio Department of Agriculture, Mediterranea University, Reggio Calabria, Italy

Juan Bautista Peris Department of Vegetal Biology, University of Valencia, Valencia, Spain

Sergey Shabala School of Land and Food, University of Tasmania, Hobart, TAS, Australia

Noomene Sleimi UR-MaNE, Faculté des Sciences de Bizerte, Université de Carthage, Tunisia

Naima Tachbibi Laboratory of Plant Biotechnologies, Faculty of Sciences, Ibn Zohr University, Agadir, Morocco

María Rosa Cárdenas Tomažič UNESCO Man and the Biosphere Programme, Division of Ecological and Earth Sciences, Paris, France

Kazuo N. Watanabe Gene Research Center, University of Tsukuba, Tsukuba City, Ibaraki, Japan

Münir Öztürk Botany Department, Science Faculty, Ege University, Bornova-Izmir, Turkey

INTRODUCTION

The world population has been increasing steadily and has reached seven billion whilst registering an increase of one billion during the last decade. One-sixth of the world population inhabits arid or/and semi-arid regions where the per capita availability of water is among the lowest in the world. Water availability has remained constant globally, however, its utilization has increased many fold due to the increase in population. Activities of humans to survive in these conditions could lead to global warming, for example, through huge expenditure of energy in the desalination of seawater for domestic purposes in the Arabian Gulf region.

Gulf Cooperation Council countries suffer from severe water scarcity and their natural resources are not sufficient for domestic usage. Therefore, using this scarce precious water for agriculture is not possible. This area is going through a period of unprecedented development and consequently the population is rising and annual water production through desalination is also increasing rapidly. Qatar is striving hard to ensure food and water security, as envisaged in Qatar National Vision 2030. Food security cannot be achieved through conventional agriculture but requires "out of the box" solutions. Halophytes are a group of plants that are naturally equipped with the mechanisms to survive under highly saline and arid conditions and produce high biomass. This high productivity could be used as fodder, forage, medicine, edible oil, and in some cases as food for humans. An "International Conference on Halophytes for Food Security in Dry Lands" was organized by the College of Arts and Sciences Qatar University from May 12–13, 2014 to address the issue of food security for Qatar and adjacent regions. The themes of the conference were: (i) halophyte ethno-botany, traditional uses, nontraditional crop development, (ii) halophyte research (ecology, bio-geography, eco-physiology, biochemistry, genetics, molecular biology; chemistry; fodder value; animal nutrition, pharmaceuticals and cosmetics, etc.) and education, (iii) food and water security, environment management, conservation and global changes, (iv) stakeholders (farmers, donors, investors, landowners, agro-industry), projects, pilot forms, network, etc. and (v) social, economic, human, and cultural aspects of scientific research.

This book addresses aspects of food security (particularly biomass production under saline conditions) that cover the themes of the conference. It also contains the communication of innovative ideas, such as research into halophyte farming with economic sustainability, as well as salt-tolerant plant utilization as a possible alternative to salt-sensitive crops. It is hoped that the information provided will not only advance vegetation science, but that it will truly generate more interdisciplinarily, networking, and awareness, and inspire farmers, and agricultural and landscaping stakeholders, to seriously engage in halophyte cash crop production in coastal and inland saline areas, especially those with an arid climate.

**M. Ajmal Khan, Munir Ozturk, Bilquees Gul,
and Muhammad Zaheer Ahmed**

CHARACTERIZATION AND FUNCTION OF SODIUM EXCHANGER GENES IN *AELUROPUS LAGOPOIDES* UNDER NaCL STRESS

Muhammad Zaheer Ahmed[1,2], Bilquees Gul[1], M. Ajmal Khan[1,3] and Kazuo N. Watanabe[2]

[1]Institute of Sustainable Halophyte Utilization, University of Karachi, Karachi, Pakistan [2]Gene Research Center, University of Tsukuba, Tsukuba City, Ibaraki, Japan [3]Centre for Sustainable Development, College of Arts and Sciences, Qatar University, Doha, Qatar

1.1 Introduction

Soil salinization is the key issue in irrigated arid and semi-arid areas that have substantial impact on plant productivity. To cope with salinity, plants have developed several adaptive mechanisms including altered growth pattern, osmotic adjustment, and ion homeostasis (Flowers and Colmer, 2008). These complex traits are extensively reported in both salt-sensitive (glycophytes) and salt-resistant (halophytes) plants (Zhu, 2001; Tester and Davenport, 2003). Moreover, recent molecular studies indicate that halophytes have better ability to alter the expression of genes linked with a wide array of plant processes which support them in surviving in saline areas (Maathuis and Amtmann, 1999; Zhu, 2001). In this scenario, there is a need to enhance knowledge about the multi-genic response of halophytes in NaCl to improve the salt tolerance of conventional crops.

Halophytes can reduce Na$^+$ toxicity in cytoplasm, minimize water deficit, manage essential mineral deficiency and reactive species damage when grown under salinity-affected soil in various ways (Blumwald, 2000; Chen et al., 2007; Cosentino et al.,

M.A. Khan, M. Ozturk, B. Gul, & M.Z. Ahmed (Eds): Halophytes for Food Security in Dry Lands.
DOI: http://dx.doi.org/10.1016/B978-0-12-801854-5.00001-7

2010). The Na^+ partitioning between the below- and aboveground biomass of plants is an important aspect for salinity resistance (Flowers and Colmer, 2008). Grasses accumulate lower amounts of Na^+ in shoots compared to dicots (Marcum, 2008). Plants exclude Na^+ from root to soil solution, regulate its loading in vascular tissues, compartmentalize in the vacuole/apoplast and excrete it from above ground epidermal bladder cells to reduce its negative effect on metabolic processes (Tester and Davenport, 2003; Flowers and Colmer, 2008; Shabala, 2013).

The movement of Na^+ into the vacuoles or toward apoplast is enabled by the action of tonoplast and plasma membrane-bound Na^+/H^+ antiporters, respectively, that use the electrochemical gradient of H^+ generated by H^+ ATPases and H^+ PPase. Knowledge is available about the sequence of Na^+/H^+ antiporters (*NHX*) (Apse et al., 1999; Shi et al., 2000; Tao et al., 2002; Zhang et al., 2008), expression and function of *NHX* genes when plants are exposed to salinity (Gaxiola et al., 1999; Oh et al., 2009). Some reports highlight the improvement in the salt tolerance of many crop plants by overexpressing *NHX* genes (He et al., 2005; Xu et al., 2010).

Poaceae is the most economically important plant family because 70% of all crops are salt-sensitive grasses. About 3.6 billion ha from 5.2 billion ha of the world's agricultural land is already salt-affected and not suitable for conventional crop farming. In contrast, the demand for food is continuously increasing and we expect to need to feed around nine billion by the end of 2050 (Millar and Roots, 2012). However, extensive efforts are underway to improve the salinity tolerance of conventional crops either through breeding or modern molecular techniques, but still no crop can tolerate half the level of salinity of seawater. In such a scenario, a major breakthrough in crop breeding for salinity tolerance is needed. Regulation of the number, size, and shape of the salt-excreting structure—trichome could be one such possibility. About 15% of halophytic grasses excrete Na^+ and Cl^- through bicellular microhairs, which are present on the leaf surface (Adams et al., 1998). *Aeluropus lagopoides* (Linn.) Trin. Ex Thw. is a salt-excreting, salinity- (1000 mmol L^{-1} NaCl; Gulzar et al., 2003) and drought-tolerant (Mohsenzadeh et al., 2006) grass. Therefore, it could be used as a model plant to improve the salinity tolerance of crops like rice, wheat, and maize (Flowers and Colmer, 2008). Detailed ecological and physiological studies on *A. lagopoides* have been carried out (Waghmode and Joshi, 1982; Sher et al., 1994; Abarsaji, 2000; Gulzar et al.,

2003). However, information related to the function of its Na^+ transport genes in salinity is lacking. Therefore, the goals of this study were: (i) to isolate the cDNA sequences of *VNHX* and *PMNHX* from *A. lagopoides*; (ii) to observe the change in the expression of both genes under saline condition; and (iii) to explore the role of both genes in the salt tolerance of *A. lagopoides*.

1.2 Materials and Methods

1.2.1 Plant Material

Tillers of *A. lagopoides* were collected from a population located in coastal areas of Karachi, Pakistan and used for the growth of new seedlings.

1.2.2 Isolation of the cDNA and Sequence Analysis of *VNHX* and *PMNHX*

One-month-old plants were treated with half-strength Hoagland culture solution containing 400 mmol L^{-1} NaCl for 2 days. Total RNA was extracted using an RNAqueous Kit (Ambion). The first strand of cDNA was synthesized from 1 μg RNA (DNA free) with the help of protocol provided with cDNA Takara RNA-PCR Kit (AMV; Ver 3.0). Polymerase chain reaction (PCR) were performed using with a pair of primers: (P_1: 5′TTC ATC TAC CTG CTC CCG CCC ATC AT3′; P_2: 5′CCA CAG AAG AAC ACG GTT AGA ATA CC3′) for *VNHX* and (P_3: 5′TTC ATC TAC CTG CTC CCG CCS ATC AT3′; P_4: 5′CCA CAG AAG AAC ACG GTT AGA ATR CC3′) for *PMNHX*, which were designed based on the conserved regions of previously reported Na^+/H^+ antiporter from other plants. PCR product was cloned through TA cloning kit (Takara) and pGEM-T vector. After cloning, plasmid was extracted and used for sequencing. The sequencing of 5′ and 3′ un-translated regions of *VNHX* was performed using P_5: 5′GTT GTG AAT GAT GCC ACG TC3′; P_6: 5′GAG AGC AGG AGA TCC CAA TC3′; P_7: 5′CCA CAG AAG AAC ACG GTT AGA ATA CC3′ and M_{13}-primer: 5′GTT TTC CCA GTC ACG AC3′. After amplification of the 3′ and 5′ regions, fragments were sequenced and assembled to provide the full-length cDNA of *VNHX*. The analysis of the *VNHX* and *PMNHX* sequences was performed by DNA-Dynamo software and NCBI program.

1.2.3 Growth Conditions and Harvest

Tillers of *A. lagopoides* were potted in plastic pots (26 cm high × 20 cm diameter) in prewashed field collected sand culture and sub-irrigated with half-strength Hoagland nutrient solution (Hoagland and Arnon, 1950) to establish for 1 month. Equal-sized plantlets were treated with different concentrations (0, 150, 300, and 600 mmol L^{-1}) of NaCl. The concentration of test solution was maintained every alternate day by distilled water to compensate for evaporation; whereas all test solutions were completely replaced after every fifth day.

Growth parameters (length of shoot and leaf, number of total and senesced leaves) were recorded initially and at the end of the experiment. Each plantlet was carefully removed from the soil after 15 days of experiment and washed thoroughly. Roots and shoots were washed and separated from each other before treating with liquid N_2. All samples were stored at $-80°C$.

1.2.4 Quantification of Gene Expression by qRT-PCR

For quantitative real-time PCR (qRT-PCR), a pair of primers were designed for *PMNHX* (PMN-F: 5′TAT CGA ATG GTG CTC GGA AGA3′; PMN-R: 5′AGC CCA GCC ACA GTA CCG ATA3′) and for *VNHX* (VNHX-F: 5′GCA GGT CCT CAA TCA GGA TG3′; VNHX-R: 5′ACT CCA AGG AAG GTG CTT GA3′) by using the gene sequence information of *A. lagopoides*. Expression of *Actin* gene was used to normalize the data. The quantitative expression data of both genes was recorded on a Light Cycler-Carousel-based System (ROCHE), while the analysis of data was performed by software 4.0. All standard curves had $R^2 \geq 0.99$.

1.2.5 Measurement of Na$^+$ in Plant Sap

The press sap method was used (Cuin et al., 2009) to determine the soluble fraction of Na$^+$ in leaves and roots of NaCl-treated plants. Sap was mixed thoroughly before preparing dilutions and used for the determination of Na$^+$ on atomic absorption spectrometer (AA-700; Perkin Elmer, Santa Clara, CA, USA).

1.2.6 Secretion of Na$^+$

Fully expanded young leaves of three plantlets were tagged from each NaCl treatment (0, 150, 300, and 600 mmol L^{-1}). All tagged leaves were prewashed 72 h before the final data collection. Leaves were rinsed with 2 mL deionized water and collected in Eppendorf tubes and the rate of Na^+ excretion was determined by atomic absorption spectrometry. The area of rinsed leaves was calculated by Image-J software version 1.45 (http://rsb.info.nih.gov/ij/) and data were expressed in μmol Na^+ cm^{-2} per day.

1.2.7 Malondialdehyde Content

Malondialdehyde (MDA) content was determined in leaf samples as an indicator of lipid peroxidation (Heath and Packer, 1986). An extinction coefficient of $155 \, mM^{-1} \, cm^{-1}$ was used to calculate the MDA content in the supernatant while absorbance was recorded at 532 and 600 nm wavelengths. The result of MDA was expressed as μg mg^{-1} FW.

1.2.8 Statistical Analyses

Statistical analysis was done by SPSS version 11.0 for Windows (SPSS, 2001). Two-way analysis of variance (ANOVA) was used to test for a significant ($P<0.05$) effect of NaCl on growth, MDA, Na^+ concentration, and expression data. A post-hoc Bonferroni test was used to test for significant differences between means. Correlation analysis was performed between different parameters of *A. lagopoides* through SPSS. Graphs were constructed with the help of SigmaPlot (11.0).

1.3 Results

1.3.1 Molecular Characterization of *VNHX* and *PMNHX*

The full-length cDNA of *VNHX* contained 2353 bp including a putative poly (A) addition signal site in the end of sequence. Whereas, the un-translated region (UTR) of 5' and 3' consisted of 337 and 393 bp respectively, the open reading frame (ORF) of 1623 bp encoded a protein of 540 amino acids with a theoretical molecular mass of 59.36 kDa (Figure 1.1A). The cDNA sequence of *VNHX* has been deposited at GenBank with the name *AlaNHX* under accession number GU199336.1. Sequence homology revealed a high degree of homology sequences of *AlNHX* (*VNHX*) and putative vacuolar Na^+/H^+ antiporter of other higher plants.

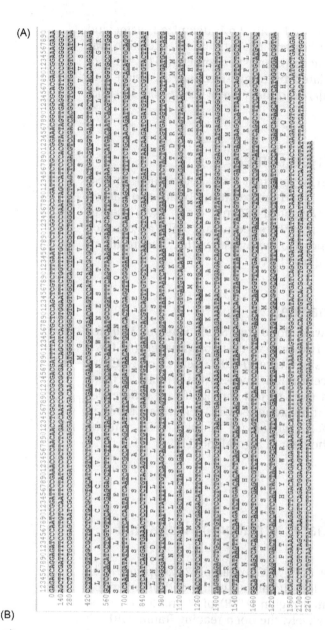

(A)

(B)

GCT GAC AGC AAG GGT AAG TGC AAT CTC TAT GAT TGT ATC ATT AAA AAT AAA GCC CAG CCA CAG TAC CGA

TAC AAT TCC AAA TGC AAG GCC CAG AGC AAC AGC CCC AAG CGC AAC TTC CGA CAA GAA CTT TAT TAT TGA

GCC AGC ATC AAA GGT TCT TCC GAG CAC CAT TCG ATA AAA TAG TTG GTA GAC GAC AAT AGC AGT CCC

Figure 1.1 Information from two isolated genes from *Aeluropus lagopoides*. (A) The cDNA and deduced amino acid sequence of *VNHX* (*AlaNHX*), and (B) cDNA sequence of *PMNHX*.

The "expressed sequence tag" (EST) of *PMNHX* contained 204 bp and showed a high degree of homology with previously reported Na^+/H^+ antiporter located on the plasma membrane of higher plants (Figure 1.1B). The cDNA sequence of *PMNHX* has been deposited at GenBank under accession number GW796824.1.

1.3.2 Growth

The number of leaves and plant height decreased significantly ($P<0.01$) with the increases in salinity. In addition, a substantial ($P<0.0001$) increase in leaf senescence was observed at 300 and 600 mmol L^{-1} NaCl (Table 1.1; Figure 1.2).

1.3.3 Peroxidation of Lipid Membrane

MDA content was unchanged at up to 300 mmol L^{-1} NaCl treatment, whereas around a 40% increase was found when plants were treated with 600 mmol L^{-1} NaCl compared to nonsaline controls (Figure 1.1).

Table 1.1 Results of One-Way ANOVA Showed the Effect of NaCl on Different Parameters of *Aeluropus lagopoides*

Parameters	df	Mean Square	$F^{significance}$
Height of shoot	3	1480.183	15.911*
# of leaves	3	1318.175	8.901*
# of yellow leaves	3	32.458	86.555***
MDA	3	2025.764	138.764***
Na^+—secretion	3	102.231	3.032*
Leaf—Na^+	3	321064.380	158.165***
Leaf—*VNHX*	3	55926.904	95.712***
Leaf—*PMNHX*	3	83948.736	76.645**
Root—Na^+	3	407017.080	293.535***
Root—*VNHX*	3	838020.333	67.553**
Root—*PMNHX*	3	9805.678	22.467**

*$P<0.05$; **$P<0.001$; ***$P<0.0001$.
Values of correlation are provided with degree of significance.

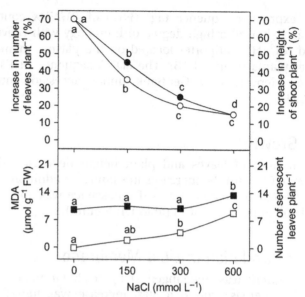

Figure 1.2 Change in the growth and biochemical parameters [filled circles: increase in number of leaves plant^{-1}; empty circles: increase in height of shoot plant^{-1}; filled squares: malondialdehyde (MDA) content; empty squares: number of senescent leaves plant^{-1}] of *Aeluropus lagopoides* treated with different NaCl concentrations (0–600 mmol L^{-1}) for 15 days ($n = 3$). Values with at least one Bonferroni letter the same were not significantly different at $P<0.05$.

1.3.4 Flux in Na$^+$

Na$^+$ content increased significantly ($P<0.0001$) in both leaves and roots of *A. lagopoides* under NaCl treatment (Table 1.1; Figures 1.3 and 1.4). Moreover, this increase was approximately tenfold higher in plants treated with 600 mmol L^{-1} NaCl than in nonsaline controls (Figures 1.3 and 1.4). In general, the amount of Na$^+$ was similar in both parts of plants except at 300 mmol L^{-1} NaCl where roots accumulated a higher amount of Na$^+$ than leaves (Figures 1.3 and 1.4).

1.3.5 Secretion of Na$^+$

Sodium excretion from the leaf surface increased significantly ($P<0.01$) with increase in NaCl concentrations up to 300 mmol L^{-1} NaCl, however, no difference was noted in the Na$^+$ secretion rate of plants exposed to 300 and 600 mmol L^{-1} NaCl (Figure 1.5).

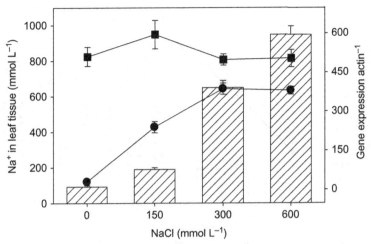

Figure 1.3 Bars represent the concentration of Na$^+$ in leaf of *Aeluropus lagopoides* treated with different NaCl concentrations (0–600 mmol L^{-1}) for 15 days ($n = 3$). Change in the expression of genes in leaves was shown by line graph (square and circle symbols were used for *VNHX* and *PMNHX* gene, respectively). Values with at least one Bonferroni letter the same were not significantly different at $P<0.05$.

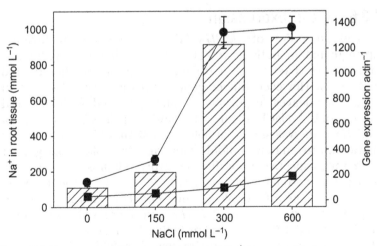

Figure 1.4 Bars represent the concentration of Na$^+$ in roots of *Aeluropus lagopoides* treated with different NaCl concentrations (0–600 mmol L^{-1}) for 15 days ($n = 3$). Change in the expression of genes in roots was shown by line graph (square and circle symbols were used for the *VNHX* and *PMNHX* genes, respectively). Values with at least one Bonferroni letter the same were not significantly different at $P<0.05$.

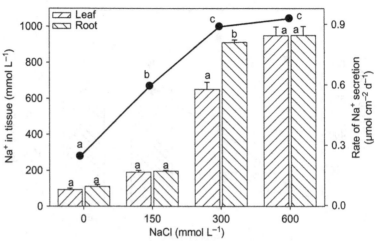

Figure 1.5 Bars represent the difference in the concentration of Na$^+$ between leaves and roots of *Aeluropus lagopoides* plants treated with different NaCl concentrations (0–600 mmol L^{-1}) for 15 days ($n = 3$). Bonferroni letters were used to show significant difference ($P<0.05$) between means of leaves and roots within each salinity level. Salt secretion rate was showed by line graph and Bonferroni letters were used to compare values among salinity concentrations.

1.3.6 Gene Expression

The expression of *AlaNHX* (*VNHX*) gene was significantly up-regulated in both leaves ($P<0.0001$; Table 1.1; Figure 1.3) and roots ($P<0.001$; Table 1.1; Figure 1.4) of plants when treated with NaCl. However, higher gene expression was observed in roots than leaves, especially in plants treated with 300 and 600 mmol L^{-1} NaCl (Figures 1.3 and 1.4). The expression of *AlaNHX* gene was similar at 300 and 600 mmol L^{-1} NaCl, where it was approximately tenfold (root) and fourfold (leaf) greater than the respective nonsaline controls (Figures 1.3 and 1.4). A negative correlation ($r^2 = 0.79$; $P<0.05$) was found between the expression of *AlaNHX* and *PMNHX* genes in leaves (Table 1.2). Expression of *PMNHX* gene increased significantly ($P<0.001$) under NaCl treatment (Table 1.1; Figures 1.3 and 1.4). *PMNHX* gene showed approximately threefold higher expression in leaves than roots (Figures 1.3 and 1.4). Gene expression did not change in leaves under salinity except at 150 mmol L^{-1} NaCl where substantial up-regulation was found (Figure 1.3). In contrast to leaves, the maximum expression of *PMNHX* gene was found in roots treated with 600 mM NaCl (Figure 1.4).

Table 1.2 Pearson Correlation Analysis Between Changes in Different Parameters of *Aeluropus lagopoides* Under NaCl

Parameters	Leaf—Na$^+$	Leaf—VNHX	Leaf—PMNHX	Root—Na$^+$	Root—VNHX	Root—PMNHX	Na$^+$—Secretion	MDA	Total leaves	Yellow Leaves
Leaf—Na$^+$	—	0.84	−0.40	0.96	0.95	0.94	0.76	0.78	−0.86	0.90
(Significance)		**	ns	**	**	**	*	*	**	**
Leaf—VNHX	0.84	—	−0.79	0.88	0.89	0.75	0.82	0.55	−0.93	0.77
(Significance)	**		*	**	**	*	*	ns	**	*
Leaf—PMNHX	−0.40	−0.79	—	−0.44	−0.48	−0.32	−0.60	−0.26	0.71	−0.41
(Significance)	ns	*		ns	ns	ns	ns	ns	*	ns
Root—Na$^+$	0.96	0.88	−0.44	—	0.98	0.85	0.80	0.60	−0.85	0.78
(Significance)	**	**	ns		**	**	*	ns	**	*
Root—VNHX	0.95	0.89	−0.48	0.98	—	0.80	0.71	0.56	−0.88	0.78
(Significance)	**	**	ns	**		**	*	ns	**	*
Root—PMNHX	0.94	0.75	−0.32	0.85	0.80	—	0.77	0.91	−0.74	0.94
(Significance)	**	*	ns	**	**		**	**	*	**
Na$^+$—Secretion	0.76	0.82	−0.60	0.80	0.71	0.77	—	0.57	−0.72	0.64
(Significance)	*	*	ns	*	*	*		ns	*	**
MDA	0.78	0.55	−0.26	0.60	0.56	0.91	0.57	—	−0.63	0.90
(Significance)	*	ns	ns	ns	ns	**	ns		ns	**
Total leaves	−0.86	−0.93	0.71	−0.85	−0.88	−0.74	−0.72	−0.63	—	−0.81
(Significance)	**	**	*	**	**	*	*	ns		*
Yellow leaves	0.90	0.77	−0.41	0.78	0.78	0.94	0.64	0.90	−0.81	—
(Significance)	**	*	ns	*	*	**	ns	**	*	

*$P<0.05$; **$P<0.001$; ns, nonsignificant.
Values of correlation are provided with degree of significance.

1.4 Discussion

Survival of salt-excreting grasses under saline conditions depends on the extent of Na^+ accumulation in cytoplasm which is the function of increase in the ability of Na^+ exclusion, excretion and sequestration into vacuoles (Ahmed et al., 2013). Sodium/hydrogen antiporter genes are considered to play an important role in controlling Na^+ flux, cytoplasmic pH and cell volume (Mahnensmith and Aronson, 1985). To better understand salt-tolerance mechanisms in *A. lagopoides* that survives successfully under highly saline conditions we cloned and characterized the cDNA of salt stress-related genes (*PMNHX* and *VNHX* (*AlaNHX*)). A full-length cDNA was isolated from *A. lagopoides* grown under saline conditions which was 2353 bp long including the predicted ORF of 1623 bp long (338–1960 bp of full-length cDNA) which encodes protein consisting of 540 amino acids. Comparison of both cDNA sequences with other proteins indicates that *AlaNHX* shares a higher identity with *AlNHX* isolated from *Aeluropus littoralis* (Zhang et al., 2008). Similarly, the EST of *PMNHX* had shown greater homology with the *SOS1* gene of *Phragmites australis* (Takahashi et al., 2009). These data allowed us to classify *PMNHX* and *AlaNHX* as new members of the plasma membrane and vacuole Na^+/H^+ antiporter family and to suggest that they might be involved in Na^+ regulation.

Growth of grasses was reduced when exposed to salinity, even if they survived in higher NaCl concentrations (Gulzar et al., 2003; Barhoumi et al., 2007; Flowers and Colmer, 2008). Similarly, *A. lagopoides* has the ability to survive in up to 1000 mmol L^{-1} NaCl but nonsaline conditions appear to be optimal for the production of plant biomass (Gulzar et al., 2003). The negative correlation between total number of leaves and Na^+ content ($r^2 = -0.86$: $P < 0.001$; Table 1.2), but a greater positive correlation between Na^+ and leaf senescence ($r^2 = 0.90$: $P < 0.001$; Table 1.2) was found. A decreasing trend in the shoot length, leaf elongation, and leaf emergence in higher salinities could be attributed to minimal Na^+ accumulation in shoots (Torrecillas et al., 2003). A delay in the emergence of new leaves and accelerated shedding of mature leaves at 600 mmol L^{-1} NaCl could be related to the specific ionic toxicity, particularly Na^+ and Cl^- (Rudmik, 1983). Halophytic grasses usually employ mature leaf shedding and decreasing leaf elongation rates that could help to reduce the Na^+ transport towards young and active plant tissues, but at the cost of reduced biomass (Munns, 2002; Flowers and Colmer, 2008). In contrast, a rapid growth reduction in 600 mmol L^{-1} NaCl-treated plants is due to oxidative stress (Sobhanian et al., 2010) indicated by higher

MDA content (40% of respective nonsaline treatment). This was further evident from a positive correlation of MDA with leaf Na ($r^2 = 0.78$: $P<0.01$; Table 1.2) and leaf yellowing ($r^2 = 0.90$: $P<0.001$; Table 1.2). MDA did not change in up to 300 mmol L^{-1} NaCl, indicating the efficient removal of toxic ions like Na^+ from the metabolically active cytoplasm (Munns and Tester, 2008; Oh et al., 2009), which was made possible through Na^+ compartmentalization inside the vacuole by *VNHX* (Cosentino et al., 2010) as we found a positive correlation between *AlaNHX* gene expression and leaf Na ($r^2 = 0.84$: $P<0.001$; Table 1.2). However, the expression of *SOS1* appeared to be unchanged during salinity stress but the higher extent of expression might be sufficient for Na^+ transport towards apoplast. The transmembrane transport of Na^+ either through *AlaNHX* or *SOS1* would be dependent on the H^+ gradient that was established by the activity of H-ATPase and H-PPase enzyme (Hedrich et al., 1989). The distribution of Na^+ varies between below- and above-ground tissues and also depends on plant species (Abogadallah, 2010; Yang et al., 2010). Most halophytic grasses accumulate a higher amount of Na^+ in roots (Marcum, 2008). The positive correlation between the Na^+ content of leaves and roots ($r^2 = 0.96$: $P<0.001$; Table 1.2) expressed by the *A. lagopoides* plant accumulated Na in both organs but a higher amount of Na^+ was found in roots when the plant was exposed to up to 300 mmol L^{-1} NaCl. This finding also validates the expression of *VNHX* gene which was threefold higher in roots than leaves. However, the expression of *PMNHX* was around twofold higher in leaves than roots. In leaves the higher expression of *PMNHX* than *VNHX* could help in the loading of Na in epidermal bladder cells for secretion through salt glands. In *A. lagopoides*, the higher uptake of Na^+ directly correlates with the secretion rate ($r^2 = 0.76$: $P<0.01$; Table 1.2). The expression of both genes (*PMNHX* and *VNHX*) and salt secretion rate were similar in seedlings treated with 300 and 600 mmol L^{-1} NaCl, suggesting Na^+ toxicity in plants treated with 600 mmol L^{-1} NaCl.

The expression of both sodium exchanger genes *PMNHX* and *VNHX* depends on tissue type and salt concentration and makes *A. lagopoides* a highly salt-resistant grass. The synchronized alteration in *PMNHX* and *VNHX* expression helps *A. lagopoides* to survive up to 300 mmol L^{-1} NaCl by successfully compartmentalizing Na^+ in apoplasts and vacuoles, respectively. The effective Na^+ secretion and shift in the biomass allocation toward roots also provide support in reducing the ion toxicity. However, plants were facing ion toxicity at 600 mmol L^{-1} NaCl due to limited Na^+ excretion and uncoordinated transcription of Na^+ transporter genes. This information will help to understand the salt-tolerance mechanisms of grasses and its use for better yields of conventional crops in saline land.

References

Abarsaji, G.A., 2000. Identification and investigation on some of eco-physiological characteristics of *Aeluropus* spp. in saline and alkaline rangelands in the north of Gorgan. Pajouhesh-Va-Sazandegi. 46, 21−25.

Abogadallah, G.M., 2010. Sensitivity of *Trifolium alexandrinum* L. to salt stress is related to the lack of long-term stress-induced gene expression. Plant Sci. 178, 491−500.

Adams, P., Nelson, D.E., Yamada, S., Chmara, W., Jensen, R.G., Bohnert, H.J., et al., 1998. Growth and development of *Mesembryanthemum crystallinum* (Aizoaceae). New Phytol. 138, 171−190.

Ahmed, M.Z., Shimazaki, T., Gulzar, S., Kikuchi, A., Gul, B., Khan, M.A., et al., 2013. The influence of genes regulating transmembrane transport of Na + on the salt resistance of Aeluropus lagopoides. Funct. Plant Biol. 40, 860−871.

Apse, M.P., Aharon, G.S., Snedden, W.A., Blumwald, E., 1999. Salt tolerance conferred by overexpression of a vacuolar Na^+/H^+ antiport in *Arabidopsis*. Science. 285, 1256−1258.

Barhoumi, Z., Djebali, W., Chaïbi, W., Abdelly, C., Smaoui, A., 2007. Salt impact on photosynthesis and leaf ultrastructure of *Aeluropus littoralis*. J. Plant Res. 120, 529−537.

Blumwald, E., 2000. Sodium transport and salt tolerance in plants. Curr. Opin. Cell. Biol. 12, 431−434.

Chen, Z., Pottosin, I.I., Cuin, T.A., Fuglsang, A.T., Tester, M., Jha, D., et al., 2007. Root plasma membrane transporters controlling K^+/Na^+ homeostasis in salt stressed barley. Plant Physiol. 145, 1714−1725.

Cosentino, C., Fischer-Schliebs, E., Bertl, A., Thiel, G., Homann, U., 2010. Na^+/H^+ antiporters are differentially regulated in response to NaCl stress in leaves and roots of *Mesembryanthemum crystallinum*. New Phytol. 186, 669−680.

Cuin, T.A., Tian, Y., Betts, S.A., Chalmandrier, R., Shabala, S., 2009. Ionic relations and osmotic adjustment in durum and bread wheat under saline conditions. Funct. Plant Biol. 36, 1110−1119.

Flowers, T.J., Colmer, T.D., 2008. Salinity tolerance in halophytes. New Phytol. 179, 945−963.

Gaxiola, R.A., Rao, R., Sherman, A., Grisafi, P., Alper, S.L., Fink, G.R., 1999. The *Arabidopsis thaliana* proton transporters, *AtNhx1* and *Avp1*, can function in cation detoxification in yeast. Proc. Natl. Acad. Sci. 96, 1480−1485.

Gulzar, S., Khan, M.A., Ungar, I.A., 2003. Effect of salinity on growth, ionic content, plant-water status in *Aeluropus lagopoides*. Commun. Soil Sci. Plant Nutr. 34, 1657−1668.

He, C., Yan, J., Shen, G., Fu, L., Holaday, A.S., Auld, D., et al., 2005. Expression of an *Arabidopsis* vacuolar sodium/proton antiporter gene in cotton improves photosynthetic performance under salt conditions and increases fiber yield in the field. Plant Cell Physiol. 46, 1848−1854.

Heath, R.L., Packer, L., 1986. Photo-peroxidation in isolated chloroplasts. Kinetics and stoichiometry of fatty acid peroxidation. Arch. Biochem. Biophys. 125, 189−198.

Hedrich, R., Kurkdjian, A., Guern, J., Flugge, U.I., 1989. Comparative studies on the electrical properties of the H^+ translocating ATPase and pyrophosphatase of the vacuolar-lysosomal compartment. EMBO J. 8, 2835–2841.

Hoagland, D.R., Arnon, D.I., 1950. The water-culture method for growing plants without soil. Calif. Agric. Exp. Stn. 347.

Maathuis, F.J.M., Amtmann, A., 1999. K^+ nutrition and Na^+ toxicity: the basis of cellular K^+/Na^+ ratios. Ann. Bot. (Lond.). 84, 123–133.

Mahnensmith, R.L., Aronson, P.S., 1985. The plasma membrane sodium-hydrogen exchanger and its role in physiological and pathophysiological processes. Circ. Res. 56, 773–788.

Marcum, K.B., 2008. Relative salinity tolerance of turfgrass species and cultivars. In: Pessarakli, M. (Ed.), Handbook of Turfgrass Management and Physiology. CRC Press, New York, NY, pp. 389–406.

Millar, J., Roots, J., 2012. Changes in Australian agriculture and land use: implications for future food security. Int. J. Agric. Sustainability. 10, 25–39.

Mohsenzadeh, S., Malboobi, M.A., Razavi, K., Farrahi-Aschtiani, S., 2006. Physiological and molecular responses of Aeluropus lagopoides (Poaceae) to water deficit. Environ. Exp. Bot. 56, 314–322.

Munns, R., 2002. Comparative physiology of salt and water stress. Plant Cell Environ. 25, 239–250.

Munns, R., Tester, M., 2008. Mechanisms of salinity tolerance. Annu. Rev. Plant Biol. 59, 651–681.

Oh, D.-H., Leidi, E., Zhang, Q., Hwang, S.-M., Li, Y., Quintero, F.J., et al., 2009. Loss of halophytism by interference with SOS1 expression. Plant Physiol. 151, 210–222.

Rudmik, T., 1983. Morphological, Anatomical, and Physiological Changes in Triglochin maritima in Response to Changes in Salinity and Nitrogen (M.S. thesis). University of Toronto, Toronto, ON, Canada.

Shabala, S., 2013. Learning from halophytes: physiological basis and strategies to improve abiotic stress tolerance in crops. Ann. Bot. (Lond.). 112, 1209–1221.

Sher, M., Sen, D.N., Mohmmed, S., 1994. Seasonal variations in sugar and protein content of halophytes in Indian desert. Ann. Arid Zone. 33, 249–251.

Shi, H.Z., Ishitani, M., Kim, C.S., Zhu, J.K., 2000. The Arabidopsis thaliana salt tolerance gene SOS1 encodes a putative Na^+/H^+ antiporter. Proc. Natl. Acad. Sci. 97, 6896–6901.

Sobhanian, H., Motamed, N., Jazii, F.R., Nakamura, T., Komatsu, S., 2010. Salt stress induced differential proteome and metabolome response in the shoots of Aeluropus lagopoides (Poaceae), a halophyte C_4 plant. J. Proteome Res. 9, 2882–2897.

SPSS, Inc, 2001. SPSS: SPSS 11.0 for Windows Update. SPSS Inc., USA.

Takahashi, R., Liu, S., Takano, T., 2009. Isolation and characterization of plasma membrane Na^+/H^+ antiporter genes from salt-sensitive and salt-tolerant reed plants. J. Plant Physiol. 166, 301–309.

Tao, X., Apse, M.P., Aharon, G.S., Blumwald, E., 2002. Identification and characterization of a NaCl-inducible vacuolar Na^+/H^+ antiporter in Beta vulgaris. Physiol. Plant. 116, 206–212.

Tester, M., Davenport, R., 2003. Na^+ tolerance and Na^+ transport in higher plants. Ann. Bot. (Lond.). 91, 503–527.

Torrecillas, A., Rodrguez, P., Sanchez-Blanco, M.J., 2003. Comparison of growth, leaf water relations and gas exchange of Cistus albidus and C. monspeliensis plants irrigated with water of different NaCl salinity levels. Sci. Hortic. (Amsterdam). 97, 353–368.

Waghmode, A.P., Joshi, G.V., 1982. Photosynthetic and photorespiratory enzymes and metabolism of 14 C-substrates in isolated leaf cells of the C_4 species of *Aeluropus lagopoides* L. Photosynthetica. 16, 17–21.

Xu, K., Zhang, H., Blumwald, E., Xia, T., 2010. A novel plant vacuolar Na^+/H^+ antiporter gene evolved by DNA shuffling confers improved salt tolerance in yeast. J. Biol. Chem. 285, 22999–23006.

Yang, M.F., Song, J., Wang, B.S., 2010. Organ-specific responses of vacuolar H^+-ATPase in the shoots and roots of C_3 halophyte *Suaeda salsa* to NaCl. J. Integr. Plant Biol. 52, 308–314.

Zhang, G.-H., Su, Q., An, L.-J., Wu, S., 2008. Characterization and expression of a vacuolar Na^+/H^+ antiporter gene from the monocot halophyte *Aeluropus littoralis*. Plant Physiol. Biochem. 46, 117–126.

Zhu, J.K., 2001. Plant salt tolerance. Trends. Plant Sci. 6, 66–71.

MULTI-TEMPORAL SOIL SALINITY ASSESSMENT AT A DETAILED SCALE FOR DISCRIMINATING HALOPHYTES DISTRIBUTION

Jorge Batlle-Sales, Juan Bautista Peris and María Ferrandis
Department of Vegetal Biology, University of Valencia, Valencia, Spain

2.1 Introduction

The salt lake of Salinas (Alicante, Spain), zone object of this study, with centroid in N38° 30.196′ W0° 53.195′, is the bottom part of an endorreic watershed where both runoff waters and subterranean water fluxes accumulate. The lake, with an extension of 2.93 km², is surrounded by several glacis and its bottom part is constituted by geological materials of the Keuper Germanic facies, with clays and gypsiferous layers dominating (IGME, n.d.), which are the source of chloride, sulfate, sodium, and magnesium ions found in the saline groundwater.

Existing historical documents (Arroyo-Ilera, 1976) report that the lake was used for salt extraction since the seventeenth century, with this activity interrupted during the eighteenth century and restarted again in the twentieth century. The semiarid climate of the area, with annual precipitation of 404 mm, average temperature of 15.6°C (Figure 2.1) and very dry summer, allows for water concentration and solute precipitation by evaporation. In 1922 works were started for desiccating the lake and in 1948 regular salt extraction by evaporation of the saline runoff and drainage waters that reached the lake was commenced. Later, the overexploitation of proximal irrigation wells depleted the groundwater level, leading to the desiccation of the lake and hence, in 1952, ceasing the salt extraction activity.

M.A. Khan, M. Ozturk, B. Gul, & M.Z. Ahmed (Eds): Halophytes for Food Security in Dry Lands.
DOI: http://dx.doi.org/10.1016/B978-0-12-801854-5.00002-9

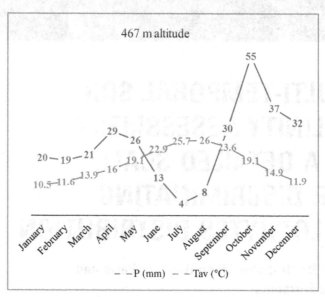

Figure 2.1 Climogram of Salinas. Mean monthly data for precipitation (mm) and temperature (°C).

Figure 2.2 Aerial photograph from 1956 showing the salt exploitation in the Salinas Lake.

In an aerial photograph from 1956 (Figure 2.2) (IGN, 1956), the existence of a residual salt layer that was later partially collected is evident. The residual soil salinity distribution was investigated by Batlle et al. (1994) and Pepiol et al. (1998), recognizing the composition and mineralogy of salt efflorescence and the distribution of salts at different depths in the soil, producing thematic maps. The soils are classified as gypsic aquisalids (Soil Survey Staff, 2014a), with massive accumulations of halite, gypsum, and calcium carbonate. Several botanical studies were realized in the zone (Rigual, 1972; Serra Laliga, 2007) and carried out in detail by Peris et al. (1999).

2.2   Objective

This study focused on the recognition of the soil conditions and salinity levels that determine the adaptation of different halophyte species, as a basis for recommendations of plant restoration of the area. The methodology adopted used measurement of soil salinity gradients in the former lake area, performing electromagnetic induction (EMI) surveys, botanical inventories, as well as soil sampling and analysis.

2.3   Methodology

2.3.1   Area of Study

The area selected for study was the south-east (SE) part of the salt lake that has undergone evident change from its appearance in the visual documents of 1956. This change includes the "soil construction" by formation of dunes and accumulation of particles transported by wind and the progressive colonization of the area by halophytes. Figure 2.3 presents the current aspect of the area: the slopes of glacis, with thick crusts of calcium carbonate,

Figure 2.3 Photograph of Salinas in May 2014, showing the glacis cultivated and the bottom part of the salt lake, partially colonized by halophytes.

are cultivated under irrigation with waters of good quality, in most cases by drip irrigation; the bottom part of the watershed, that constitutes the former salt lake, showing bare saline soil areas colonized progressively by halophytes. In the SE border of the lake the "soil construction" is an active process by formation of dunes by wind transport of soil particles. At the times of salt exploitation, the soil surface was sealed by a salt layer, formed by evaporation of the brine. After ceasing of salt mining, the salt crust was taken as a residual product. The soluble salts at the soil surface started to slowly leach downwards into the soil. The formation of crusts in the surface, by quick evaporation of soil solutions that arrived by capillary ascent, further impedes the arrival of solutions due to capillary disruption, giving "fluffy" micro-relief, with a soft and loose surface layer of several millimeters that is highly erodible by wind. In the study area the soil particles of this layer are transported by the winds of dominant orientation from WNW–ESE. Seeds of the most salinity-tolerant halophytes (mainly *Arthrocnemun macrostachyum*) enter the soil cracks and, after growth, reduce the wind speed, thereby promoting the deposition of particles transported in suspension after the plant, hence starting the dune formation process visible in Figure 2.4.

This "soil construction" by the frequent windy conditions in the zone gives additional opportunities for colonization to other plant species that are less salt-tolerant. The upper part of the "constructed soil" has high permeability and the soluble salts can be leached more effectively from a layered and compacted soil. It is very important to assess the existing salinity conditions that allow the plants to germinate in order to understand how the colonization process proceeds.

Figure 2.4 Formation of dunes with soil particles transported by wind, in May 2014.

2.3.2 Design of the Soil Survey and Vegetation Inventories

A geographical information system (GIS), based on ETRS89 ellipsoid and UTM projection, was implemented with the following information available for the area: aerial geo-referenced orthophoto from 2012 (30 cm of resolution) (ICV, 2012), Lidar EDM (ICV, 2009), geological maps (IGME, n.d.), and previous information about soils (Batlle et al., 1994; Pepiol et al., 1998) and plants (Peris et al., 1999), as well as our own unpublished information. The software used was Quantum GIS (2011). The photointerpretation of the aerial orthophoto suggested the existence of "vegetation bands" arranged along a topographical gradient, until the center of the lake, with bare soil. A series of survey transects were planned perpendicular to this topographical gradient, from an altitude of 474.5 to 472.7 m (lowest part of the salt lake). Four survey campaigns were completed on October 24, 2013 (first EMI survey), November 5, 2013 (second EMI survey with botanic inventories), May 7, 2014 (intensive EMI survey and soil sampling in six selected areas) and July 24, 2014 (intensive EMI survey in the six selected areas). Points of measurement (dots) and position of soil sampling (PI, PII, PIII, PIV, PV, and PVI) are represented in Figure 2.5.

2.3.3 Plants Inventory

During the EMI survey of November 5, 2014, sampling of plants was done and 81 botanical inventories were annotated according to the Braun-Blanquet methodology (Braun-Blanquet, 1964),

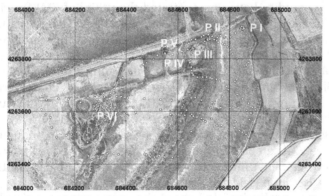

Figure 2.5 Points of EMI measurement and situation of the six soil profiles (PI–PVI) into the six monospecific areas selected for intensive EMI measurement.

recording the total area in square meters, percentage of area coverage and the values of ECa provided by the EMI device for two dipoles arrangements (EMv and EMh).

Six monospecific botanical inventories were selected for more detailed EMI and soil sampling research: *Suaeda vermiculata* (one area, PI), *Suaeda vera* (one area, PII), *Sarcocornia fruticosa* (two areas, PIII and PVI) and *A. macrostachyum* (two areas, PIV and PV).

2.3.4 Soil Sampling and Analysis

Composite soil samples were taken by soil augering at depths of 0–30, 30–60, and 60–90 cm, in each of the six selected points of monospecific plant inventories, PI–PII–PIII–PIV–PV–PVI (Figure 2.5). Field tests were performed for soil texture by hand, including presence of solid carbonates (effervescence with 6 M HCl), presence of chlorides (reaction with a solution 0.05 M of $AgNO_3$). Undisturbed samples taken with steel cylinders of 100 cm^3 were used for determining gravimetric water content at field conditions as well as for determining the soil bulk density. At the laboratory, soil samples were analyzed for total carbonates and soil saturated paste extract was prepared, extracted and analyzed for anion and cation concentration, ECe and pHe, according to standard methods 4F2 (Soil Survey Staff, 2014b). Sample color was coded according to the Munsell Color Book, page 10YR.

2.3.5 Electromagnetic Soil Salinity Survey, Geostatistics EMI Calibration and Mapping

Many studies have demonstrated the utility of EMI for characterizing the spatial variability of salt-affected soils (Farifteh et al., 2006; Corwin and Lesch, 2005a,b; Doolittle and Brevik, 2014), discriminating different types of salt-affected soils (Arriola-Morales et al., 2009), and for detecting temporal changes in soil salinity in irrigated areas (Batlle-Sales et al., 2000; Herrero et al., 2011; Akramkhanov et al., 2014), but only a few have used EMI in rapport to halophyte research (He et al., 2014).

The measurement of the apparent soil electrical conductivity (ECa) by EMI can be performed very efficiently in a quick way and can provide better information than ECe (Amakor et al., 2014). In many cases crop yield can be predicted from EMI surveys.

The physical principle behind the measurements using EMI is as follows: when a magnetic field reaches an electrical conductor, it induces an electrical current that, in turn, promotes a secondary magnetic field. The EMI instruments induce a primary magnetic field into the soil and measure the secondary magnetic field induced. From the comparison of the two magnitudes an "apparent" or "bulk" soil electrical conductivity (ECa) can be derived. For this research we used a standard EM38 instrument (GEONICS Ltd., 2014) that provides measurements of either the quad-phase (conductivity) or in-phase (magnetic susceptibility) component data, as selected by the operator, without contact with the soil. The instrument can be rotated to collect data in either the horizontal or vertical dipole mode and provides approximately depths of exploration of 1.5 and 0.75 m in the vertical and horizontal dipole modes, respectively. The main soil components capable of conducting electricity are ionic solutions and certain soil solids (soluble salts and clays). Hence the measured ECa is highly dependent on soil moisture, solution salinity and texture, bulk density, and soil temperature, among other factors. A review of soil properties influencing the ECa can be found in Friedman (2005).

A model of the electrical conductivity of mixed soil/water volumes was developed by Rhoades et al. (1999) idealizing the soil system as a two-pathway conductance model, highlighting the contribution to total electrical conductivity of solutions in large and small pores, as well as of the solid phase in the soil. A simplified formula for saline soils was derived by Rhoades et al. (1999) for relating the ECa measured by EMI to several soil conditions influencing the measurement.

$$\text{ECa} = \left(\frac{(\theta\text{s} + \theta\text{ws})^2 \cdot \text{ECs}}{\theta\text{s}} \right) + (\theta\text{w} - \theta\text{ws}) \cdot \text{ECwc} \qquad (2.1)$$

where ECa is the electrical conductivity of the bulk soil, ECs is the surface conductance of soil solids without indurated layers, ECwc is the specific electrical conductivity of the continuous conductance element, θs is the volumetric water content of the surface conductance of the soil, θws is the volumetric soil water content in small pores, and θw is the total volumetric soil water content. A detailed description of the physical principles and of the two-ways conductance model can be found in Lesch (2005) and Corwin and Lesch (2005b,c). This equation is used by the ESAP software (Lesch et al., 2000) to make a calibration of the EM measurements against measured soil data and converting EMv and EMh into estimated electrical conductivity of soil saturated paste extract (ECe) using data obtained from an EMI survey grid or transect. We used the version 2.35 of ESAP and made calibration of the EMI data for ECe prediction at depths 0−30,

30−60, and 60−90 cm, using data from six soil profiles. After calibration of the measurements of the EMI survey against ECe, several maps of soil salinity at different maps can be derived.

The data obtained from the soil salinity survey using the EM38 meter were analyzed with the PAST software (Hammer et al., 2001) for basic univariate statistics, checking the hypothesis of normal distribution of the variables, computing bivariate correlation and performing data comparison. The VESPER software (Minasny et al., 2005) was used for examining the data autocorrelation, computing the variograms of the EM38 raw and calibrated signal data, as well as for producing the kriged maps of predicted ECe, at different depths.

An intensive measurement with EMI was performed in two epochs in each of the six selected areas with plant inventories to obtain a mean value of ECa for each area and to put into evidence the variance of ECa values at a micro-scale. A set of 20 ECa measurements was obtained in May and in July, in points at distances ranging from 2 to 5 m, depending on the area extent to be covered. Soil temperature was recorded for the topsoil (0−30 cm) with a penetration thermometer, after thermal equilibration.

2.4 Results and Discussion

2.4.1 EMI Survey

The graphical exploratory analysis of the signals from the EMI device reveal that both signals are highly correlated, as presented in the biplot in Figure 2.6, with values of EMv>EMh in all cases, suggesting a "normal salinity profile." The histograms and normal probability plot suggest that both signals approach a normal statistical distribution. The univariate data of EMv and EMh are presented in Table 2.1 and the statistical tests of normality in Table 2.2, suggesting that the data of both measurements (EMv and EMh) may be distributed normally with 95% confidence.

Both signals present similar distribution of salinity values, with maximum values of ECa in the center of the lake and a gradual diminution of the ECa values towards the right side of the area that is two meters higher. The results are consistent with the conceptual model of downslope transport of salts by surface water and groundwater. Salinity is not homogeneous in the bottom part of the area and several "hotspots" of salinity can be identified.

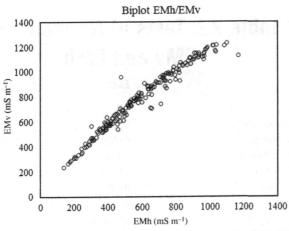

Figure 2.6 Biplot of the EMI signals of the general survey, in vertical (EMv) and horizontal (EMh) dipole orientation (units mS m^{-1}).

Table 2.1 Univariate Data of EMv and EMh

	EMv	EMh
N	179	179
Min	235	140
Max	1232	1170
Sum	140,538	106,837
Mean	785.1285	596.8547
Standard error	18.65326	17.54532
Variance	62,282	55,103.06
Standard deviation	249.5636	234.7404
Median	778	577
25th percentile	587	405
75th percentile	987	772
Skewness	−0.098293	0.2309455
Kurtosis	−0.8990609	−0.7990932
Geometric mean	740.4308	546.816
Coefficient of variation	31.78634	39.32957

Table 2.2 Tests of Normality for EMv and EMh

	EMv	EMh
N	179	179
Shapiro-Wilk W	0.9747	0.9788
p (normal)	0.00243	0.007787
Jarque-Bera JB	6.425	6.461
p (normal)	0.04025	0.03953
p (Monte Carlo)	0.0352	0.0406
Chi^2	6.8994	5.0223
p (normal)	0.0086223	0.025022
Chi^2 OK (N>20)	Yes	Yes
Anderson-Darling A	1.015	1.04
p (normal)	0.01106	0.009588

2.4.2 Vegetation Inventories and Relation with Salinity Gradients

Figure 2.7 presents the vegetation inventories, grouped by associations and a topographic profile indicating the form of relief as well as the topographic position in which each association appears. In the most saline soils (indicated as 5 in the scheme) the dominant association is *Cynomorio coccinae–Arthrocnemetum macrostachyi*, in the area of dunes formation (indicated as 3 in the scheme) dominate the association *Parapholi incurvae–Frankenietum pulverulentii* and in the less saline part (indicated as 1 in the scheme) dominates the association *Atriplici glaucae–Suaedetum verae*. The associations *Suaedo splendentis–Salicornietum ramosissimae* and *Cistancho luteae–Sarcoconietum fruticosii* (indicated in the scheme as 2 and 4, respectively) appear in areas of intermediate salinity with respect to 1 and 3–5.

2.4.3 Soil Properties

The soils of the study area consist of a surficial crust followed by a series of consecutive C layers without development of structure (massive), presenting clayed texture (by hand) and no clear horizon differentiation. Soluble salts, carbonates, and

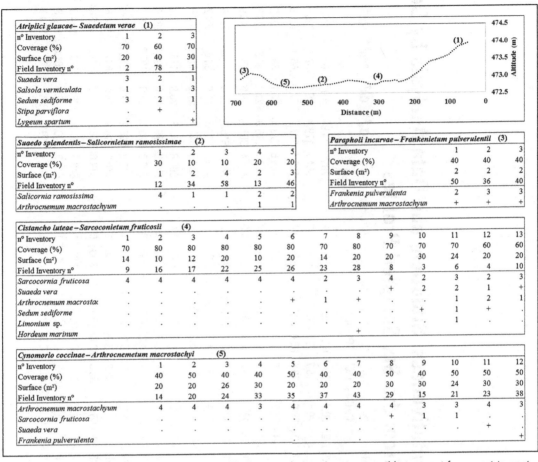

Atriplici glaucae– Suaedetum verae (1)

n° Inventory	1	2	3
Coverage (%)	70	60	70
Surface (m²)	20	40	30
Field Inventory n°	2	78	1
Suaeda vera	3	2	1
Salsola vermiculata	1	1	3
Sedum sediforme	3	2	1
Stipa parviflora	.	+	.
Lygeum spartum	-	.	+

Suaedo splendentis– Salicornietum ramosissimae (2)

n° Inventory	1	2	3	4	5
Coverage (%)	30	10	10	20	20
Surface (m²)	1	2	4	2	3
Field Inventory n°	12	34	58	13	46
Salicornia ramosissima	4	1	1	2	2
Arthrocnemum macrostachyum	.	.	.	1	1

Parapholi incurvae– Frankenietum pulverulentii (3)

n° Inventory	1	2	3
Coverage (%)	40	40	40
Surface (m²)	2	2	2
Field Inventory n°	50	36	40
Frankenia pulverulenta	2	3	3
Arthrocnemum macrostachyun	+	+	+

Cistancho luteae– Sarcoconietum fruticosii (4)

n° Inventory	1	2	3	4	5	6	7	8	9	10	11	12	13
Coverage (%)	70	80	80	80	80	80	70	80	70	70	70	60	60
Surface (m²)	14	10	12	20	10	20	14	20	20	30	24	20	20
Field Inventory n°	9	16	17	22	25	26	23	28	8	3	6	4	10
Sarcocornia fruticosa	4	4	4	4	4	4	2	3	4	2	3	2	3
Suaeda vera	.	.	.	.	.	.	.	.	+	2	2	1	+
Arthrocnemum macrostac	.	.	.	.	.	+	1	+	.	.	1	2	1
Sedum sediforme	.	.	.	.	.	.	.	.	.	+	1	+	.
Limonium sp.	.	.	.	.	.	.	.	.	.	.	1	.	.
Hordeum marinum	.	.	.	.	.	.	.	+	.	.	.	.	.

Cynomorio coccinae– Arthrocnemetum macrostachyi (5)

n° Inventory	1	2	3	4	5	6	7	8	9	10	11	12
Coverage (%)	40	50	40	40	50	40	40	50	40	50	50	50
Surface (m²)	20	20	26	30	20	20	20	30	30	24	30	30
Field Inventory n°	14	20	24	33	35	37	43	29	15	21	23	38
Arthrocnemum macrostachyum	4	4	4	3	4	4	4	4	3	3	4	3
Sarcocornia fruticosa	.	.	.	.	.	.	.	+	1	1	.	.
Suaeda vera	.	.	.	.	.	.	.	.	.	.	+	.
Frankenia pulverulenta	.	.	.	.	.	.	.	.	.	.	.	+

Figure 2.7 Vegetation inventories and position with the associations in a topographic transect from west to east.

gypsum are present in the entire profile. The data from soil samples are presented as Table 2.3, with indication of the vegetation area in which they were taken. Profiles PI and PII, situated in the most elevated part of the toposequence, are "normal saline profiles" with ECe increasing downwards as a consequence of salt leaching, with sulfates dominating over chlorides in solution. Profiles PIII, PIV, PV, and PVI, with similar concentrations of sulfates and chlorides in solution, are "inverted salinity profiles" due to there being no possibility of leaching, because of their lowest topographic position and the salt precipitation in the surface by capillary ascent of soil solutions. The levels of ECe at 0–30 cm depth, in which many plants have important root development, are so high that only halophytes

Table 2.3 Soil Representativeness, Situation, and Data Analysis by Depths from the Six Sampled Profiles

Soil Profile	Area	Centroid Position	Depth (cm)	Color	% $CaCO_3$	w (%)	% SP	ECe (dS/m)	pHe	Alkalinity	Cl^-	SO_4^{2-}	Ca^{2+}	Mg^{2+}	Na^+	K^+	SAR
PI	I	N38° 30.266' W0° 52.808'	0–30	10YR 5/2	30.2	3.7	6.9	2.65	7.07	1.8	28.2	74.9	5.0	6.0	83.4	2.7	35.6
			30–60	10YR 4/1	20.4	16.7	56.2	40.10	6.57	1.2	304.7	179.7	26.9	82.1	289.4	6.9	39.2
			60–90	10YR 5/1	29.9	23.4	85.4	39.80	6.29	1.4	293.4	287.7	17.0	66.0	421.6	8.4	65.4
PII	II	N38° 30.245' W0° 52.893'	0–30	10YR 6/1	35.2	12.0	30.0	25.00	7.00	1.1	298.6	323.5	23.1	39.7	298.2	2.0	53.2
			30–60	10YR 7/1	34.7	20.0	65.0	40.00	7.40	0.8	389.3	315.6	21.7	40.3	476.3	3.1	85.5
			60–90	10YR 7/1	43.2	25.0	80.0	50.00	6.80	1.0	415.4	301.1	19.6	65.1	548.9	3.9	84.3
PIII	III	N38° 30.226' W0° 52.942'	0–30	10YR 7/1	38.5	11.7	53.7	77.30	7.12	1.2	581.1	516.2	22.0	154.0	996.6	10.3	106.2
			30–60	10YR 7/1	52.4	23.4	74.2	57.30	7.21	1.4	406.2	510.3	21.0	96.0	897.2	9.7	117.3
			60–90	10YR 8/1	58.4	28.5	75.0	68.80	6.85	1.0	496.5	323.1	21.0	119.0	743.6	4.5	88.9
PIV	IV	N38° 30.221' W0° 53.010'	0–30	10YR 7/1	42.6	17.6	64.4	94.50	6.81	1.4	806.8	649.5	26.9	136.1	1379.2	9.2	152.8
			30–60	10YR 8/1	60.1	34.2	86.9	76.20	7.82	0.8	575.5	479.1	23.0	109.0	983.6	7.1	121.1
			60–90	10YR 7/1	56.4	35.7	84.0	83.00	7.31	0.8	603.7	323.1	23.0	107.0	936.5	6.4	116.2
PV	V	N38° 30.231' W0° 53.014'	0–30	10YR 7/1	31.6	18.8	60.9	98.60	8.09	1.0	846.3	516.2	22.0	203.0	1214.7	8.2	114.5
			30–60	10YR 7/1	48.8	35.2	78.6	60.30	7.00	1.0	440.1	510.3	18.0	83.0	934.6	6.8	131.5
			60–90	10YR 7/1	55.8	39.9	82.6	68.20	7.03	0.8	519.0	598.9	21.0	82.0	1261.4	9.1	175.8
PVI	VI	N38° 30.107' W0° 53.196'	0–30	10YR 7/1	46.4	26.6	67.4	83.70	7.31	1.0	648.8	602.3	28.9	130.1	984.3	9.3	110.4
			30–60	10YR 8/1	75.9	33.3	82.4	73.80	7.21	1.2	552.9	340.8	25.0	101.0	814.3	6.1	102.6
			60–90	10YR 8/1	53.5	40.4	86.3	88.70	7.23	1.2	722.1	309.6	28.9	120.1	890.1	4.2	103.1

can survive in such conditions. The only exception is the 0–30 cm layer of the profile PI, with low salinity. The SAR values are elevated but do not present a risk for soil structure stability because the values of pH are close to neutrality and the ECe is very high.

2.4.4   Maps of Predicted ECe (Calibrated)

The EMI signals were calibrated in reference to the measured soil data for depths of 0–30, 30–60, and 60–90 cm, using a stochastic model based on the dual-pathway conductivity model, giving predicted values of ECe at every point where EMI measurements were taken. After normalizing each population of data (by depth), experimental variograms were computed and spherical functions were fitted (Table 2.4).

Maps of ECe and their respective maps of error of prediction were produced by ordinary kriging (Figure 2.8). The ECe pattern shows minimum values in the more elevated parts of the area and maximum values in the bottom part, with hotspots of salinity.

2.4.5   Spatial Variation of Soil Salinity ECa at the Micro-Scale

Two intensive EMI surveys (more than 20 measurements per area) were performed in the six selected areas during the dates of May 7, 2014 and July 24, 2014. Measurements were interpolated by inverse distance algorithm for producing the maps and are shown in Figure 2.9 (data from May). ECa showed microvariations at the metric scale within an apparently uniform area. When sampling soils, a particular point must be selected for augering or pit opening. However, there is the risk of introducing a bias in the ECe values, along with the inconvenience

Table 2.4 Variogram Parameters for Predicted ECe at Different Depths

Variograms	Model	Nugget	Sill	Range	RMSE
ECe 0–30 cm	Spherical	280.7	639.3	409.2	102.3
ECe 30–60 cm	Spherical	141.5	221.9	329.5	34.8
ECe 60–90 cm	Spherical	221.6	422.3	332.2	66.3

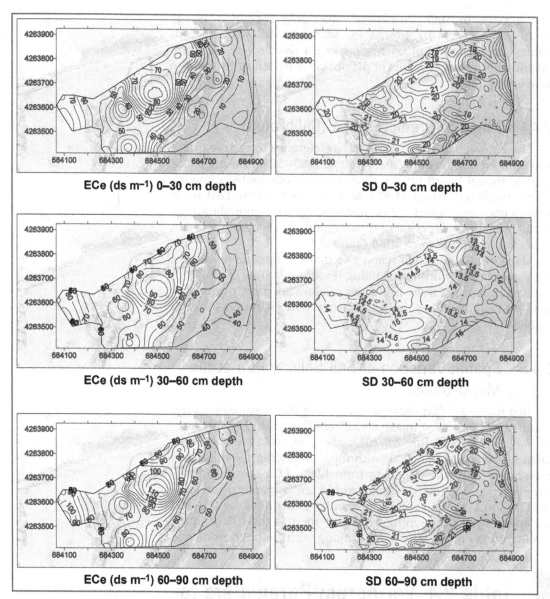

Figure 2.8 Maps of kriged predicted ECe for depths of 0–30, 30–60, and 60–90 cm (units dS m^{-1}) with the corresponding maps of error for predicted values for each depth mapped (in standard deviation units).

of the small volume of samples used for analysis as well as the long time needed for sampling. Soil sampling should be guided by a prior knowledge of ECa variations within the area of interest, measured by EMI in a quick, cheap and non-destructive way.

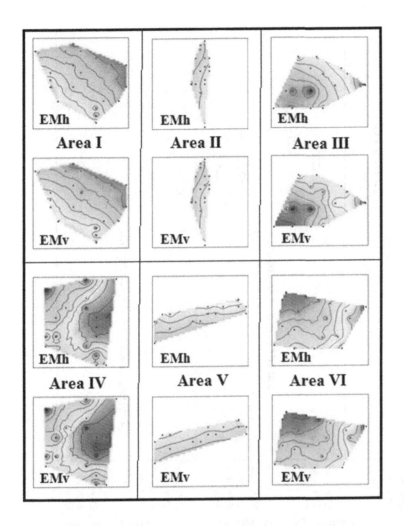

Figure 2.9 Detailed maps of interpolated EMI signals (EMv and EMh) for the six selected monospecific inventories.

2.4.6 Multi-Temporal EMI Measurement in Selected Vegetation Inventories

The EMI measurements made in each area in May and July of 2014 are grouped in Figure 2.10 as "box and whisker" diagrams, with indication of the plant species present. It is evident that the different plants appear in a certain range of ECa values, without overlapping and with clear discrimination of the mean ECa value for each species. For each plant the mean values of May and July differ slightly, with the highest values in July. The test of means equality supports that the means are different. The quotients of means in July and May

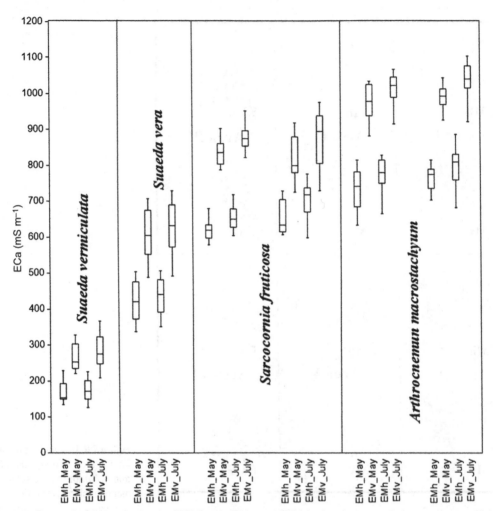

Figure 2.10 Box and whisker plot for the EMI signals (EMv and EMh) in the six areas of plant mono-specificity, for two dates (May and July, 2014).

in each of the selected vegetation areas (Table 2.5) give results ranging from 1.026 to 1.064, with inter-areas averaged values of 1.050 and 1.048. Those quotients, very close in the ECa of the bottom and upper part of the soil, suggest that the difference of the means is due to the different soil temperature in May (average 19.8°C) and July (average 24.0°C), rather than to saline soil solution movements, because those increments in the averaged ECa for EMv and EMh do not mark a

Table 2.5 Comparison of Signals for Two Dates

Quotients	Area I	Area II	Area III	Area IV	Area V	Area VI	Interareas Average	CV
EMh_July/EMh_May	1.026	1.041	1.058	1.074	1.051	1.048	1.050	1.539
EMv_July/EMv_May	1.057	1.033	1.051	1.064	1.037	1.047	1.048	1.123

difference between top and bottom soil. The deeper part of the soils shows an inertia to temperature changes, tending to equilibrate at depths >50 cm with the monthly average temperature with a time delay, whereas the soil temperature at the surface changes through the day. Due to the composite nature of the soil, with materials of different thermal capacity and heath conductivity, soil atmosphere, water, soluble, and insoluble solids, is it difficult to find a correction factor of ECa for different temperatures (Friedman, 2005), as exists for solutions (Richards, 1954). However, the quotients are not far from the value of 1.095 obtained by dividing the correction factor applied for reporting EC of solutions at 19.8°C (factor 1.117) and 24.0°C (factor 1.020) to the reference temperature of 25°C.

2.5 Conclusions

The use of EMI for discriminating soil salinity gradients and their correspondence with zonation of halophytes can efficiently guide subsequent soil sampling, providing ECa information for higher soil volume support than the samples, with very high spatial and temporal resolution. EMI signals calibrated against soil profile ECe were used for producing maps of soil salinity for the study area, at different depths, allowing us to establish the correspondence of the plants present and the predicted soil salinity. The soil salinity conditions for several halophyte species have been discriminated: areas monospecific for *S. vermiculata*, *S. vera*, *S. fruticosa*, or *A. macrostachyum* were discriminated by EMI according to increasing values of ECa and ECe, providing the range of salinity variation and showing the spatial variability of soil salinity inside the plant community.

References

Akramkhanov, A., Brus, D.J., Walvoort, D.J.J., 2014. Geostatistical monitoring of soil salinity in Uzbekistan by repeated EMI surveys. Geoderma. 213, 600–607.

Amakor, X.N., Jacobson, A.R., Cardon, G.E., Hawks, A., 2014. A comparison of salinity measurement methods based on soil saturated pastes. Geoderma. 219–220, 32–39.

Arriola-Morales, J., Batlle-Sales, J., Valera, M.A., Linares, G., Acevedo, O., 2009. Spatial variability analysis of soil salinity and alkalinity in an endorreic volcanic watershed. Int. J. Ecol. Dev. 14, 17.

Arroyo-Ilera, R., 1976. La laguna de Salinas (Alicante) y su desecación. Cuadernos de Geografía. 18, 37–48.

Batlle, J., Abad, A., Bordás, V., Pepiol, E., 1994. Soil transformations in salt-stressed lagoon ecosystem. Presented at the 15th World Congress of Soil Science, Acapulco. México, pp. 262–277.

Batlle-Sales, J., Hurtado, A., Batlle-Montero, E., 2000. Cartografía quasi-tridimensional de cambios multitemporales en la salinidad del suelo, mediante la medida del campo electromagnético inducido y geoestadística. In: Quintero-Lizaola, R., Reyna-Trujillo, T., Corlay-Chee, L., Ibáñez-Huerta, A., García-Calderón, N.E. (Eds.), La Edafología Y Sus Perspectivas Al Siglo XXI. Tomo II. Colegio de Postgraduados. Universidad Nacional Autonoma de México, Universidad Autónoma Chapingo México, Mexico, pp. 667–677.

Braun-Blanquet, J., 1964. Pflanzensoziologie, Grundzüge der Vegetationskunde. third ed. Springer, Wien.

Corwin, D.L., Lesch, S.M., 2005a. Characterizing soil spatial variability with apparent soil electrical conductivity: I. Survey protocols. Comput. Electron. Agric. 46, 103–133.

Corwin, D.L., Lesch, S.M., 2005b. Apparent soil electrical conductivity measurements in agriculture. Comput. Electron. Agric. 46, 11–43.

Corwin, D.L., Lesch, S.M., 2005c. Characterizing soil spatial variability with apparent soil electrical conductivity: Part II. Case study. Comput. Electron. Agric. 46, 135–152.

Doolittle, J.A., Brevik, E.C., 2014. The use of electromagnetic induction techniques in soils studies. Geoderma. 223–225, 33–45.

Farifteh, J., Farshad, A., George, R.J., 2006. Assessing salt-affected soils using remote sensing, solute modelling, and geophysics. Geoderma. 130, 191–206.

Friedman, S.P., 2005. Soil properties influencing apparent electrical conductivity: a review. Comput. Electron. Agric. 46, 45–70, GEONICS Ltd., 2014. EM38.

Hammer, Ø., Harper, D.A.T., Ryan, P.D., 2001. PAST: paleontological statistics software package for education and data analysis. Palaeontol. Electron. 4, 9.

He, B., Cai, Y., Ran, W., Jiang, H., 2014. Spatial and seasonal variations of soil salinity following vegetation restoration in coastal saline land in eastern China. CATENA. 118, 147–153.

Herrero, J., Netthisinghe, A., Hudnall, W.H., Pérez-Coveta, O., 2011. Electromagnetic induction as a basis for soil salinity monitoring within a Mediterranean irrigation district. J. Hydrol. 405, 427–438.

ICV, 2009. Lidar MDE 1mt 2009 (Sistema de referencia ETRS89H30. Formato ERS) (Digital Elevation Model).

ICV, 2012. Ortofoto 1:5.000 Comunidad Valenciana año 2012 (Sistema de referencia ETRS89H30. Formato ECW) (Aerial photograph).

IGME, n.d. Cartografia 1:50.000 MAGNA 50 (2a serie). Hoja 845 YECLA. (Map).

IGN, 1956. Vuelo Americano 1956–1957. (Aerial photograph).

Lesch, S.M., 2005. Sensor-directed response surface sampling designs for characterizing spatial variation in soil properties. Comput. Electron. Agric. 46, 153–179.

Lesch, S.M., Rhoades, J.D., Corwin, D., 2000. The ESAP Version 2.01r user manual and tutorial guide. Research Report #146. George E. Brown, Jr., Salinity Laboratory, Riverside, CA.

Minasny, B., McBratney, A.B., Whelan, B.M., 2005. VESPER version 1.62. Australian centre for precision agriculture, McMillan Building A05, The University of Sydney, NSW 2006.

Pepiol, E., Batlle-Sales, J., Bordás, V., 1998. Geostatistic study of salt distribution in "Laguna de Salinas", Alicante, Spain. Proceedings of GeoENV II-Geostatistics of Environmental Applications 10. Kluwer Academic Publishers, The Netherlands, pp. 441–452.

Peris, J.B., Stübing, G., Cirujano, S., Medina, L., 1999. Estudio de la vegetación de los PP.NN. del Sur de Alicante. Real Jardín Botánico de Madrid. CSIC.

Quantum GIS, 2011. Development team, 2012. Quantum GIS Geographic Information System. Open Source Geospatial Foundation Project.

Rhoades, J.D., Chanduvi, F., Lesch, S., 1999. Soil salinity assessment. Methods and interpretation of electrical conductivity measurements, FAO Irrigation and drainage paper. Food and Agriculture Organization of the United Nations, Rome.

Richards, L.A., 1954. Diagnosis and improvement of saline and alkali soils. In: Richards, L.A., et al., (Eds.), Agriculture Handbook. United States Salinity Laboratory Staff, U.S. Department of Agriculture, Washington, DC.

Rigual, A., 1972. Flora y vegetación de la provincia di Alicante. nte. Inst. Estud. Alicantinos. Alicante.

Serra Laliga, L., 2007. Estudio crítico de la flora vascular de la provincia de Alicante: aspectos nomenclaturales, biogeográficos y de conservación. Monografías del Real Jardín Botánico de Madrid. CSIC.

Soil Survey Staff, 2014a. Keys to Soil Taxonomy. Twelfth ed. USDA-Natural Resources Conservation Service, Washington, DC.

Soil Survey Staff, 2014b. Kellogg Soil Survey Laboratory Methods Manual. Soil Survey Investigations Report No. 42, Version 5.0. U.S. Department of Agriculture, Natural Resources Conservation Service.

NUTRITIONAL VALUE OF *CHENOPODIUM QUINOA* SEEDS OBTAINED FROM AN OPEN FIELD CULTURE UNDER SALINE CONDITIONS

Meryem Brakez[1], Salma Daoud[1], Moulay Chérif Harrouni[2], Naima Tachbibi[1] and Zahra Brakez[3]

[1]*Laboratory of Plant Biotechnologies, Faculty of Sciences, Ibn Zohr University, Agadir, Morocco* [2]*Hassan II Agronomic and Veterinary Institute, Agadir, Morocco* [3]*Laboratory of Cell Biology & Molecular Genetics, Faculty of Sciences, Ibn Zohr University, Agadir, Morocco*

3.1 Introduction

The degradation of soil and water by salinization is a worldwide problem. The loss of soil and water quality, the explosion of consumption, and fluctuations in food prices are great enemies of food security.

According to the United Nations (UN), the world now has a population of 7 billion people, nearly a billion more than in 2000 (6.1 billion). Beyond the purely demographic aspect, this number encourages researchers to think about an essential issue for the future of humanity: food and food security.

Morocco, a Mediterranean country with mainly arid and semiarid climates, has gone through years of drought and still continues to suffer from it (Anonymous, 2009). The country has conventional water resources that are unequally distributed in space and time and that do not meet the needs of local populations, especially for agriculture (Anonymous, 2006). Furthermore, more land is affected by salinization due to climate change and lack of good-quality water (Anonymous, 2008).

M.A. Khan, M. Ozturk, B. Gul, & M.Z. Ahmed (Eds): Halophytes for Food Security in Dry Lands.
DOI: http://dx.doi.org/10.1016/B978-0-12-801854-5.00003-0

The combination of these factors constitutes the major cause for crop losses in Morocco (Daoud, 2004). Nationally, about 160,000 ha of land are affected by salinity (Ftouhi, 1981; Badraoui and Debbarh, 2003). Covering an area of nearly 8.7 million hectares, cereals are by far the main crops in the agricultural production system in Morocco. However, studies conducted on the most consumed cereals in Morocco (barley, wheat, and durum wheat and triticale) showed that water salinity in the range of $0.74-11.69$ mS cm^{-1} caused a reduction in the yield potential of these cereals estimated at about 30–50% (Lahlou, 1999). Thus, the most obvious solution, to insure food security and rehabilitation of land and water degraded by salinity, is to use plants that can grow and produce under saline conditions. The pseudo-cereal and facultative halophyte *Chenopodium quinoa* is among the halophyte species with agronomic and nutritional value that could match or even exceed the cultivation of cereals around the world, including Morocco.

Chenopodium quinoa, commonly known as quinoa, is a plant native to the Andes, mainly grown for human consumption. Quinoa has great plasticity and flexibility. It can grow from sea level to 4000 m altitude in extremely diverse conditions (drought, salinity, cold, wind) (Jacobsen, 2003; Jacobsen et al., 2007). Quinoa is a highly salt-tolerant species. Some varieties can tolerate concentrations similar to those of seawater (about 500 mM NaCl; Adolf et al., 2010). Quinoa seeds, under nonsaline conditions, appear to be an important and complete food thanks to their high protein content (11–20%) as well as their amino acid composition and quality. Seeds are also rich in minerals, especially potassium and iron (Abugoch, 2009). In addition to the high protein content, quinoa seeds exhibit an interesting lipid content ranging between 5.5% and 8.5% (Koziol, 1992; Wright et al., 2002; De Bruin, 1963; Dini et al., 1992). The fatty acid composition is mostly represented by oleic and linoleic acids (Jahaniaval et al., 2000). Thanks to the high nutritional value of quinoa seeds, the UN general assembly declared 2013 the "International Year of Quinoa" to pay tribute to the traditional practices of the Andean people, who have managed to preserve quinoa to feed the present and future generations.

Currently, we witness a growing interest in quinoa seeds as sources of food and for the quinoa plant as fodder in the United States and Europe. Given this importance, quinoa is well positioned to be grown in Morocco as well.

The aim of this study was to evaluate the yield and the qualitative aspect of *C. quinoa* seeds obtained under saline conditions in an open field.

3.2 Materials and Methods

3.2.1 Site Experiment

The experiment was carried out in an open field during Fall 2011–2012 at the botanical garden at the Hassan II Institute of Agronomy and Veterinary Medicine (IAV) in Agadir, Morocco. The IAV is located in the Souss area in south-western Morocco. The plain of Souss is bounded in the north by the High Atlas Mountains, in the west by the Atlantic Ocean and in the south by the Anti-Atlas Mountains. Land is relatively flat, and soils are loamy to sandy. Agriculture relies partly on surface water collected and stored in dams, but mainly on underground water pumped from the aquifer with a subsequent intrusion of seawater along the coast. The climate is arid but moderated by the maritime influence.

3.2.2 Experimental Design and Irrigation Protocol

The study was conducted using the Hualhuas cultivar of quinoa provided by Prof. Koyro of the Justus Liebig Institute of Plant Ecology, Giessen, Germany. The cultivated area was 110 m^2. The soil was prepared to contain eight lines of 11 m in length, in the form of ridges, on which quinoa seeds were sown. The physical and chemical soil characteristics for 0.0–0.4-m soil layer are reported in Table 3.1. Two treatments were performed: 0 mM NaCl as a control and 125 mM NaCl which gave the maximum growth of quinoa in controlled conditions (Jacobsen, 2003; Koyro et al., 2008; Hariadi et al., 2011). Control was tapwater and the saline solution was prepared by adding NaCl to tapwater in a tank of 1 m^3. The irrigation was supplied by using the drip irrigation system. During the experiment, temperatures ranged from 13.6°C (mean minimum temperature) to 23.9°C (mean maximum temperature), and relative humidity ranged between 46% and 88%.

At physiological maturity, the crop was hand-harvested. The plants were cut at the base and laid out on a tarpaulin sheet. The panicles, separated from the rest of the plant, were dried and then the dried seeds were removed. Seed yield was determined on three plants per line. Biochemical parameters

Table 3.1 Physical and Chemical Characteristics of the Soil Layer 0.0–0.40 m at the Beginning of the Experiment

Depth (m)	0.0–40
Physical Characteristics	
Coarse sand (%)	17.86
Fine sand (%)	35.61
Coarse silt (%)	20.18
Fine silt (%)	14.55
Clay (%)	11.8
Total limestone (%)	21.9
Active limestone (%)	7.8
Chemical Characteristics	
EC (mS m^{-1})	0.11
K$^+$ (ppm)	194.92
Na$^+$ (ppm)	18.17
Ca^{2+} (ppm)	2405.26
Mg^{2+} (ppm)	338.22
pH	8.9

EC, electrical conductivity at 25°C.

were determined from six plants harvested from each treatment and each line. The protein content determination was performed by the method of Lowry et al. (1951). The carbohydrate and starch contents have been determined by the methods described by Dubois et al. (1956) and Allefrey and Northcote (1977) (cited by Ben Dkhil and Denden, 2010), respectively. Lipids were extracted with hexane in Soxhlet and their composition was determined using GPC. Polyphenol content was performed according to the method described by Shetty and Kwok (1995) and modified by Ben Dkhil and Denden (2010).

Flavonoid content was determined according to Miliauskas et al. (2004). The content of vitamin C was estimated using the method cited by Borah et al (2011). Mineral composition was determined by the method of Munter (1980). K$^+$ and Na$^+$

content were determined with a flame photometer 410 (FLM3b Radiometer Copenhagen) and those of Ca^{2+}, Mg^{2+}, Fe, Cu, Zn, and Mn were measured using an atomic absorption spectrophotometer (Perkin Elmer 3110). Mineral content was calculated on a dry weight basis.

All results were analyzed with STATISTICA 6. The methods used in this study are the analysis of variance (ANOVA) with one factor and LSD test to determine significant differences ($P < 0.05$) between treatments.

3.3 Results and Discussion

3.3.1 Yield Data

The yield of quinoa seeds per plant was increased in 125 mM NaCl treatment compared to the control treatment without salinity. However, there was no significant difference for seed yield per plant among the two treatments (Figure 3.1). This result is similar to that obtained by Pulvento et al. (2012) where they found no significant difference in the yield of quinoa variety "Titicaca" subjected to $22\ mS\ m^{-1}$ (about 200 mM NaCl) in an open field experiment in southern Italy.

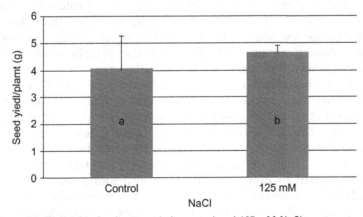

Figure 3.1 The yield of quinoa seeds in control and 125 mM NaCl treatments. Bars with the same letters do not differ significantly at $P < 0.05$.

3.3.2 Composition

The biochemical analysis of quinoa seeds shows that 125 mM NaCl treatment induced higher and/or similar values

Table 3.2 Biochemical Composition of Quinoa Seeds Obtained from Nonsaline and Saline Conditions

	Control	125 mM NaCl
Protein (%)	21.63 ± 8.66	23.88 ± 2.29
Starch (%)	46.28 ± 0.43	49.02 ± 1.04
Carbohydrates (%)	47.12 ± 0.01	44.98 ± 0.25
Vitamin C (mg g^{-1})	0.03 ± 0.002	0.03 ± 0.001
Polyphenols (mg g^{-1})	1.22 ± 0.07	1.12 ± 0.04
Flavonoids (mg g^{-1})	0.12 ± 0.06	0.13 ± 0.04
Lipids (%)	6.57 ± 1.32	7.98 ± 0.66

for the analyzed elements. However, there was no significant difference in the composition of the seed samples between quinoa grown under saline and nonsaline irrigation (Table 3.2).

We noted that the protein content of quinoa seeds ranged from 21.6% to 23.8% of dry weight, respectively, in control and 125 mM NaCl treatments (Table 3.2). Koziol (1992) and Schilk and Budenheim (1996) reported that, generally, protein content in quinoa seeds (11–22% of dry weight) is higher than those reported for cereals (7–13%). Pulvento et al. (2012) reported that saline treatment did not affect the protein content of "Titicaca" seeds compared to the nonsaline one.

The carbohydrate content ranged from 44.9% to 47.1% in nonsaline and saline treatments, respectively (Table 3.2). Pulvento et al. (2012) showed a content of 54–57% of carbohydrates in "Titicaca" seeds with no significant difference between the saline and nonsaline treatments. In controlled conditions, Koyro et al. (2008) obtained between 58.4% and 68.9% carbohydrate content in quinoa seeds. They also found that 100 and 200 mM NaCl treatments did not affect carbohydrate content compared to the control (0 mM NaCl). Starch content in quinoa seeds ranged from 46% to 49% of dry weight (Table 3.2). Starch is the most abundant carbohydrate in quinoa seeds; it typically represents about 32–69.2% of dry weight (Ando et al., 2002; Lindeboom et al., 2004; Wright et al., 2002; De Bruin, 1964; Ruales, 1992; Schlick, 2000).

Quinoa seeds appear to be a good source of vitamin C compared to wheat, rice, and barley, which do not contain vitamin C (Nikkhah, 2012). In this study, vitamin C content in quinoa seeds was about 0.03 mg g^{-1} (3 mg 100 g^{-1}) both in control and 125 mM NaCl treatments (Table 3.2). Koziol (1992) reported a value of 4 mg 100 g^{-1} of vitamin C in quinoa seeds obtained from the field in nonsaline conditions. However, in hydroponic conditions, vitamin C was not detected (Schlick, 2000).

The polyphenol content of quinoa seeds ranged between 1.12 and 1.22 mg g^{-1} (112 and 122 mg 100 g^{-1}) (Table 3.2). This is similar to the levels of polyphenols reported for durum and wheat (137 mg 100 mg^{-1} and 134 mg 100 g^{-1}, respectively) (Repo-Carrasco et al., 2010) and quite large compared to other grains such as barley, corn, millet, and oats, where the polyphenol content ranged between 25 and 60 mg 100 g^{-1} (Mattila et al., 2005). In nonsaline conditions, quinoa seeds exhibited amounts of polyphenols ranging from 30.3 to 530 mg 100 g^{-1} (Valencia-Chamorro, 2003; Repo-Carrasco et al., 2010; Alvarez-Jubete et al., 2010; Chlopicka et al., 2012).

Flavonoid content was 0.12 mg g^{-1} in control compared to 0.13 mg g^{-1} in 125 mM NaCl, with no significant differences between treatments (Table 3.2). In nonsaline conditions, Chlopicka et al. (2012) reported a flavonoid content of 92 μg g^{-1} in quinoa seeds, and Repo-Carrasco et al. (2010) reported levels of flavonoids between 0.36 and 0.73 mg g^{-1} in the seeds of different varieties of quinoa.

The lipid content in quinoa seeds increased from 6.57% in control to 7.98% in 125 mM NaCl treatment, however there was no significant difference between treatments (Table 3.2). Under natural growth conditions (without salinity) levels between 5.5% and 8.5% of lipid in quinoa seeds have been reported (Koziol, 1992; Wright et al., 2002; De Bruin, 1963; Dini et al., 1992). Pulvento et al. (2012) found that salinity conditions had no effect on the lipid content of "Titicaca" seeds, which they estimated to range between 5.7% and 5.8%.

Mineral analysis showed that potassium, iron, and manganese are the most abundant ions in quinoa seeds (Figure 3.2). Mineral content in quinoa is generally higher than that reported for most cereal crops (De Bruin, 1963; Koziol, 1992; Repo-Carrasco et al., 2003; Dini et al., 2005; Vega-Galvez et al., 2010).

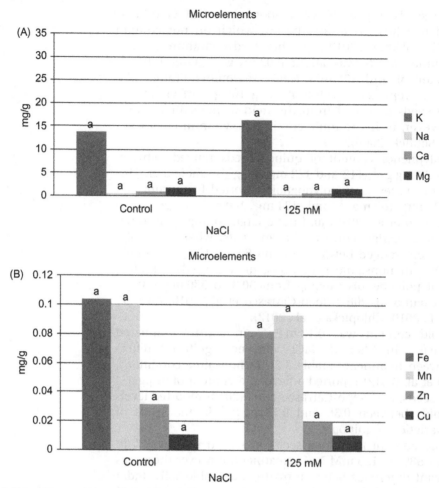

Figure 3.2 Salinity effect on (A) macroelements and (B) microelements contents (mg g^{-1} dry weight) in quinoa seeds. Bars with the same letters are not significantly different at $P < 0.05$.

3.4 Conclusion

Our study shows that quinoa seeds are rich in essential nutritive components (proteins, sugars, fatty acids, minerals, and antioxidants), which are of great benefit to human health. The study also shows that treatment of quinoa in the field with 125 mM NaCl improves the nutritional value of seeds. Therefore, quinoa appears to be a particularly interesting option for human consumption in salt-affected lands.

Given its economic quality and its tolerance to salinity and varying climate contexts further studies are needed to elucidate the effect of salt stress on the nutritional quality of seeds of other quinoa varieties to identify the best variety to cultivate.

References

Abugoch, L.E., 2009. Quinoa (*Chenopodium quinoa* Willd.): composition, chemistry, nutritional, and functional properties. Adv. Food Nutr. Res. 58, 1–31.

Adolf, V.I., Jacobsen, S.E., Shabala, S., 2010. Salt tolerance mechanisms in quinoa (*Chenopodium quinoa* Willd). Environ. Exp. Bot. 92, 43–54.

Alvarez-Jubete, L., Wijngaard, H., Arendt, E.K., Gallagher, E., 2010. Polyphenol composition and *in vitro* antioxidant activity of amaranth, quinoa buckwheat and wheat as affected by sprouting and baking. Food Chem. 119, 770–778.

Allefrey, J.M, Northcote, D.H., 1977. The effects of the axis and plant hormones on the mobilization of storage materials in the groundnut (*Arachis hypogaea*) during germination. New Phytol. 78.

Ando, H., Chen, Y., Tang, H., Shimizu, M., Watanabe, K., Miysunaga, T., 2002. Food components in fractions of quinoa seed. Food Sci. Technol. Res. 8 (1), 80–84.

Anonymous, 2006. Plate-forme du débat national sur l'eau, MATEE, Novembre (2006), pp. 48.

Anonymous, 2008. Ministère de l'Energie, des Mines, de l'Eau et de l'Environnement; Secrétariat d'Etat chargé de l'Eau et de l'Environnement (MEMEE—SEEE). Eaux souterraines saumâtres au Maroc, potentialités en tant que ressources alternatives, Direction de la recherche et de la planification de l'eau, pp. 21.

Anonymous, 2009. Plan d'Action National de Lutte Contre la Désertification Agadir: La lutte contre l'avance du désert dans cinq préfectures de la Région Souss Massa Drâa, Comment s'y prendre? pp. 12.

Badraoui, M., Debbarh, A., 2003. Irrigation et environnement au Maroc: situation actuelle et perspectives. In: Marlet, S., Ruelle, P. (Eds.), Vers une Maitrise Des Impacts Environnementaux de l'Irrigation. Cederom du CIRAD, Montpellier, France.

Ben Dkhil, B., Denden, M., 2010. Salt stress induced changes in germination, sugars, starch and enzyme of carbohydrate metabolism in *Abelmoschus esculentus* (L.) Moench seeds. Afr. J. Agric. Res. 5 (6), 408–415.

Borah, A., Yadav, R.N.S., Unni, B.G., 2011. Evaluation of antioxidant activity of different solvent extracts of *Oxalis corniculata* L. J. Pharm. Res. 5 (1), 91–93.

Chlopicka, J., Pasko, P., Gorinstein, S., Jedryas, A., Zagrodzki, P., 2012. Total phenolic and total flavonoid content, antioxidant activity and sensory evaluation of pseudocereal breads. LWT Food Sci. Technol. 46, 548–555.

Daoud, S., 2004. Effet de l'irrigation a l'eau de mer sur les paramètres agronomiques et physiologiques de quelques espèces halophytes. Thèse de doctorat présentée à la Faculté des Sciences, Université Ibn Tofail, Kenitra, Maroc.

De Bruin, A., 1963. Investigation of the food value of quinoa and canihua seed. J. Food Sci. 29, 872–876.

De Bruin, A., 1964. Investigation of the food value of quinoa and canihua seed. J. Food Sci. 29, 872–876.

Dini, A., Rastrelli, L., Saturnino, P., Schettino, O., 1992. A compositional study of *Chenopodium quinoa* seeds. Nahrung. 36, 400–404.

Dini, I., Tenore, G.D., Dini, A., 2005. Nutritional and antinutritional composition of Kancolla seeds: an interesting and underexploited andine food plant. Food Chem. 92, 125–132.

Dubois, M., Gilles, K.A., Hamilton, J.K., Rebers, P.A., Smith, F., 1956. Colorimetric method for determination of sugars and related substances. Anal. Chem. 38, 350–356.

Ftouhi, A., 1981. Les sols salés dans les ORMVA. Mémoire de fin d'étude ENA, Meknès.

Hariadi, Y., Marandon, K., Tian, Y., Jacobsen, S.-E., Shabala, S., 2011. Ionic and osmotic relations in quinoa (*Chenopodium quinoa* Willd) plants grown at various salinity levels. J. Exp. Bot. 62, 185–193.

Jacobsen, S.-E., 2003. The worldwide potential for quinoa (*Chenopodium quinoa* Willd.). Food Rev. Int. 19 (1–2), 167–177.

Jacobsen, S.E., Monteros, C., Corcuera, L.J., Bravo, L.A., Christiansen, J.L., Mujica, A., 2007. Frost resistance mechanisms in quinoa (*Chenopodium quinoa* Willd.). Eur. J. Agron. 26, 471–475.

Jahaniaval, F., Kakuda, Y., Marcone, M.F., 2000. Fatty acid and triacylglycerol compositions of seed oils of five amaranthus accessions and their comparison to other oils. J. Am. Oil Chem. Soc. 77 (8), 847–852.

Koyro, H.-W., Eisa, S.S., Leith, H., 2008. Influence of salinity on biomass production, yield, composition of reserves in the seeds, water and solute relations. In: Lieth, H., et al., (Eds.), Mangroves and Halophytes: Restoration and Utilisation, vol. 43. Springer, Dordrecht, pp. 133–145.

Koziol, M., 1992. Chemical composition and nutrtional evaluation of quinoa (*Chenopodium quinoa* Willd.). J. Food Compost. Anal. 5, 35–68.

Lahlou, A., 1999. La production de récoltes au moyen de l'exploitation des eaux saumâtres. Publication de l' organisation islamique pour l'éducation, les sciences et la culture 3–62.

Lindeboom, N., Chang, P., Tyler, R., 2004. Analytical, biochemical and physicochemical aspects of starch granule size, with emphasis on small granule starches: a review. Starch/Starke. 56, 89–99.

Lowry, O.H., Rosenbrough, N.J., Farr, A.L., Randall, R.J., 1951. Protein measurement with the folin phenol reagent. J. Biol. Chem. 193, 265–275.

Mattila, P., Pihlava, J.M., Hellstrom, J., 2005. Contents of phenolic acids, alkyland alkenylresorcinols, and avenanthramides in commercial grain products. J. Agric. Food Chem. 53, 8290–8295.

Miliauskas, G., Venskutonis, P.R., Van-beek, T.A., 2004. Screening of radical scavenging activity of some medicinal and aromatic plant extracts. Food Chem. 85, 231–237.

Munter, C.R., 1980. Preparation of Plant Material. Research Analytical Laboratory. University of Minnesota, St paul, MN.

Nikkhah, A., 2012. Barley grain for ruminants: a global treasure or tragedy. J. Anim. Sci. Biotechnol. 3, 22.

Pulvento, C., Riccardi, M., Lavini, A., Lafelice, G., Marconi, E., d'Andria, R., 2012. Yield and quality characteristics of quinoa grown in open field under different saline and non saline irrigation regimes. J. Agron. Crop Sci. 198, 254–263.

Repo-Carrasco, R., Espinosa, C., Jacobson, S.-E., 2003. Nutritional value and use of the Andean crops quinoa (*Chenopodium quinoa*) and Kaniwa (*Chenopodium pallidicaule*). Food Rev. Int. 19 (1–2), 179–189.

Repo-Carrasco, R.V., Hellstrom, J.K., Pihlava, J.-M., Mattila, P.H., 2010. Flavonoids and other phenolic compounds in Andean indigenous grains: quinoa (*Chenopodium quinoa*), kaniwa (*Chenopodium pallidicaule*) and kiwicha (*Amaranthus caudatus*). Food Chem. 120, 128–133.

Ruales, J., 1992. Development of an Infant Food from Quinoa (*Chenopodium quinoa*, Willd): Technological Aspects and Nutritional Consequences (Ph.D. thesis). University of Lund, Lund, Sweden.

Schlick, G., Bubenheim, D.L., 1996. Quinoa: candidate crop for NASA's controlled ecological life support systems. In: Janick, J. (Ed.), Progress in New Crops. ASHS Press, Arlington, VA, pp. 632–640.

Schlick, G., 2000. Nutritional Characteristics and Biomass Production of *Chenopodium quinoa* Grown in Controlled Environments (Master's thesis). Paper 2112.

Shetty, K., Curtis, O.F., Levin, R.E., Withowsky, R., Ang, W., 1995. Prevention of vitrification associated with *in vitro* shoot cultures of oregano (*Origanum vulgare*) by Pseudomonas spp. J. Plant. Physiol. 147, 447–451.

Valencia-Chamorro, S.A., 2003. Quinoa. In: Caballero, B. (Ed.), Encyclopedia of Food Science and Nutrition, vol. 8. Academic Press, Amsterdam, MA, pp. 4895–4902.

Vega-Galvez, A., Miranda, M., Vergara, J., Uribe, E., Puenteb, L., Martınezc, E.A., 2010. Nutrition facts and functional potential of quinoa (*Chenopodium quinoa* willd.), an ancient Andean grain: a review. J. Sci. Food Agric.1–7.

Wright, K., Pike, O., Fairbanks, D., Huber, C., 2002. Composition of *Atriplex hortensis*, sweet and bitter *Chenopodium quinoa* seeds. J. Food Sci. 67 (4), 1380–1383.

HALOPHYTES AND SALINE VEGETATION OF AFGHANISTAN, A POTENTIAL RICH SOURCE FOR PEOPLE

Siegmar-W. Breckle

Department of Ecology, University of Bielefeld, Bielefeld, Germany

4.1 Introduction

At least 6% of the world's land surface area is salt-affected land. This amounts to about $9.5 \times 10^6 \, km^2$ according to UNEP figures (Flowers and Yeo, 1995) or $8.0 \times 10^6 \, km^2$ according to FAO data (Munns, 2005). Large areas are naturally salt-affected, but irrigation under arid climatic conditions carries a large threat of secondary salinization (Breckle, 1989).

Halophytes, which can complete their whole life-cycle with sea-water conditions, have evolved various adaptive mechanisms by becoming halo-succulent, by accumulating inorganic salts, or by synthesis of osmolytes. It is even questionable whether there are obligate halophytes in the strict physiological sense (Hedenström and Breckle, 1974). All the adaptations of halophytes are tools for survival, not for high productivity. Their productivity is low for biochemical, physiological (see Figure 4.2) and ecophysiological reasons (Waisel, 1972; Ungar, 1991), as well as anatomical and morphological adaptations their survival on sea-water salinity is maintained but this inhibits high crop yields (Munns, 1993). However, degraded land saline agriculture, even with low crop yields, may be advisable purely as an alternative to bare wasteland.

For saline- and alkaline-degraded land only a few halophytic species can be used for phytomelioration. This requires special techniques for propagation, and for planting seedlings and saplings (Breckle et al., 2011; Wucherer et al., 2005). There are many applications but very few for food production.

M.A. Khan, M. Ozturk, B. Gul, & M.Z. Ahmed (Eds): Halophytes for Food Security in Dry Lands.
DOI: http://dx.doi.org/10.1016/B978-0-12-801854-5.00004-2

Under an arid climate sustainable agriculture with high production of crops per surface area is always only achievable with nonsaline conditions. In the long run it may pay to spend additional costs to maintain sustainable irrigation and leaching systems to keep the salinity of soils low. The important rule "No irrigation without drainage" also points to the fact that it pays in the long run more to invest in desalinization technology systems (inverse osmosis, energy sources from high radiation in deserts, photovoltaic devices, etc.), to keep soils low in salt, since fresh water is always indispensable for human welfare. But on several marginal areas and waste sites saline vegetation is also used sustainably for various purposes.

In Afghanistan traditional knowledge is widespread. For centuries villagers have used and collected plants for fodder, fuel, food, medicines, and other purposes. Afghanistan is a very mountainous country. Only in the lowlands do saline sites play a major role and only from distinct sabkha flats are some halophytic species used. We give an overview of those sources from saline sites.

4.2 Methods

During my former 3-year stay in Afghanistan and shorter travels, I have explored many vegetation types and visited many villages and bazaars. My own observations and ecological studies on saline sites during several excursions are compared with details from the few literature sources. Herbarium samples collected are now in Gött. Samples for ion analysis from green plant parts were taken, and fresh and dry weight noted. Cations were analyzed by atomic absorption, chloride by titration, sulfate by reduction and reaction with methylene blue spectrophotometrically (for details see Breckle, 1986; Lötschert and Köhm, 1977; Mirazai and Breckle, 1978). Many details on basic literature on flora and vegetation are given in Breckle (1983, 2002b, 2007), Breckle and Rafiqpoor (2010), Breckle et al. (2013), and Freitag (1971, 1991).

4.3 Results

4.3.1 Vegetation

There are several endorrheic lakes and salt flats in the lowlands of Afghanistan, surrounding the central mountain ranges of Hindu Kush and Kohe Baba. Some of them have become dry during recent decades mainly due to lower water input, since

Figure 4.1 Salt crusts along irrigation channels, Bamyan-Valley, central Afghanistan. (Photograph: SWBr, 1966).

the tributaries have been increasingly used for irrigation in upstream areas. Sophisticated irrigation systems are maintained with a network of small channels (Figure 4.1).

Saline flats have become more prominent despite inter-annual fluctuations of water balance, and thus lake levels in former decades have also been tremendous. In parallel, fluctuation of salinity of remnant water bodies has been and is great and soil salinity around lakes or in dried sea floors is also very variable. However, for all lakes or basins, a reliable recent history of water balance data has not been recorded.

Examples of the main saline lake and salt flats are given in Table 4.1.

The vegetation around saline flats and salt lakes in Afghanistan is often a mosaic of halophytic, gypsophilous and psammophytic vegetation (Breckle, 1983, 1986; Mirazai and Breckle, 1978). Often, salt flats are surrounded by loessic pene-plains and may also include clay pans and clay funnels (takyr). The halophytic vegetation of saline flats is often almost bare of vascular plants, especially when soil salinity is very high and salt crusts develop on top, even more so, when alkalinity is high, which is often the case under the very strong continental climatic conditions in north Afghanistan and in high dry

Table 4.1 Salt Lakes and Saline Flats in Afghanistan

Name	Province	Latitude °N; Longitude °E	Altitude m a.s.l.	Approximate Size (km²)	Salinity
Dasht-e Nawor	Ghazni	33.7°N; 67.8°E	3130	380	Alkaline
Ab-e Istada	Ghazni	32.5°N; 67.9°E	1970	210	Low
Dasht-e Shorak	N Mazar	36.8°N; 67.1°E	310	220	Moderate
Namaksar	Herat/Iran	34.0°N; 60.7°E	600	340 (+90 Iran)	Moderate
Ishkin-e Am	Nimroz	31.2°N; 62.0°E	480	550	High
Hamun-e Hilmend	Nimroz/Iran	31.6°N; 61.8°E	470	480 (+120 Iran)	High
Hamun-e Saberi (Hamun-e Puzak)	Farah/Iran	31.6°N; 61.3°E	470	570 (+500 Iran)	High
Hamun-e Gavdizirin	Nimroz	29.7°N; 62.2°E	460	1700	Moderate

Figure 4.2 *Halocnemum strobilaceum* (Chenopodiaceae) an extremely salt-resistant subshrub in the salt deserts of Iran and Central Asia, (A) left Touran Biosphere reserve, eastern Iran; (B) Aralkum, Kazakhstan. (Photographs: SWBr, 1977, 2003).

mountain valleys in central Afghanistan. With a soil pH of 9–10, only a few plants can survive for example, *Camphorosma*.

In the lowlands, high-saline sabkhas are characterized by very few isolated stands of *Haloxylon salicornicum* subshrubs, occasionally *Halocnemum strobilaceum* (Figure 4.2A and B)

Figure 4.3 Saxaul (*Haloxylon aphyllum*), at the north coast of northern Aral Sea, Kazakhstan. (Photograph: SWBr, 2003).

Figure 4.4 *Tamarix* tree, northwest Afghanistan. (Photograph: Ian Hedge, 1968).

occurs, and sporadically they contain some others, mostly chenopodiaceous shrubs and subshrubs (e.g., *Haloxylon aphyllum* (Figure 4.3), *Salsola arbuscula*, *Cornulaca monacantha*, *Seidlitzia rosmarinus*, *Anabasis setifera*, *Ephedra scoparia*, and some *Tamarix* species (Figure 4.4)) and a number of xero-halophytic annuals (Freitag et al., 2010), (e.g., *Tribulus macroptera, Atriplex dimorphostegia, Halocharis* sp., *Fagonia* sp.). Similar vegetation types may also occur on gravelly, very dry sites and land with lower salinity. In this case, some hemicryptophytes such as *Aristida plumosa, Astragalus* spp., and other therophytes such as *Plantago ciliata, Adonis* spp. and others, are mixed.

A detailed study of halophilic phytosociological units is still lacking. It may be difficult since the cover percentages are very low, the salinity in soil is temporarily and locally very variable and most saline areas are a complex mosaic of sandy, loessic and clayey more-or-less saline spots, even salinity or alkalinity may be predominantly on the top or middle or in lower horizons of soil profiles. From year to year vegetation is thus very variable.

On rather undisturbed sites it is possible to recognize distinct belts of various halophytic and pseudohalophytic vegetation up to the nonsaline desert or semi-deserts surrounding them. Those belts from lower to higher sites are normally governed by decreasing salinity in soils as well as by increasing drought stress. Accordingly, plant functional types and dominant lifeform of plants change along this catena (Breckle, 1983; Mirazai and Breckle, 1978; Frey et al., 1985).

4.3.2 The Halophytic Flora and Useful Halophytes

There are many halophytic species which exhibit a broad spectrum of races more or less adapted to saline sites. Halophytic ecotypes are normally smaller in comparison to their nonsaline ecotypes. The nomenclatorial distinctions as different subspecies are not made in all cases and in some is difficult to assess. Examples of some halophilic subspecies are known from *Heliotropium dasycarpum* ssp. *transoxanum, Silene vulgaris, Polygonum aviculare, Peganum harmala, Tribulus terrestris*, and *Juncus gerardii*. However, cultivation experiments or even cross-breeding tests have rarely or not been done at all.

Studies on the flora and vegetation of Afghanistan show that the Afghan flora is very rich, with almost 5000 species (Breckle et al., 2013). This parallels the rather dissected mountainous country with a high geo-diversity (Breckle and Rafiqpoor, 2010).

Examples of often-collected medicinal plants include *Glycyrrhiza* species, sometimes exported for example, to Pakistan, and less commonly small enterprises produce licorice products. *Glycyrrhiza* species are sometimes found on slightly saline loess slopes, they have very deep and far-reaching stolones, which are dug. *Glycyrrhiza* is not cultivated, thus collecting from the wild is still important.

Figure 4.5 Sheep and goats grazing on a slightly saline swamp area, Iran, close to Shiraz. (Photograph: SWBr, 2004).

Another example is *Ferula assa-foetida* and related species. These produce assa-foetida, which has been used as a medical drug for centuries in this region. *Ferula* and *Dorema* species are also found in the desert areas where salinity is very variable. The most common, often collected but now mostly cultivated, drug is *Cannabis sativa*. As a ruderal plant it is common everywhere on nutrient-rich waste grounds. Again slightly saline soils are not a problem, though growth is limited and yield is low.

Examples of medicinal plants from saline sites which are used in households are very limited in contrast to other rangeland species in the mountains. Some are used as herbal tea plants, for example, *P. aviculare* ecotypes, or *Potentilla anserina*. *Ephedra* spp. are used as an anticoughing herb.

A more general use of wastelands and saline sites is as rangeland plants (Figure 4.5). Goats, sheep, camels, horses, donkeys, and cows of nomadic people graze along their migration routes throughout the year. Grazing is selective, with the result that spiny, poisonous and unpalatable species are less grazed, which results in a lot of weedy and unpalatable species in many areas. In the last few decades grazing pressure on saline sites has also strongly increased.

Collecting all kind of plants for fuel is rather common in all parts of Afghanistan (Figure 4.6), and this is certainly the most striking impact factor on Afghan vegetation. All woody parts, even from dwarf shrubs and semi-shrubs, and also large annuals are collected and dried, often piled up on the roofs of houses for winter. This stored material is also partly used for fodder. Similarly, woody plants and subshrubs from saline areas

Figure 4.6 Large loads of desert plants for fuel and fodder, mainly from *Artemisia* and *Salsola* semi-desert and sebkha lowlands, near Andkhoi, north Afghanistan. (Photograph: Paap, 2011).

Figure 4.7 *Peganum harmala* (Zygophyllaceae), a multi-purpose poisonous, alkaloid-containing medical plant in south-west and Central Asia, from north Afghanistan. (Photograph: Ian Hedge).

are used. The highest wood biomass is from *H. aphyllum* (Figure 4.3) and *Haloxylon persicum* (on sand dunes), however, both are rare in Afghanistan. They may have been more common in former times, as was *Calligonum*. But even also dry annuals (*Salsola, Petrosimonia, Salicornia* (Figure 4.7), *Suaeda*, etc.) are collected. Herbs such as dry *Rheum* leaves or branches from *P. harmala* (Figure 4.8) are ideal for starting fires in the kitchen, and again they are also used for fodder.

It remains an open question whether saline sites along sebkhas may be used as cash crop areas with adapted halophytes or as sources for potential future crops. In any case they are a rich floristic mosaic with a high biodiversity conservation value.

Figure 4.8 *Salicornia europaea* s.l., at the coastline of northern Aral Sea. (Photograph: SWBr).

4.4 Discussion

For centuries nomadic people in Afghanistan have used summer pastures in the mountains but semi-deserts, including saline flats, in the lowlands during winter. The continental winter rain climate favors this traditional land-use. In winter, salinity decreases and ephemerals germinate and are a rich source for fodder (and fuel).

Biotic factors like grazing certainly are influencing the distribution of halophytes in saline habitats (Ungar, 1998), and grazing and overgrazing can be observed in all regions of Afghanistan. From Europe it is reported that *herbivory* may both increase and decrease species richness in salt marshes (Adam, 1990; Bakker and de Vries, 1992). Both mowing and grazing was found to cause a change in the relative cover of species and species composition of communities on salt marshes in the Netherlands (Bakker, 1978). Whether this is also the case in semi-deserts and saline flats in Afghanistan is not clear. Grazing causes gaps to open in the salt marsh vegetation and allows both annuals and perennials from the low marsh to establish on the higher marsh. The salinity content of the soil decreased more in the control area than in the grazed and mowed areas (Bakker and Ruyter, 1981).

Jerling and Andersson (1982) reported that cattle were selective in their grazing of species in a Baltic seashore meadow. Selective herbivory by feral horses is also suggested to be a significant factor in determining the competitive success of grasses *Spartina alterniflora* and *Distichlis spicata* in Maryland salt marshes (Furbish and Albano, 1994). This is certainly the case also with sheep and goat grazing, as well as by the camels of Afghan nomadic people; the present vegetation in many parts, as well as in saline flats, is spiny, poisonous or by other means almost unpalatable after the long selective grazing and overgrazing history in Afghanistan.

Grazing favors halophytes on less saline sites. Heavily grazed salt marshes of northern Germany (10 sheep ha^{-1}) were dominated by the perennial grass *Puccinellia maritima* and annuals *Suaeda maritima* and *Salicornia europaea* (Kiehl et al., 1996). Two perennial species, *Halimione portulacoides* and *Aster tripolium*, became rare on these marshes when they were exposed to heavy grazing. After 4 years without grazing, the cover of *Festuca rubra*, *A. tripolium*, and *H. portulacoides* increased, and cover of *Plantago maritima*, *Suaeda depressa*, and *S. europaea* decreased. Kiehl et al. (1996) concluded that grazing caused an upward shift of zonation boundaries between lower-, mid-, and upper-salt marsh vegetation. Increased salinity and soil compaction, because of grazing and trampling, opened up the higher areas of the marshes in Germany and the Netherlands for invasion by lower marsh species (Bakker, 1985; Kiehl et al., 1996). Gaps produced in marsh vegetation by grazing animals were the most significant factor determining the establishment of annual species since the more competitive perennial and less halophytic species were absent (Jensen, 1985; Ellison, 1987; Gibson and Brown, 1991; Bakker and de Vries, 1992; Kiehl et al., 1996).

Grasses are the main potential fodder crops for saline soils, as Gulzar et al. (2007) pointed out. Some of the most productive species yield similar biomasses as conventional crops on seawater irrigation—this was checked with *Salicornia bigelovii*. They may outperform conventional crops in yield and water use efficiency at lower salinities, as was claimed by Glenn et al. (1999).

The lack of *fuel* in many villages in Afghanistan, even more with the nomadic people, forces the use of all kinds of plant material for fuel. Saxaul wood is very famous for its value and good burning quality for kebab stoves.

The use of the rich ash of halophytes (potash) for soap production or other purposes is not yet reported but in former times was observed by us. In villages oven ash is regularly used also as a fertilizer in gardens.

Medicinal plants have long been studied in Afghanistan. Aitchison (1890) gave a long list of the uses of various plants. Volk (1955, 1961) and Pelt et al. (1965) listed all known medicinal plants and drugs from the bazaars, comprising only a few halophytes. From their lists our observations are complemented (see Table 4.2). Many plant species are rich in secondary

Table 4.2 Halophytic Plant Species of Afghanistan

Species	Plant Family	Halophyte Type	Known Uses, Applications
Ephedra strobilacea	Ephedraceae	P	u, m
Amaranthus albus	Amaranthaceae	P	f
Apium graveolens	Apiaceae	P	m, s
Artemisia maritima	Asteraceae	P	u, s
Epilasia hemilasia	Asteraceae	P	x
Pulicaria arabica	Asteraceae	P	x
Saussurea salsa	Asteraceae	E	x
Sonchus maritimus	Asteraceae	P	f
Taraxacum bessarabicum	Asteraceae	P	f
Heliotropium arguzioides	Boraginaceae	E	x
Heliotropium dasycarpum	Boraginaceae	E	x
Crambe kotschyana	Brassicaceae	P	f
Hornungia procumbens	Brassicaceae	P	x
Lepidium cartilagineum	Brassicaceae	E	s
Lepidium perfoliatum	Brassicaceae	P	f
Silene vulgaris	Caryophyllaceae	P	f
Spergularia marina	Caryophyllaceae	E	x
Spergularia media	Caryophyllaceae	E	x
Anabasis setifera	Chenopodiaceae	P	f, u
Atriplex dimorphostegia	Chenopodiaceae	P, (R)	f
Atriplex leucoclada	Chenopodiaceae	P, (R)	f
Bassia eriophora	Chenopodiaceae	P	f
Bienertia cycloptera	Chenopodiaceae	E	f, u
Camphorosma monspeliaca	Chenopodiaceae	P	f, u
Caroxylon nitrarium	Chenopodiaceae	E	f, u
Caroxylon scleranthum	Chenopodiaceae	E	f, u
Climacoptera lanata	Chenopodiaceae	E	x
Climacoptera longipistillata	Chenopodiaceae	E	x
Climacoptera longistylosa	Chenopodiaceae	E	x
Climacoptera turcomanica	Chenopodiaceae	E	x
Cornulaca monacantha	Chenopodiaceae	E	f, u
Gamanthus commixtus	Chenopodiaceae	E	x
Gamanthus gamocarpus	Chenopodiaceae	E	x
Halimocnemis mollissima	Chenopodiaceae	E	f
Halocharis hispida	Chenopodiaceae	P	x
Halocharis sulphurea	Chenopodiaceae	P	x
Halocharis violacea	Chenopodiaceae	P	x
Halocnemum strobilacea	Chenopodiaceae	E	f, u

(Continued)

Table 4.2 (Continued)

Species	Plant Family	Halophyte Type	Known Uses, Applications
Halogeton glomeratus	Chenopodiaceae	E	x
Halostachys belangeriana	Chenopodiaceae	E	f, u
Haloxylon ammodendron	Chenopodiaceae	P	f, u
Haloxylon salicornicum	Chenopodiaceae	E	f, u
Kali tragus	Chenopodiaceae	E	f, u, s
Kalidium caspicum	Chenopodiaceae	E	f, u
Kaviria tomentosa	Chenopodiaceae	E	f, u
Oxybasis rubra	Chenopodiaceae	E	f
Petrosimonia sibirica	Chenopodiaceae	E	f
Salicornia perennans etc. (Figure 4.8)	Chenopodiaceae	E	f, d
Salsola rosmarinus	Chenopodiaceae	P	f, u
Spinacia turkestanica	Chenopodiaceae	P	f, u, d
Several *Suaeda*-species	Chenopodiaceae	E	f, u
Xylosalsola richteri	Chenopodiaceae	P	f, u
Cressa cretica	Convolvulaceae	E, R	x
Prosopis farcta	Fabaceae	P	f, u
Alhagi maurorum	Fabaceae	P	f, u
Halimodendron halodendron	Fabaceae	P	f, u
Medicago sativa	Fabaceae	P	f
Melilotus indicus	Fabaceae	P	f, u
Trifolium fragiferum	Fabaceae	P	f
Frankenia pulverulenta	Frankeniaceae	E, R	f, u
Centaurium pulchellum	Gentianaceae	P	m
Plantago coronopus	Plantaginaceae	P	x
Plantago maritima ssp. *salsa*	Plantaginaceae	P	x
Limonium reniforme	Plumbaginaceae	E, R	x
Polygonum aviculare	Polygonaceae	P	f, m
Portulaca oleracea	Portulacaceae	P	f, d
Glaux maritima	Primulaceae	E, R	x
Ranunculus sceleratus	Ranunculaceae	P	m
Potentilla anserina	Rosaceae	P	f, m
Populus euphratica	Salicaceae	P	u, f
Populus pruinosa	Salicaceae	P	u, f
Reaumuria halophila	Tamaricaceae	E, R	f, u
Tamarix hispida	Tamaricaceae	E, R	u, f
Tamarix ramosissima	Tamaricaceae	E, R	u, f
Tamarix sp. (Figure 4.4)	Tamaricaceae	P, R	u
Nitraria schoberi	Zygophyllaceae	E, R	f, u
Fagonia bruguieri	Zygophyllaceae	P	f
Peganum harmala (Figure 4.7)	Zygophyllaceae	P	m, f

(Continued)

Table 4.2 (Continued)

Species	Plant Family	Halophyte Type	Known Uses, Applications
Tribulus terrestris	Zygophyllaceae	P	f
Zygophyllum eurypterum	Zygophyllaceae	P	f, u
Bolboschoenus maritimus	Cyperaceae	P, (H)	f
Cyperus fuscus	Cyperaceae	P	f
Cyperus rotundus	Cyperaceae	P	f, d
Schoenoplectus litoralis	Cyperaceae	P, (H)	f
Schoenoplectus tabernaemontani	Cyperaceae	P, (H)	f
Scirpoides holoschoenus	Cyperaceae	P, (H)	f
Juncus gerardii	Juncaceae	P, (H)	x
Aeluropus lagopoides	Poaceae	E, R	x
Aeluropus littoralis	Poaceae	E, R	x
Aeluropus macrostachyus	Poaceae	E, R	x
Crypsis aculeata	Poaceae	E	x
Crypsis schoenoides	Poaceae	E	x
Desmostachya bipinnata	Poaceae	P	f
Phragmites australis	Poaceae	P, (H)	f, u
Puccinellia distans	Poaceae	P, R	f
Zannichellia palustris	Potamogetonaceae	P, H	x

Approximate indicator values for salinity (see Breckle, 1985, Ellenberg et al., 1991) are S = 7—9 in E, 3—6 in P, but in P also ecotypes are known with less than S = 3.
Halophyte types: E, Euhalophyte; P, Pseudohalophyte; H, Hydrophyte; R, Salt-recreting halophyte.
Known uses: m, medicinal plants; f, fodder; u, fuel; d, food; s, spice; x, no use known yet.

compounds. They are often used for distinct medicinal purposes. The potential of halophytes in this respect has only started to be studied (Buhmann and Papenbrock, 2013), however, it can be deducted from other phytochemical properties which are often taxon-specific. It needs suitable conditions for cultivation. Besides good agricultural fields, normally only moderate- or low-salinity sites can be used to start such halophyte cultivations (Dagar et al., 2011). It will need more detailed cultivation field experiments to promote a suitable and sustainable utilization of halophytic crop plants (Debez et al., 2011).

It is known that *C. sativa* ssp. *indica*, the source of marihuana, is slightly or moderately salt-tolerant and as a ruderal plant is dependent on a good nitrogen supply. *Papaver somniferum*, the source of opium, is more drought-, but less salt-tolerant than *Cannabis*. Neither is suitable for salinized sites. For a sound economy both of these drug plants should be replaced by good cash crop plants. Under special conditions the melioration of abandoned saline soils can be achieved with valuable crops like *Glycyrrhiza glabra*, suitable for licorice production, as was shown by Khushiev et al. (2005) in north Kazakhstan, other halophytic crops out of 587 halophytic species in China are discussed by Kefu et al. (2002, 2011) and by Ksouri et al. (2012) for food, medical, and nutraceutical applications. The list of medicinal plants used in the Tajik and Afghan Pamirs, given by Kassam et al. (2010) and the treatise by Shawe (2007), does not contain any real halophytes, except *Amaranthus* and *P. harmala*, which are known as pseudo-halophytes from some moderately saline sites. The herbal drug and the seeds of *P. harmala* (Figure 4.8) are widely used in Central Asia. In Afghanistan the seeds (ispand) are used not only as an anthelminticum, but also against headache, rheuma, lumbago, and also dizziness (Volk, 1955). It is a very traditional drug with ritual applications, for example, against bad mood and the evil eye together with talismans.

By far the greatest potential of the rich Afghan halophytic flora is to be seen in their rich biodiversity. Nature conservation started rather early when the country was a kingdom, partly to serve as hunting grounds. UN officials raised detailed plans for various national parks and reserves, including sebkha areas (Petocz and Larsson, 1977). However, a functioning nature protection system is still lacking, due to the war situation and the historical unrest over recent decades. A few national parks have been established however, and a long list of future reserves is proposed (Breckle and Rafiqpoor, 2010), including Dasht-e Nawor or Hamun-e Puzak.

Desertification in arid regions is a common threat. Overexploitation and overgrazing have not stopped at sebkha areas, and additionally a huge part of agricultural land has become saline due to incorrect irrigation practices and the poor quality of irrigation water. Rehabilitation of degraded saline flats is only possible by using suitable halophytic plant species for phytomelioration. It is a costly process, but in the long run it can lead to better conditions (Breckle, 1982, 1990, 2002a; Norman et al., 2013). One striking example is the huge saline seafloor of the former Aral sea, now called the Aralkum (Breckle et al., 2001; Breckle, 2013; Breckle and Wucherer, 2011; Wucherer et al., 2011). Saxaul (*H. aphyllum*) has been shown in large recultivation areas to be suitable, on moderately saline soils, to regenerate quickly and to cover solonchak soil if furrows are used where sand is blown in. In former times in the deserts of Afghanistan saxaul and other chenopod shrubs (*S. arbuscula*, *Halostachys caspica*, *Halothamnus subaphyllus*, *S. rosmarinus*) were certainly more abundant, but they have been overexploited for centuries. These should be used for afforestation, especially on saline sites. Saxaul wood can be used for fuel after only a few years of growth. On very saline sites, even with salt crusts, *H. strobilaceum* is useful for phytomelioration, however, reproduction by cuttings is not as successful as with saxaul (Breckle et al., 2011).

4.5 Conclusions

They have a rather wide variety of uses and plant products, but for the overall economy of villagers and as food source, the many halophytic plant species in Afghanistan are negligible. They still play only a marginal role and only a small proportion is used. But they bear increasing future prospects, as in general already Aronson (1985) has pointed out. A greater diversification of the uses of natural products will be essential for a broader and sound economic basis much less dependent on the deadly drug production in Afghanistan. The use of saline areas and halophytes can add to a new multipurpose agriculture and can give hope for new incomes. This is discussed by Sardo and Hamdy (2005), with Rozema and Flowers (2008) listing many interesting applications, and Kefu et al. (2011) indicating the rich gene pool being a precious resource for the future; however, a breakthrough of seawater agriculture is not yet visible.

Acknowledgments

The help and hospitality of many people and institutions in various countries in Central Asia is greatly acknowledged. Many thanks to DAAD for sponsoring exchange students and an expedition to the Pamirs in collaboration with GTZ-CCD-Project. Many thanks to the University of Bonn and the University of Bielefeld, and to unknown reviewers for their great assistance. The financial help of BMBF in Germany for Aralkum Research is greatly acknowledged.

References

Adam, P., 1990. Saltmarsh Ecology. Cambridge University Press, New York, NY.

Aitchison, J.E.T., 1890. Notes on the products of western Afghanistan and north-eastern Persia. Trans. Bot. Soc. Edinb. 18, 1–228.

Aronson, J., 1985. Economic halophytes—a global review. In: Wickens, G.E., Goodin, J.R., Field, D.V. (Eds.), Plants for Arid Lands. Royal Botanic Gardens, Kew, pp. 177–188.

Bakker, J.P., 1978. Changes in a salt-marsh vegetation as a result of grazing and mowing—a five-year study on permanent plots. Vegetatio. 38, 77–87.

Bakker, J.P., 1985. The impact of grazing on plant communities, plant populations and soil conditions on salt marshes. Vegetatio. 62, 391–398.

Bakker, J.P., de Vries, Y., 1992. Germination and early establishment of lower salt-marsh species in grazed and mown salt marsh. J. Veg. Sci. 3, 247–252.

Bakker, J.P., Ruyter, C., 1981. Effects of five years of grazing on a salt-marsh vegetation. Vegetatio. 44, 81–100.

Breckle, S.-W., 1982. The significance of salinity. In: Spooner, B., Mann, H.S. (Eds.), Desertification and Development: Dryland Ecology in Social Perspective. Acad. Press, London, pp. 277–292.

Breckle, S.-W., 1983. Temperate deserts and semideserts of Afghanistan and Iran. In: West, N.E. (Ed.), Temperate Deserts and Semideserts. Ecosystems of the World (Ed. Goodall, D.W.), vol. 5. Elsevier, Amsterdam, MA, pp. 271–319.

Breckle, S.-W., 1985. Die Siebenbürgische Halophyten-Flora—Ökologie und ihre pflanzengeographische Einordnung. Siebenbürgisches Arch. 3 (20), 53–105.

Breckle, S.-W., 1986. Studies on halophytes from Iran and Afghanistan. II. Ecology of halophytes along salt-gradients. Proc. R. Bot. Soc. Edinb. 89B, 203–215.

Breckle, S.-W., 1989. Role of salinity and alkalinity in the pollution of developed and developing countries. Internat. In: Öztürk, M.A. (Ed.), Symposium on the effect of pollutants to plants in developed and developing countries in Izmir/Turkey 22.-28.8.1988, pp. 389–409.

Breckle, S.-W., 1990. Salinity tolerance of different halophyte types. In: Bassam, N.El. et al. (Eds.), Genetic aspects of plant mineral nutrition. Plant Soil. vol. 148, pp. 167–175.

Breckle, S.-W., 2002a. Salinity, halophytes and salt affected natural ecosystems. In: Läuchli, A., Lüttge, U. (Eds.), Salinity: Environment–Plants–Molecules. Kluwer Acad.Publ., Dordrecht, pp. 53–77.

Breckle, S.-W., 2002b. Salt deserts in Iran and Afghanistan. In: Böer, B., Barth, H.-J. (Eds.), Sabkha Ecosystems. Kluwer, Netherlands, pp. 109–122.

Breckle, S.-W., 2007. Flora and vegetation of Afghanistan. Basic Appl. Dryland Res. (BADR online). 1 (2), 155–194.

Breckle, S.-W., 2013. From aral sea to aralkum—an ecological disaster or halophytes' paradise. Prog. Bot. 74, 351–398.

Breckle, S.-W., Rafiqpoor, D.M., 2010. Field Guide Afghanistan—Flora and Vegetation. Scientia Bonnensis, Bonn, pp. 864.

Breckle, S.-W., Wucherer, W., 2011. Halophytes and salt desertification in the Aralkum Area. In: Breckle, S.-W., Dimeyeva, L., Wucherer, W., Ogar, N.P. (Eds.), Aralkum—A Man-Made Desert. The Desiccated Floor of the Aral Sea (Central Asia). Ecological Studies, vol. 218, pp. 271–299 (Chapter 12).

Breckle, S.-W., Scheffer, A., Wucherer, W., 2001. Halophytes on the dry sea floor of the Aral Sea. In: Breckle, S.-W., Veste, M., Wucherer, W. (Eds.), Sustainable Land-Use in Deserts. Springer, Heidelberg, pp. 139–146.

Breckle, S.-W., Dimeyeva, L., Wucherer, W., Ogar, N.P. (Eds.), 2011. Aralkum—A Man-Made Desert. The Desiccated Floor of the Aral Sea (Central Asia). Ecological Studies, vol. 218. Springer, Berlin.

Breckle, S.-W., Hedge, I.C., Rafiqpoor, M.D., 2013. Vascular Plants of Afghanistan—An Augmented Checklist. Scientia Bonnensis, Bonn, p. 598.

Buhmann, A., Papenbrock, J., 2013. An economic point of view of secondary compounds in halophytes. Funct. Plant Biol. 40, 952–967.

Dagar, J.C., Minhas, P.S., Kumar, M., 2011. Cultivation of medicinal and aromatic plants in saline environments. CAB Rev. Perspect. Agric. Vet. Sci. Nutr. Nat. Resour. 6 (009).

Debez, A., Huchzermeyer, B., Abdelly, C., Koyro, H.-W., 2011. Current challenges and future opportunities for a sustainable utilization of halophytes. In: Öztürk, M., et al., (Eds.), Sabkha Ecosystems. Tasks for Vegetation Science. Springer Science, Berlin, pp. 9–77.

Ellenberg, H., Weber, H.E., Düll, R., Wirth, V., Werner, W., Paulißen, D., 1991. Zeigerwerte von Pflanzen in Mitteleuropa. Scr. Geobot. 18, 1–248.

Ellison, A.M., 1987. Effects of competition, disturbance, and herbivory on *Salicornia europaea*. Ecology. 68, 576–586.

Flowers, T.J., Yeo, A.R., 1995. Breeding for salinity resistance in crop plants: where next? Aust. J. Plant. Physiol. 22, 875–884.

Freitag, H., 1971. Die natürliche Vegetation Afghanistans. Beiträge zur Flora und Vegetation Afghanistans. Vegetatio. 22 (4–5), 255–344.

Freitag, H., 1991. The distribution of some prominent Chenopodiaceae in SW Asia and their phytogeographical significance. Flora Vegetatio Mundi. 9, 281–292.

Freitag, H., Hedge, I.C., Rafiqpoor, M.D., Breckle, S.-W., 2010. Flora and vegetation geography of Afghanistan. In: Breckle, S.-W., Rafiqpoor, M.D. (Eds.), Field Guide Afghanistan—Flora and Vegetation. Scientia Bonnensis, Bonn, pp. 79–115.

Frey, W., Kürschner, H., Stichler, W., 1985. Photosynthetic pathways and ecological distribution of halophytes from four littoral salt marshes (Egypt/Sinai, Saudi Arabia, Oman and Iran). Flora. 177, 107–130.

Furbish, C.E., Albano, M., 1994. Selective herbivory and plant community structure in a mid-Atlantic salt marsh. Ecology. 75, 1015–1022.

Gibson, C.W.D., Brown, V.K., 1991. The effect of grazing on local colonization and extinction during early succession. J. Veg. Sci. 2, 291–300.

Glenn, E.P., Brown, J.J., Blumwald, E., 1999. Salt tolerance and crop potential of halophytes. CRC Crit. Rev. Plant Sci. 18, 227–255.

Gulzar, S., Khan, M.A., Liu, X., 2007. Seed germination strategies of *Desmostachya bipinnata*: a fodder crop for saline soils. Rangeland Ecol. Manage. 60, 401–407.

Hedenström, H.V., Breckle, S.-.W., 1974. Obligate halophytes? a test with tissue culture methods. Z. Pflanzenphysiol. 74, 183–185.

Jensen, A., 1985. The effect of cattle and sheep grazing on salt-marsh vegetation at Skallingen, Denmark. Vegetatio. 60, 37–48.

Jerling, L., Andersson, M., 1982. Effects of selective grazing by cattle on the reproduction of *Plantago maritima*. Holarct. Ecol. 5, 405–411.

Kassam, K.-A., Karamkhudoeva, M., Ruelle, M., Baumflek, M., 2010. Medicinal plant use and health sovereignty: findings from the tajik and Afghan pamirs. Hum. Ecol. 38, 817–829.

Kefu, Z., Hai, F., Ungar, I.A., 2002. Survey of halophyte species in China. Plant Sci. 163, 491–498.

Kefu, Z., Song, J., Feng, G., Zhao, M., Liu, J., 2011. Species, types, distribution, and economic potential of halophytes in China. Plant Soil. 342 (1–2), 495–509.

Khushiev, H., Noble, A., Abdullaev, I., Toshbekov, U., 2005. Remediation of abandoned saline soils using *Glycyrrhiza glabra*: a study from the hungry steppe of central Asia. Int. J. Agric. Sustainability. 3, 102–113.

Kiehl, K., Eischeid, I., Gettner, S., Walter, J., 1996. Impact of different sheep grazing intensities on salt marsh vegetation in northern Germany. J. Veg. Sci. 7, 99–106.

Ksouri, R., Ksouri, W.M., Jallali, I., Debez, A., Magné, C., Hiroko, I., 2012. Medicinal halophytes: potent source of health promoting biomolecules with medical, nutraceutical and food applications. Crit. Rev. Biotechnol. 32 (4), 289–326.

Lötschert, W., Köhm, H.-J., 1977. Characteristics of tree bark as an indicator in high-immission areas. Oecologia. 27, 47–64.

Mirazai, N.A., Breckle, S.-W., 1978. Untersuchungen an afghanischen Halophyten. I. Salzverhältnisse in Chenopodiaceen Nord-Afghanistans. Bot. Jahrb. Syst. 99, 565–578.

Munns, R., 1993. Physiological processes limiting plant growth in saline soils: some dogmas and hypotheses. Plant Cell Environ. 16, 15–24.

Munns, R., 2005. Genes and salt tolerance: bringing them together. New Phytol. 167, 645–663.

Norman, H.C., Masters, D.G., Barrett-Lennard, E.G., 2013. Halophytes as forages in saline landscapes, interactions between plant genotype and environment change their feeding value to ruminants. Environ. Exp. Bot. 92, 96–109.

Pelt, J.-M., Hayon, J.C., Younus, M.C., 1965. Plantes medicinales et drogues de l'Afghanistan. Bull. Soc. Pharm. Nancy. 66, 16–61.

Petocz, R.G., Larsson, J.Y., 1977. National Parks and Utilization of Wildlife Resources, Afghanistan. Ecological Reconnaissance of Western Nuristan with Recommendations for Management, 9. UNDP, FAO, Kabul, Field Doc pp. 63.

Rozema, J., Flowers, T., 2008. Crops for a salinized world. Science. 322, 1478–1480.

Sardo, V., Hamdy, A., 2005. Halophytes—a precious resource. In: Hamdy, A., El Gamal, F., Lamaddalena, N., Bogliotti, C., Guelloubi, R. (Eds.), Options Méditerranéennes: Sér. B. Etudes et Recherches, vol. 53, pp. 119–128.

Shawe, K., 2007. The Medicinal Plant Sector in Afghanistan. FAO Report, Shawe Medic Plants Mission, Kabul, p. 46.

Ungar, I.A., 1991. Ecophysiology of Vascular Halophytes. CRC Press, Boca Raton, FL.

Ungar, I.A., 1998. Are biotic factors significant in influencing the distribution of halophytes in saline habitats? Bot. Rev. 64, 176–199.

Volk, O.H., 1955. Afghanische drogen. Planta Med. 3, 171–178.

Volk, O.H., 1961. A survey of Afghan medicinal plants. Pak. J Sci. Ind. Res. 4, 232–238.

Waisel, Y., 1972. Biology of Halophytes. Acad Press, New York & London.

Wucherer, W., Veste, M., Herrera Bonilla, O., Breckle, S.-W., 2005. Halophytes as useful tools for rehabilitation of degraded lands and soil protection. In: Proceedings of the First International Forum on Ecological Construction of the Western Beijing, Beijing, pp. 87–94 (English); pp. 169–175 (Chinese).

Wucherer, W., Breckle, S.-W., Kaverin, V.S. Dimeyeva, L.A., Zhamantikov, K., 2011. Phytomelioration in the Northern Aralkum. In: Breckle, S.-W., Dimeyeva, L., Wucherer, W., Ogar, N.P. (Eds.), Aralkum—A Man-Made Desert. The Desiccated Floor of the Aral Sea (Central Asia). Ecological Studies, vol. 218, pp. 343–386.

COMPARISON OF SEED PRODUCTION AND AGRONOMIC TRAITS OF 20 WILD ACCESSIONS OF *SALICORNIA BIGELOVII* TORR. GROWN UNDER GREENHOUSE CONDITIONS

Cylphine Bresdin[1], Edward P. Glenn[1] and J. Jed Brown[2]
[1]*Environmental Research Laboratory of the University of Arizona, Tucson, AZ, USA* [2]*Center for Sustainable Development, College of Arts and Sciences, Qatar University, Doha, Qatar*

5.1 Introduction

Salicornia bigelovii has potential as a seawater-irrigated crop that can yield: fiber, feed, oilseed, and food for human consumption (Glenn et al., 1991, 1999, 2013; Jaradat and Shahid, 2012; Weber et al., 2007; Shahid et al., 2013; Zerai, 2010). There is a need to develop environmentally sound agronomic and crop management techniques for seawater agriculture in conjunction with selective breeding of *S. bigelovii* (Brown et al., 2014; Jaradat and Shahid, 2012; Shahid et al., 2013) to produce high-yielding varieties appropriate to field conditions where seawater, brackish or other low-quality water can be used for irrigation (Glenn et al., 1992, 1994, 1998, 1999, 2009, 2013; Grattan et al., 2008; Jordan et al., 2009; Masters et al., 2007; Rozema and Flowers, 2008; Zerai et al., 2010). We investigated *S. bigelovii* as a potential oilseed and biomass crop as part of an arid land biofuels program under development in the UAE. We evaluated 20 wild accessions of *S. bigelovii*

M.A. Khan, M. Ozturk, B. Gul, & M.Z. Ahmed (Eds): Halophytes for Food Security in Dry Lands.
DOI: http://dx.doi.org/10.1016/B978-0-12-801854-5.00005-4

over two annual crop cycles under greenhouse conditions in Tucson, Arizona, USA. Our primary focus was on seed yield. As a secondary focus, we selected for other desirable agronomic characteristics including heat tolerance, large flower spikes with synchronized flowering and seed with high oil content, with the aim of selecting founder lines for a selection and breeding program.

5.2 Materials and Methods

5.2.1 Relevant Biology of *S. bigelovii*

Salicornia bigelovii Torr. (Amaranthaceae), a small annual succulent halophyte native to coastal salt marshes of North America and the Caribbean, is usually less than 60 cm tall under natural conditions and has leafless succulent segmented stems that may display lateral branching. Both stems and branches terminate in long upright spikes that are comprised of three-flowered cymes bearing perfect flowers in response to photoperiod (Munz, 1974; York et al., 2000). In its native range, *S. bigelovii* begins flowering in June and the spike continues to elongate from the apex, producing new flowers over 30–60 days, causing a population of *S. bigelovii* to contain flowers and seeds at all stages of maturation during the majority of the flowering season (Zerai et al., 2010). Stigmas appear about 10 days before anthers emerge, allowing out-crossing to occur, and at anthesis, self-fertilization will occur in the absence of out-crossing (Zerai et al., 2010). Each flower produces a single seed held onto the spike by a persistent fleshy calyx until the flower spike senesces and dries in the autumn when it detaches from the spike (Sheperd et al., 2005).

5.2.2 Source of Wild Accessions

Seventeen *S. bigelovii* accessions were collected in 2011 from wild populations in Puerto Rico, the Atlantic coast of the United States, and the US portion of the Gulf of Mexico (Table 5.1). Two additional accessions were collected from the Florida Keys in 2012. Seed previously produced from an accession that had been collected from Galveston, Texas, USA, and held in cold storage, was also included in these trials.

5.2.3 Experimental Design

Plants of each annual crop were grown in 4-L pots containing a 1:1 mix of soil and medium sand in a greenhouse in

Table 5.1 Locality Data for Accessions and Number of Plants Planted in 2012 and 2013

No. Planted 2012	2013	ID	Name	State	Location
129	114	BC	Boca Chica	TX	N25 59.873 W97 09.607
123	118	BCW	Boca Chica West	TX	N25 59.221 W97 11.339
37		B	Brewster	MA	N41 45.601 W70 06.930
1		C	Chincoteaque	VA	N37 56 13.95 W75 24 35.05
128	104	CCB	Corpus Christi Bridge	TX	N27 37 54.01 W97 14 13.09
122	197	EI	Elmer Island	LA	N29 10.742 W 90 04.197
128	104	G	Galveston	TX	N 29.269 W 95.0065
122	94	GI	Grand Isle	LA	N29 14.709 W89 59.240
122	118	MI	Mustang Island	TX	N27 37 16.85 W97 12 38.82
16		PR	Puerto Rico	PR	N17 57.125 W67 11.745
48		Q	Quimby	VA	N37 33 53.12 W75 44 32.13
123	131	Sp	SP 2	TX	N26 10.071 W97 10.560
		Sp	SP 3	TX	N26 09.390 W97 10.435
53	130	SpN	SP north	TX	N26 11.889 W97 10.863
29		WL	Web Landing	VA	N37 24 7.28 W755 2 0.98
		WL	Well2	MA	N41 54.372 W69 59.940
16		WF	Well Fleet	MA	N41 53.780 W70 00.547
3		WB	West Barnstable	MA	N41 42.959 W70 22.046
	45	FIB	Big Coppitt Key	FL	N24 35.836 W081 39.311
	45	FIL	Little Torch Key	FL	N24 39.985 W081 23.615

Tucson, Arizona, USA, an arid location with generally clear skies. A block design was used to account for a positive 6°C gradient from the cooling pad end to the fan end of the evaporatively cooled greenhouse. For each crop cycle, accessions were randomized using rand function (Microsoft Excel) into blocks that consisted of five columns and ten rows, 50 pots per block. Pots were spaced 0.30 m apart within columns and rows on four parallel steel-mesh tables oriented along the temperature gradient, giving each plant an area of 0.093 m^2. This spacing allowed plants to form closed canopies, allowing yields to be expressed on an area basis as well as on a per plant basis, for comparison with open-field yields. Crop 1 (2012) consisted of 1200 plants. Crop 2 (2013) consisted of 1100 plants.

5.2.4 Greenhouse Procedures and Crop Observations

For Crop 1, the first 18 accessions listed in Table 5.1 were sown on March 26, 2012 in seeding trays and irrigated with municipal water by overhead spray three times a day. Poor germinating accessions: *West Barnstable, South Padre 2, South Padre North, Web Landing, Well Fleet, Quimby, Brewster,* and *Elmer Island* were reseeded on April 10–12, 2012. *South Padre 2* and *South Padre 3* were combined and labeled as *South Padre,* and *Well 2* and *Well Fleet* were combined under the label of *Well Fleet.*

Crop 2 was sown following the procedure used for Crop 1, but was sown on March 19, 2013 and consisted of the top nine performing accessions from Crop 1 plus two new accessions from Florida. Seedlings were transplanted on May 16 through June 16 for 2012 and May 12–21 for 2013. During 2012, overhead spray was maintained for the transplants with an additional manual spray of saline water, 10 ppt NaCl plus $MgSO_4 \cdot 7H_2O$, soluble fertilizer to supply 25 mg L^{-1} N, P, K, and fungicide (Ridomil, BASF, Inc., Florham Park, New Jersey). Individual pots were put into place on tables and drip irrigation began June 4, 2012. Crop 2 seedlings were transplanted into pots with the same soil mixture as the previous year and directly put on drip with synthetic sea salt (Crystal Sea Marinemix, Marine Enterprises International, Baltimore, MD) at 10 ppt with the same fertilizer and fungicide additive used for Crop 1. The irrigation regimen was three times per day at a rate to keep the soil moist and allow some leaching without depleting salts. Salinity of leachate was 9–11 ppt.

Individual plant height was measured every other week during 2012 until flowering became prominent.

5.2.5 Harvest and Processing

Crop 1 plants were individually harvested in early November 2012 when most of the spikes had formed seeds but before seeds began to drop from spikes. They were cut at the base, placed in paper bags, weighed and allowed to dry for 2 months. Plants, except the Atlantic accessions, were processed individually in batches by accession. Individual dried plants were weighed, passed through a hammer mill and the milled content was then hand-sifted and winnowed in front of a greenhouse exhaust fan to separate seed from chaff. Seeds were put into a paper coin envelope and stored in a cold-room. Small accessions from the Atlantic region (*Quimby*, *Chincoteaque*, *Well Fleet*, *Web Landing*, *West Barnstable*, *Brewster*, and *Puerto Rico*) were pooled per accession for processing.

Crop 2 plants were harvested in early December 2013 and processed the same as the Crop 1 plants with the exception that small plants less than 65 g dry weight were pooled and processed as one plant.

5.2.6 Seed Purity and Proximate Analysis

Ten randomly chosen envelopes per accession were individually analyzed for seed purity for Crop 1. A sample (seed plus chaff) was weighed, seed and chaff were separated by hand, then seeds were weighed and counted. Seed weight was divided by sample weight to obtain purity. Weight per seed was obtained by dividing seed weight by seed count. The average number of seeds per plant per accession was calculated by dividing seed number by sample weight and then multiplying by weight of envelope contents. The mean purity of the ten assays was taken to represent the accession and applied to all seed envelopes (individual plants) of that accession.

For Crop 2 purity assays, a sample of about 0.6 mL of seed from ten randomly chosen envelopes per accession was mixed and weighed. Chaff and seed from a sample of the mixed seed were separated and each component was weighed. Seed weight divided by the sum of seed weight and chaff was taken as accession purity. Eight groups of 25 seeds were weighed and weight was divided by 25 to estimate individual seed weight for that accession. Harvest index (HI) was calculated on a per accession level as total weight of seed yield divided by total dry above-ground biomass before separating seeds from plant.

Mean seed yield was multiplied by 10.6 to obtain seed yield per square meter in order to approximate accession yields on an area basis. The average number of seed per square meter was obtained by dividing seed yield per square meter by average seed weight.

Cleaned seed used for purity calculations from Crop 1 and Crop 2 were combined per accession and sent to Litchfield Analytical Service (Litchfield, Michigan, USA) for total fat content by ethanol extraction and a proximate feed analysis was done on the biomass waste from Crop 2.

5.2.7   Environmental Measurements

We used a plant growth photometer (International Light Inc., Newburyport, MA, USA) to measure light transmission into the greenhouse at irregular intervals over each crop cycle. Air temperatures were measured with maximum–minimum thermometers installed at the pad end and the fan end of the array of plants. Maximum daily temperatures recorded at approximately weekly intervals from July 25, 2012 to August 24, 2012 were used to establish the temperature gradient during the period of flowering and seed filling for Crop 1.

5.2.8   Statistical Methods

For analysis, data from individual blocks were consolidated into nine complete blocks running parallel to the temperature data in the greenhouse. Crop 1 and Crop 2 were analyzed by two-way analysis of variance (ANOVA) with accession (seed source) and block along the temperature gradient as categorical variables and seed yield, biomass yield, and HI as dependent variables. When ANOVAs were significant ($P < 0.05$), means were separated using the Holm-Sidak Test for comparing multiple means. Statistical procedures were carried out using Sigma Plot software (Systat, Inc., Santa Cruz, CA).

5.3   Results

5.3.1   Survival and Growth

An early display of pale yellow shoot tips during Crop 1 suggested that there was either a reduction of light in the greenhouse or that there was a mineral deficiency. Blue and red light were reduced in the greenhouse compared to outside by about 38% and far red was reduced by about 54%. Yellow tips receded when we substituted a synthetic sea salt compound (Instant Ocean, Union Pet Group, Inc., Cincinnati, OH, USA) for the in-house saline mix made from Epsom salts. No yellow shoot tips were evident during the course of Crop 2. Of 1200 established transplants in Crop 1, 996 (83%) survived the 2012 growing season and produced seed; 14 plants had no seed. Of 1100 (84%) established transplants in Crop 2, 922 survived the 2013 growing season and produced seed, whereas 11 plants had no seed. Crop 2 produced seed with a higher mean weight per seed than Crop 1. Nonseed-bearing plants tended to be either small plants from the Atlantic coast or lanky plants grown at the cool end of the greenhouse.

Table 5.2 Data for 2012 and 2013 Accessions for Mean Biomass, Seed Yield, Purity, Seed Weight and Total Seed Produced

	Biomass (g DW)	Seed (g)	Purity (%)	Seed (g m^{-2})	Wt Seed^{-1}	# Seed	Total Seed (g)	Total # Seed
2012								
BC	157.910	22.084	77.4	234.094	4.78E − 04	4.62E + 04	2429.273	5.08E + 06
BCW	154.491	20.240	78.5	214.549	4.62E − 04	4.38E + 04	2287.173	4.95E + 06
CCB	163.418	21.624	79.4	229.214	5.30E − 04	4.08E + 04	2378.639	4.49E + 06
EI	168.000	17.506	68.6	185.565	3.80E − 04	4.61E + 04	1698.097	4.47E + 06
G	123.627	22.996	75.9	243.759	3.35E − 04	6.87E + 04	2782.532	8.31E + 06
GI	146.462	21.905	80.3	232.191	4.29E − 04	5.10E + 04	2278.100	5.31E + 06
MI	136.640	18.595	65.1	197.111	4.43E − 04	4.19E + 04	2324.424	5.24E + 06
Sp	173.651	23.664	75.7	250.837	4.43E − 04	5.34E + 04	2579.364	5.82E + 06
SpN	156.240	21.159	67.6	224.285	4.81E − 04	4.40E + 04	1079.107	2.24E + 06
Mean	153.382	21.086	74.3	223.512	4.42E − 04	4.84E + 04	19836.709	4.59E + 07
2013								
BC	145.680	20.261	95.7	214.766	3.60E − 04	5.63E + 04	2066.618	5.74E + 06
BCW	134.663	18.640	91.2	197.587	4.60E − 04	4.05E + 04	1994.508	4.34E + 06
CCB	120.796	19.029	94.3	201.708	5.20E − 04	3.66E + 04	1769.701	3.40E + 06
EI	130.950	17.746	89.3	188.109	5.40E − 04	3.29E + 04	1082.512	2.00E + 06
G	107.476	19.966	94.6	211.643	4.40E − 04	4.54E + 04	1357.708	3.09E + 06
GI	135.984	20.645	88.9	218.839	4.80E − 04	4.30E + 04	1403.873	2.92E + 06
MI	150.055	19.575	93.9	207.493	4.80E − 04	4.08E + 04	1918.335	4.00E + 06
Sp	125.550	20.240	94.6	214.545	5.40E − 04	3.75E + 04	2226.407	4.12E + 06
SpN	146.927	20.360	88.8	215.815	5.20E − 04	3.92E + 04	2280.307	4.39E + 06
Mean	133.120	19.607	92.4	207.834	4.82E − 04	4.13E + 04	16099.968	3.40E + 07

Crop 1 produced a total of about 20 kg seed with an estimate of 4.59×10^7 total seeds and a mean weight of 0.442 mg per seed (Table 5.2). ANOVA of accessions for Crop 1 was significant for seed yield, biomass, and HI (Table 5.3). Grand Isle had a higher HI than South Padre, Corpus Christie Bridge, and Boca Chica West in 2012. Crop 2 produced a total of about 16 kg seed with an estimate of 3.40×10^7 total seeds and a mean weight of 0.482 mg per seed (Table 5.2). ANOVA of accessions for Crop 2

Table 5.3 Results of Two-Way ANOVAs for Seed Yield, Biomass and Harvest Index (HI) for Crops 1 and 2 *Salicornia bigelovii* Grown in a Tucson, Arizona, USA.

Crop/Dependent Variable	DF Accession, Block, Residual	F Accession, Block	P Accession, Block
Crop 1			
Seed yield	8, 9, 879	7.61, 10.0	<0.001, <0.001
Biomass	9, 8, 882	17.3, 14.1	<0.001, <0.001
HI	9, 8, 865	19.9, 32.3	<0.001, <0.001
Crop 2			
Seed yield	8, 7, 795	2.50, 3.09	0.003, 0.011
Biomass	8, 7, 795	3.76, 24.5	<0.001, <0.001
HI	8, 7, 779	2.96, 9.59	0.003, <0.001

Accession (Seed Source) and Block Along a Salinity Gradient were the Categorical Variables.

was also significant for biomass, seed yield, and HI (Table 5.3). Although the ANOVA for HI by accessions was significant ($P = 0.003$) for Crop 2, the means separation test did not find differences among any of the accessions at $P < 0.05$ for Crop 2. There were no significant differences between mean seed yield per plant between crops ($P > 0.05$).

The two Florida accessions were not included in Crop 2 analysis because they performed poorly. The Florida accessions produced a total of 403.4 g of seed from 42 plants or 9.6 g per plant, which is 50% less than the mean yield for Crop 1 accessions and the other accessions in Crop 2. Seed yield, biomass, and HI of individual accessions for Crop 1 and Crop 2 are in Figures 5.1 and 5.2, respectively.

Differences among accessions tended to be more pronounced for Crop 1 than for Crop 2. *Quimby*, from Massachusetts on the Atlantic Coast, had only half the seed and biomass yield as accessions from the Gulf of Mexico, and was not included in Crop 2. All of the Texas accessions performed well, yielding about 19–23 g of seed per plant (Table 5.3).

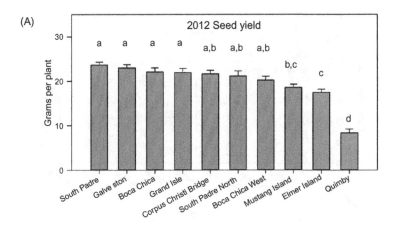

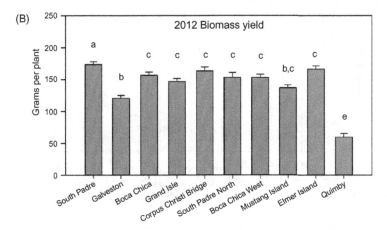

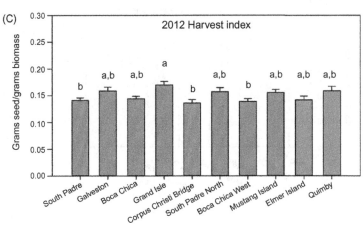

Figure 5.1 Mean seed yield (A), biomass yield (B), and harvest index (C) of Crop 1 accessions of *Salicornia bigelovii* grown in a greenhouse in Tucson, Arizona, USA in 2012. Error bars are standard errors of mean. Different letters over bars indicate means were significantly different at $P < 0.05$.

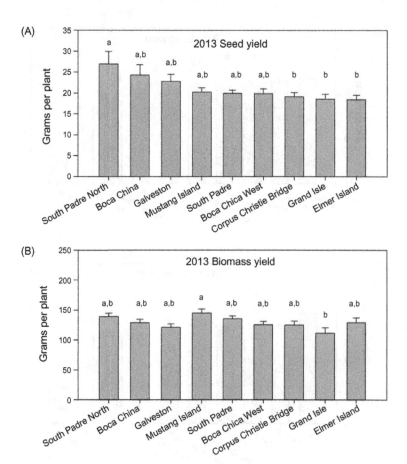

Figure 5.2 Mean seed yield (A), biomass yield (B) and harvest index (C) of Crop 2 accessions of *Salicornia bigelovii* grown in a greenhouse in Tucson, Arizona, USA in 2013. Error bars are standard errors of mean. Different letters over bars indicate means were significantly different at $P < 0.05$.

5.3.2 Temperature Effects

During the period of flowering and seed filling, maximum daily temperatures ranged from 33.9–38.9°C at the pad end to 40.6–43.8°C at the fan end of the greenhouse. For both Crop 1 and Crop 2, the effect of blocking for temperature was significant for seed yield, biomass, and HI (Figure 5.3). Seed yields peaked between 39°C and 42°C (Figure 5.3A) for all accessions except Quimby, for which yield decreased across the gradient (not shown). Biomass decreased over the whole temperature range (Figure 5.3B) and HI was maximal at 42.5°C (Figure 5.3C). Scatterplots of height versus temperature for Crop 1 showed that height increased with decreasing temperature (Figure 5.4).

5.3.3 Oil Content

Seed oil content as percent dry weight of combined Crops 1 and 2 for the nine best-performing accessions ranged from 22.3% to 25.9% while oil content for *Quimby* was 15.9%. Proximate analysis shows the biomass waste to be high in digestible nutrients and judged by the ash and sodium content, also high in salts (Table 5.4).

5.4 Discussion

Mean seed weight from greenhouse-grown accessions was lower than observed for accessions subjected to repeated harvest and mechanical threshing in previous field trials (Glenn et al., 1998). We attribute this to mass selection under repeated harvesting and threshing, in which the smaller seeds tend to be culled out at each harvest (Zerai et al., 2010).

Over two crop cycles, more than 1918 plants survived from 2300 transplanted, showing that greenhouse conditions were favorable for evaluating the yield potential of these accessions. The greenhouse had a 2-year average seed yield of 216.3 g m^{-2} from individually processed plants. The best-performing accessions came from the Gulf of Mexico and yielded the equivalent of 188–244 g m^{-2} of seeds and 1200–1800 g m^{-2} of biomass, similar to open-field yields under seawater irrigation (Glenn et al., 1991). Field trials of many of these same accessions were grown in outdoor plots in the United Arab Emirates with saline water irrigation, and similarly, Gulf of Mexico accessions performed best (Brown, unpublished data). Field yields of seed from other seed oil

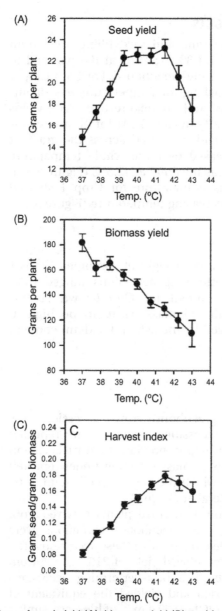

Figure 5.3 Means across accessions of seed yield (A), biomass yield (B), and harvest index (C) versus maximum day temperatures during the flowering and seed-filling stages of growth for *Salicornia bigelovii* grown in a greenhouse in Tucson, Arizona, USA. Error bars are standard errors of means.

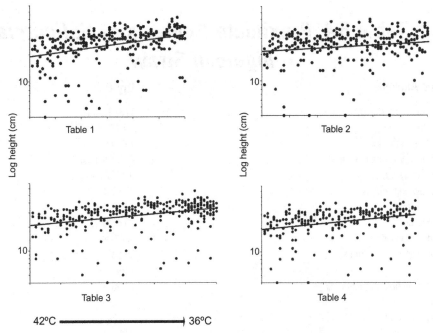

Figure 5.4 2012 Height (cm) of individual plant/row/table at time of last measure showing that the correlation between height and temperature favors a cooler temperature.

crops: hemp, canola, mustard, sesame, and poppy are in the range of 115–180 g m^{-2} (FAOSTAT, 2012). The oil content of these greenhouse-grown plants was slightly lower than reported for field-grown *S. bigelovii*, but still within the range of other seed oil crops (Glenn et al., 1991). Hence, this study confirms the high yield potential of this species under brackish water irrigation.

There was an inverse trend where plants that produced small seeds produced more seeds than plants that produced larger seeds. This is relevant to the goals of a breeding program because these seeds are small and difficult to process compared to some other oilseed crops. *Corpus Christi Bridge*, the tallest accession, and the *South Padre*s produced the largest seeds both years. *Elmer Island* showed the greatest improvement in HI and average seed weight from Crop 1 to Crop 2. *Quimby* from Virginia was not included in Crop 1 analysis or grown in Crop 2 because it was a poor-performing accession, but it had distinctive characteristics. It displayed little vegetative growth,

Table 5.4 Proximate Analysis of *Salicornia bigelovii* Straw

Feed Analysis	Dry Basis
Metabolizable energy	1731 Kcal kg^{-1}
Digestible energy	2.1 Mcal kg^{-1}
Net energy for lactation	1.06 Mcal kg^{-1}
Net energy for maintenance	0.90 Mcal kg^{-1}
Net energy for gain	0.20 Mcal kg^{-1}
Effective net energy	37.43%
Crude protein	4.86%
Crude fiber	20.47%
Crude carbohydrates	30.25%
Digestible carbohydrates	39.55%
Fat	1.9%
Total digestible nutrients	47.96%
Ash	33.23%
Phosphorus (P)	0.22%
Calcium (Ca)	0.25%
Potassium (K)	2.46%
Magnesium (Mg)	0.48%
Sodium (Na)	10.72%

but when the photoperiod and ambient temperatures began to change in August it grayed and with no further vegetative growth it produced large flower spikes that produced a low number of large black seeds; 0.933 mg per seed. Hence, this germplasm should be retained in a future breeding program.

5.5 Conclusion

Seed from S. *bigelovii* accessions collected from the Atlantic coast of the United States, the US portion of the Gulf of Mexico coast and *Puerto Rico* were grown and evaluated in a common garden experiment under greenhouse conditions in Tucson, Arizona, USA. Seed harvested from the highest-performing accessions from Crop 1 were used to produce Crop 2. Seed weights were larger for Crop 2 than Crop 1, showing the potential of improving the crop through selective breeding and mass selection. Accessions from the Gulf of Mexico tolerated the high greenhouse temperatures encountered in this study, but temperatures above 41°C reduced seed yield and should be avoided if possible. Seeding earlier in the season could allow plants to establish before extreme ambient temperatures (40–47°C) are reached and maintained for extended periods in the United Arab Emirates and similar climates. A successful strategy for hot climates could be to sow seed early (e.g., October) and harvest in April or May to avoid heat stress in summer, as was practiced in Eritrea (Zerai, 2010).

Continued cultivation of crops from individual top seed producers of each accession should be carried out in the greenhouse and concurrently in the field to establish whether sufficient genetic variation within accessions exists and to create founder lines of top producers, and to develop environmentally sound management techniques. In addition to improving seed yield and seed size, breeding goals should address improving heat sensitivity, selecting for early-flowering (day neutral) plants and selection for shorter but more numerous flower spikes, to reduce the long period from flowering to maturation as new flowers are produced along elongating spikes.

5.6 Acknowledgment

This research was funded by a grant from Masdar Institute's Sustainable Bioenergy Research Consortium, Abu Dhabi, UAE.

References

Brown, J.J., Glenn, E.P., Smith, S.E., 2014. Feasibility of halophyte domestication for high-salinity agriculture. In: Khan, M.A., et al., (Eds.), *Sabkha Ecosystems: Volume IV: Cash Crop Halophyte and Biodiversity 73 Conservation*, Tasks for Vegetation Science 47. Springer Science + Business Media Dordrecht, 2014.

FAOSTAT, 2012. <http://fao.org/site/291/default.aspx>. (Last visited August 2014).

Glenn, E., Hodges, C., Lieth, H., Pielke, R., Pitelka, L., 1992. Growing halophytes to remove carbon from the atmosphere. Environment. 34, 40–43.

Glenn, E., Brown, J., Blumwald, E., 1999. Salt tolerance and crop potential of halophytes. Crit. Rev. Plant Sci. 18, 227–255.

Glenn, E.P., O'Leary, J.W., Watson, M.C., Thompson, T.L., Kuehl, R.O., 1991. *S. bigelovii* Torr.: an oilseed halophyte for seawater irrigation. Science. 251, 1065–1067.

Glenn, E.P., Olsen, M., Frye, R., Moore, D., 1994. Use of halophytes to remove carbon from the atmosphere: results of a demonstration experiment. ElectricPower Research Institute, TR-103310, Research Report 8011-03, Palo Alto, CA, 6 chapters, various pages.

Glenn, E.P., Brown, J., O'Leary, J.W., 1998. Irrigating crops with seawater. Sci. Am. 279, 56–61.

Glenn, E.P., Mckeon, C., Gerhart, V., Nalger, P.L., 2009. Deficit irrigation of a landscape halophyte for reuse of saline waste water in a desert city. Landscape Urban Plan. 89, 57–64.

Glenn, E.P., Anday, T., Chaturvedi, R., Martinez-Garcia, R., Pearlstein, S., Soliz, D., et al., 2013. Three halophytes for saline-water agriculture: an oilseed, a forage, a grain. Environ. Exp. Bot. 92, 110–121.

Grattan, S.R., Benes, S.R., Peters, D.W., Diaz, F., 2008. Feasibility of irrigating pickelweed (*S. bigelovii* Torr.) with hyper-saline drainage water. J. Environ. Qual. 37, 149–156.

Jaradat, A.A., Shahid, M., 2012. The dwarf saltwort (*Salicornia bigelovii* Torr.): evaluation of breeding populations. ISRN Agron.article ID 151537.

Jordan, F.L., Yocklic, M., Morino, K., Seaman, R., Brown, P., Glenn, E.P., 2009. Consumptive water use and stomatal conductance of *Atriplex lentiformis* irrigated with industrial brine in a desert irrigation district. Agric. For. Meteorol. 149, 899–912.

Masters, D., Benes, S.R., Norman, H., 2007. Biosaline agriculture for forage and livestock production. Agric. Ecosyst. Environ. 119, 234–248.

Munz, P., 1974. A Flora of Southern California. University of California Press, Berkeley, CA.

Rozema, J., Flowers, T., 2008. Crops for a salinized world. Science. 322, 1478–1480.

Shahid, M., Jaradat, A.A., Rao, N.K., 2013. Use of marginal water for *Salicornia bigelovii* Torr. planting in the United Arab Emirates. In: Shahid, S.A., Abdelfattah, M.A., Taha, F.K. (Eds.), Developments in Soil Assessment and Reclamation: Innovative Thinking and Use of Marginal Soil and Water Resources in Irrigated Agriculture. Springer, Dordrecht, pp. 451–462.

Sheperd, K.A., Macfarlane, T.D., Colmer, T.D., 2005. Morphology, anatomy and histochemistry of Salicornioideae (Chenopodiaceae) fruits and seeds. Ann. Bot. (Lond.). 95, 917–933.

Weber, D., Ansari, R., Gul, B., Khan, M.A., 2007. Potential of halophytes as sources of edible oil. J. Arid. Environ. 68, 315–321.

York, J., Lu, Z., Glenn, E.P., John, M.E., 2000. Daylength affects floral initiation in *S. bigelovii* Torr. Plant Biol. 2000, 41–42 (abstract).

Zerai, D.B., Glenn, E.P., Chatervedi, R., Lu, Z., Mamood, A.N., Nelson, S.G., et al., 2010. Potential for the improvement of *S. bigelovii* through selective breeding. Ecol. Eng. 36, 730–739.

6

CARBON MITIGATION: A SALT MARSH ECOSYSTEM SERVICE IN TIMES OF CHANGE

Isabel Caçador[1], Bernardo Duarte[1], João Carlos Marques[2] and Noomene Sleimi[3]

[1]Marine and Environmental Sciences Centre, Faculty of Sciences of the University of Lisbon, Lisbon, Portugal [2]Marine and Environmental Sciences Centre, Faculty of Sciences and Technology, University of Coimbra, Coimbra, Portugal [3]UR-MaNE, Faculté des Sciences de Bizerte, Université de Carthage, Tunisia

6.1 Salt Marshes: Key Ecosystems

Estuarine systems represent important ecosystems in the coastal landscape, providing important ecosystem services such as water quality improvement, fishery resources, habitat and food for migratory and resident animals, and recreational areas for human populations. In the past 40 years, estuarine conservation was recognized as a priority at the national and international levels through several acts as the Ramsar Convention, 1977 (Munari and Mistri, 2008), and Water Framework Directive, 2000. Coastal, estuarine, and transitional waters have been affected by man's activities all over the world. Historically, developing human civilizations have often been concentrated in coastal areas where access to water promoted trade and commerce. As a consequence, human alteration of natural ecosystems is profound in coastal areas and a central theme of environmental management is to develop a policy to balance socioeconomic growth and environmental protection (Borja and Dauer, 2008). In the last 250 years, industrial activity has increased with a concomitant increase in fossil fuel usage (Houghton, 1999) and consequent atmospheric CO_2 increase. This has recognized consequences on climate change, namely increasing the global

M.A. Khan, M. Ozturk, B. Gul, & M.Z. Ahmed (Eds): Halophytes for Food Security in Dry Lands.
DOI: http://dx.doi.org/10.1016/B978-0-12-801854-5.00006-6

surface temperature (Bluemle et al., 1999; IPCC, 2007), and can put at risk the estuarine systems, such as salt marshes.

Salt marshes occupy the transition zone between terrestrial and marine ecosystems and are characterized by high productivity, which is considered essential in maintaining the detritus-based food chain supporting estuarine and coastal ecosystems (Marinucci, 1982). Salt marshes are key areas for the estuarine system, namely for primary production and nutrient regeneration (Caçador et al., 2009), thus becoming one of the most productive ecosystems on the planet (Lefeuvre et al., 2003). Estuarine wetlands, as salt marshes, constitute good carbon sinks having simultaneously reduced rates of greenhouse gas emissions (Magenheimer et al., 1996), with a carbon sequestration capacity per unit area of about one order of magnitude higher than other wetland systems (Bridgham et al., 2006). Salt marshes are usually located in estuarine systems and their primary production allows for a greater reduction of CO_2 in the atmosphere and incorporation into organic tissues through photosynthesis (Sousa et al., 2010a,b). Wetlands represent the largest carbon pool with a capacity of 770 Gt of carbon, overweighing the total carbon storage of farms and rainforests (Han et al., 2005).

6.2 Salt Marsh Sediments: Sinks or Sources?

In Mediterranean wetlands one can distinguish essentially two areas: the mudflat extension, prone to submersion and erosion, and the salt marsh, strongly colonized by dense networks of a variety of halophyte species. Wetlands, in particular salt marsh sediments, are very organic (Richert et al., 2000), being a good support for the microbial activity, providing large amounts of belowground litter as well as organic compound exudates by living plants (Duarte et al., 2007). The amount of organic carbon (C) transferred yearly to sediments and the decomposition rate of organic matter are reflected in the vertical profiles of C concentrations in sediments. Depth variation of C differed considerably between vegetated and nonvegetated sediments (Figure 6.1). Carbon concentrations in nonvegetated sediments are rather low, even at the surface, and decrease gradually with depth. On the other hand, vegetated sediments exhibited a subsurface C enrichment in layers of higher root biomass. This indicates unequivocally a marked influence of the halophyte vegetation, not only by introducing low-molecular-weight organic compounds (Duarte et al., 2007)

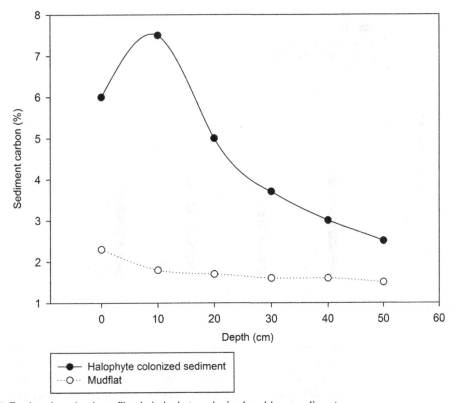

Figure 6.1 Total carbon depth profiles in halophyte colonized and bare sediments.

but also dead carbon-rich necromass (Duarte et al., 2008, 2010). Although locally this pattern is unequivocal, there are significant geographic patterns to be considered in terms of carbon retention (Figure 6.2). Typically warmer systems (Mediterranean-like) are more productive, thus providing higher carbon sequestration rates and consequent elevated litter production amounts (Duarte et al., 2013a). On the other hand, colder systems (Northern Atlantic-like) tend to present low productivity rates, thus influencing the carbon retention rates negatively (Duarte et al., 2013a).

Intrinsically connected to these halophyte-driven inputs are the microbial communities that play a key role on marsh biogeochemistry. Sediment microbial communities play an important role in these processes, being on the base of the ecosystem as they are the major decomposers of organic matter releasing nutrients into more phytoavailable forms (Ravit et al., 2003). This kind of sediment is often waterlogged and shows low levels of oxygen, which has adverse effects on plant growth (Richert et al., 2000). However, salt marsh plants are well known

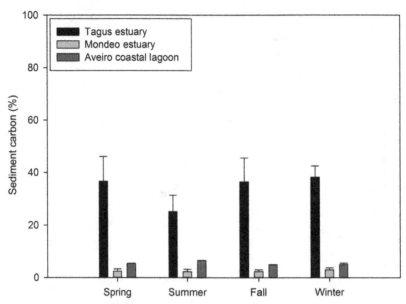

Figure 6.2 *Spartina maritima* sediment carbon stock in three different coastal systems.

for pumping oxygen from the atmosphere to the deeper sediment, turning the redox conditions of the root zone oxidative and as a consequence stimulating aerobic microbial activity (Ludemann et al., 2000). These kinds of plant–microflora interaction are quite variable depending not only on the plant species but also on the plant growth period and physiological state. As a primary and common product of this decomposing activity, CO_2 is released from the sediment.

Sediment CO_2 efflux rates suffer seasonal variations along the year intrinsically connected with three factors: organic matter quality (C:N ratio), pH, and temperature (Figure 6.3). The temperature versus respiration plots display the typical Gaussian curve of an enzymatic mechanism, with an optimal peak of activity around 18°C, decreasing towards extreme temperature. This temperature optimum coincides with the temperatures verified during a large part of the year in this system (Duarte et al., 2013a). Mangrove forest sediments also have a respiratory activity pattern and present a similar feedback to temperature, affecting CO_2 effluxes in a nonlinear way, with a peak in respiration rates around 25–27°C (Lovelock, 2008). Inevitably, temperature also affects indirectly other highly correlated parameters, such as the sediment water content. Orchard and Cook (1983) found that respiration increases with soil moisture and that rewetting soils caused a rapid increase in soil CO_2 efflux. Salt marsh sediments

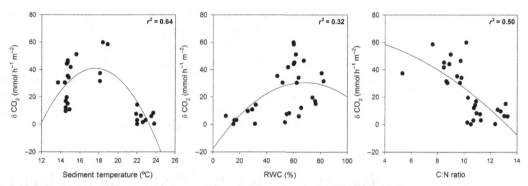

Figure 6.3 Sediment temperature, moisture and organic matter quality (C:N ratio) influence on sediment respiration rates.

can have water contents up to 80%, describing again a weak, but visible, Gaussian distribution with a peak of sediment respiration around 70%. Therefore, oxygen seems to be a limiting factor in moisture content above this optimum value. Salt marsh vegetation also has some influence on sediment O_2 dynamics. Although salt marsh sediments are periodically flooded, plants are able to colonize these sediments, partly due to their ability to diffuse oxygen through their roots into the sediment (Ludemann et al., 2000; Lillebø et al., 2006; Duarte et al., 2008; Caetano et al., 2011). This way, if oxygen is not a limiting factor, increases in sediment water content may be beneficial for sediment respiration. Also, pH is known to control not only enzymatic activities but also microbial growth (Madigan et al., 2009). Recently, some evidence has suggested the role of sediment pH in shaping the activity of extracellular phosphatase isoforms, indicating the decomposing efficiency of phosphorous-based respiratory substrates (Freitas et al., 2014). Other studies, focusing mainly on forest and agriculture soils pointed towards a shift in the soil pH driven by litter and necromass introduction (Prilha and Smolander, 1999; Smolander and Kitunen, 2002). Given the relatively high soil respiration rate at slightly acidic sediment pH, it is likely that a range of sediment pH exists where microbial activity and respiration could be high (Lee and Jose, 2003). The organic matter quality, instead of its total amount, also presents a good correlation with the sediment CO_2 fluxes. Costa et al. (2007) found that the microbial communities inhabiting the sediments have preference for the degradation of carboxylic acids, carbohydrates, and phenolic compounds, and only a small part of the microbial community showed any relation with nitrogen-based organic molecules (amino acids and amines). Also, other studies found that nitrogen (N) addition to soils, and thus a decrease in the C/N ratio, leads to an enhancement of the soil respiration. Gallardo and Schlesinger (1994) found an increase in soil respiration when nitrogen was added experimentally to forest soils in central North Carolina. Similar results were reported in a temperate forest in Germany (Brumme and Beese, 1992). Low tidal nutrient inputs into salt marshes point toward a N-limited sediment system, whereas N-rich organic matter favors the decomposition processes and thus increases sediment respiration.

6.3 Halophytes: An Efficient Carbon Pump

Alongside the sediments, the most important biological carbon sink in tidal wetlands is the halophytes; salt marsh plants vary carbon-harvesting efficiencies. This is evident from the stable isotope signature of the halophytes (Figure 6.4). C_4 plants, like the ones belonging to the genus *Spartina*, have a highly efficient photosynthetic mechanism to fix carbon both in the mesophyll and bundle sheath cells, thus having a lower (less negative) $\delta^{13}C$ signature. On the other hand, C_3 plants with the less efficient carbon-harvesting mechanism accumulate carbon at lower rates and thus present more negative

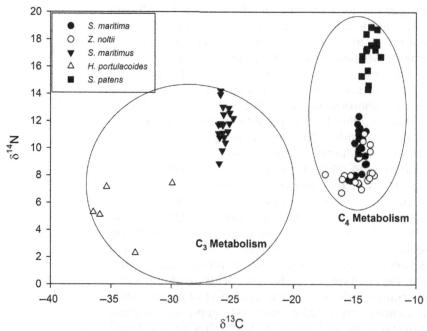

Figure 6.4 $\delta^{13}C$ and $\delta^{15}N$ stable isotope signatures of several C_3 and C_4 halophytes.

$\delta^{13}C$ signatures. Also, Middelburg et al. (1997) showed that in Georgia and Waarde marshes, the C_4 plants had a range of $\delta^{13}C$ between $-12\permil$ and $-17\permil$. The $\delta^{13}C$ of the atmospheric CO_2 is approximately $-7\permil$ (Zhou et al., 2006) and because of the anthropogenic burning of fossil fuels this value tends to be more negative, similar to a C_3 plant (approximately $-26\permil$) (Lloyd and Farquhar, 1994). This isotope approach illustrates very well the actual ability of the halophytes to harvest and accumulate carbon with their biomass, but can also help to understand the origin of the organic carbon trapped in the sediments. Baeta et al. (2009) showed that the sedimentary organic matter is mostly composed of the dead material derived from the halophyte coverage alongside a mix of terrestrial sources, micro- and macro-algae. This shows that part of the carbon exported by these species remains in the sediment, mostly driven from root senescence, reinforcing that the estuary acts as a carbon sink. On the other hand the aboveground senescence can act as a source of carbon too, since the major part of carbon is exported to the water column (Table 6.1). The relationship between the carbon metabolism (here illustrated by its stable isotope signature) and the carbon standing stocks, is very evident in the standing biomass role as a carbon sink, with the C_3 species (*Halimione portulacoides*, *Sarcocornia perennis*, and *Juncus maritimus*) presenting low-stocking ability when compared with the fast-growing halophytic grasses like *Spartina maritima*.

Table 6.1 Net Primary Production (NPP, Carbon kg m^{-2} y^{-1}) and Losses (Carbon kg m^{-2} y^{-1}) of Four Distinct Halophytes

		NPP	Losses
Halimione portulacoides	Aerial	0.04 ± 0.01	0.03 ± 0.01
	Root	0.09 ± 0.03	0.02 ± 0.01
S. perennis	Aerial	0.03 ± 0.01	0.02 ± 0.01
	Root	0.01 ± 0.01	0.01 ± 0.01
Juncus maritimus	Aerial	0.01 ± 0.00	0.00 ± 0.00
	Root	0.01 ± 0.00	0.00 ± 0.00
Spartina maritima	Aerial	0.12 ± 0.01	0.05 ± 0.01
	Root	1.72 ± 0.03	0.91 ± 0.04

6.4 Out-Welling Carbon

While the dead belowground biomass stays mostly in the sediment (Duarte et al., 2008), the aboveground detritus on the sediment surface experiences periodic tidal inundation and is transported to the adjacent water bodies (Duarte et al., 2010; Caçador et al., 2009). During dieback processes, for example, there is an enormous increase in the amount of aboveground necromass generated by the marsh due to increased senescence rates, and thus the out-welling phenomenon acquires an increased dimension enhanced by wetland degradation (Dagg et al., 2007). Nevertheless this detrital production is highly dependent on the plant species, which in European elevated marshes presents high diversity compared to the American low flat marshes (Beeftink, 1977; Lefeuvre et al., 1994; Caçador et al., 2009). The exportation of particulate organic matter, in the form of plant detritus, to support offshore consumers and water column nutrient regeneration, has been considered an important functional role of wetlands providing strong arguments for their conservation (Snedaker, 1978; Macintosh, 1981; Caçador et al., 2009; Duarte et al., 2010, 2014c). This characteristic was first postulated as out-welling by Odum (1968). Some years later, Odum (1980) reviewed the status of his hypothesis and concluded that out-welling from estuarine wetlands, while generally in existence, greatly depends on the physical characteristics of the wetland. Considering these hydrodynamic features and the nature of the plant detritus, in the present conditions these exposed salt marshes act as detritus exporters but to other areas of the estuarine domain, with a low contribution to the outer oceanic shelf (Figure 6.5). Only extreme events like storm surges or floods, acting similarly to spring tides, enhance this detritus exportation to the ocean. During these events there is a higher input of detrital organic carbon to the ocean, thereby increasing the secondary production of this area (Deegan et al., 2000). On the other hand, with sea level rise (SLR), the exports would increase even during neap tides, resulting in high organic load of the adjacent coastal waters. Two outcomes evolve from this: (i) enhanced fueling of secondary production in the coastal shelf and (ii) excessive N load and possible eutrophication of the coastal area. These two consequences (positive and negative) acquire greater importance if the full extension of the marsh ecosystem along with the cumulative plant-derived detritus pool is considered (Caçador et al., 2009, 2013).

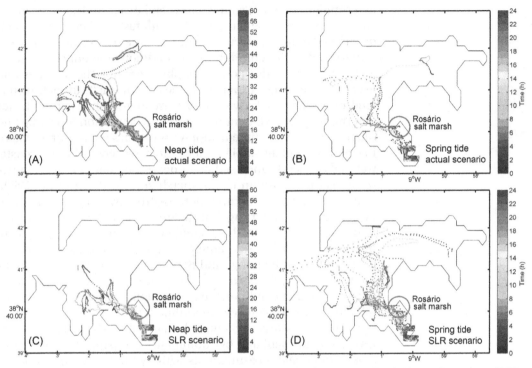

Figure 6.5 Detrital halophyte particles trajectories during neap (A,C) and spring (B,D) tides under present (A,B) and future SLR (C,D) conditions, at Tagus estuary (Portugal).

6.5 Hydrological Control of Carbon Stocks

Due to the highly dynamic and sensitive nature of salt marshes, it is important to consider that the forces that drive their decay are the same processes that drive their growth (Townend et al., 2011). The tidal dynamics contributes very strongly to salt marsh heterogeneity, simultaneously affecting chemical and physical factors like the extent and frequency of tidal flooding, sediments transport, salinity, and nutrient concentrations. Tidal drivers represent a set of factors that induce species to develop resistance mechanisms, thus adapting to the environment during their evolution (Vernberg, 1993). As such, the physical processes represent some of the factors lined by tide that are important to consider, especially while evaluating the biomass stock of pioneer species. *Spartina maritima* productivity depends on a complex interaction of physical and chemical factors, varying not only seasonally but also from marsh to marsh (Ibañez et al., 2000; Sousa et al., 2010a,b). In salt marshes, the nature of any tidal asymmetry is significant for determining sediment supply (Townend et al., 2011). According to Dias and Sousa (2009), in estuaries with fast tidal currents, flood-dominated currents result in net sediment transport into the estuary and under ebb-dominant conditions the opposite occurs and net seaward transport takes place, eroding the marshes and leading to sediment export into the nearby shelf. In fact, in this kind of estuary tidal asymmetry is frequently the shaping factor determining net sediment transport and deposition (Castaing and Allen, 1981). This sediment supply constitutes an important factor for salt marsh plant stability, since the persistence of these coastal wetlands depends upon sediment deposition and thus the control of the vertical position of the marsh surface (Reed et al., 1999). This takes us to the "competition-to-stress hypothesis" (Guo and Pennings, 2012), inferring that the upstream limits of plant distri-butions are determined by competition, and the downstream limits by abiotic stress, namely flooding. According to Bertness et al. (1992) and Pennings and Bertness (2001) competitive domi-nants (e.g., *H. portulacoides*) are typically unable to survive in physically harsh conditions (e.g., frequent flooding), while stress-tolerant, but competitively subordinate plants (e.g., *S. maritima*) grow in more stressful habitats because they are displaced from less stressful habitats by dominant competitors. Combining *S. maritima* field measurements with hydrological models points toward an expected increase in organic matter content and may result in higher future nutrient availability, which may even lead to an increase in the belowground/aboveground biomass ratio. On the other side, the increase in sediment moisture as a result of SLR effects and the consequent decrease in biomass is also a pos-sibility, which is in accordance with Sharpe and Baldwin (2012) who also claim that future SLR will reduce plant biomass. Consequently, if the first effect is dominant the *S. maritima* devel-opment in response to SLR will help to offset the effects of higher local dynamics, contributing to increase the marginal stability in both salt marshes and reinforce their coastal defense natural function. Otherwise, the reduction in the species abundance will accelerate the marginal instability and therefore coastal erosion (Valentim et al., 2013).

Considering this new information provided by the hydro-logical models, *S. maritima* will be negatively affected by increasing inundation periods and frequency, decreasing its biomass and thus its coverage area (Valentim et al., 2013). This way it becomes important to understand the photo-biological causes underlying this projected drawback in the *S. maritima* population under SLR scenarios. As observed in previous studies focusing on *S. maritima* communities (Silva et al., 2005), despite the mainly terrestrial characteristics of this species, there was no inhibition of photosynthetic activity in submerged individuals. These authors found that comparing both communities (submerged and air-exposed), in the terrestrial environment *S. maritima* scavenges about four times as much carbon as during submersion (Silva et al., 2005). This species presents a high adaptation capacity to submersion. Although after short periods of submersion an evident reduction in primary PSII photochemistry could be observed mainly due to its inability to deal with the absorbed light, after prolonged periods it displays effective mechanisms to dissipate the excessive light energy and for detoxification of the accumulated ROS, increasing the photosynthetic efficiency (Figure 6.6). All these aspects point to an evident photochemical plasticity of this species towards prolonged submersion periods like those expected under SLR, allowing it to maintain its photosynthetic activity even during prolonged submersion periods, and thus surviving the near-future climate change scenarios (Duarte et al., 2014a). Nevertheless, these ecophysiological counter-measures are made at severe energy costs, and thus have negative implications on the biomass production of the tussocks, in agreement with the previous findings that pointed to a reduction in *S. maritima* biomass under SLR.

6.6 Global Warming and Carbon Stocks

IPCC (2007) reports show that global temperature increased approximately 0.3°C per decade from 1979 to 2005. Salt marshes reduce the rates of greenhouse gas emissions (Magenheimer et al., 1996) and have a carbon sequestration capacity per unit area approximately one order of magnitude higher than other wetland systems (Bridgham et al., 2006), with the potential to sequester carbon continuously over thousands of years (Brevik and Houmburg, 2004). Ecological models have been used to clarify the effects of anthropogenic impacts on a global scale by integrating processes related to the biota of the ecosystem in its conceptual structure (Fragoso et al., 2009). Accordingly, an ecological model focuses on the objects of interest for a considered and well-defined problem. There can be many different ecological models of the same ecosystem, as the model version is selected according to the goals of modeling (Jørgensen and Fath, 2011). This is the case of the primary productivity models applied to halophytic vegetation (Figure 6.7). These models have as a basis, the modulation of the primary productivity of the halophytes and their relationships with the light environment,

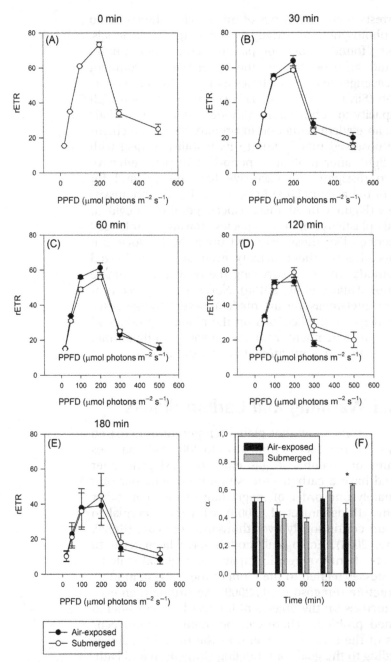

Figure 6.6 Photosynthetic electron transport rates (rETR; A,B,C,D,E) and photosynthetic efficiency (F) in submerged and air-exposed individuals.

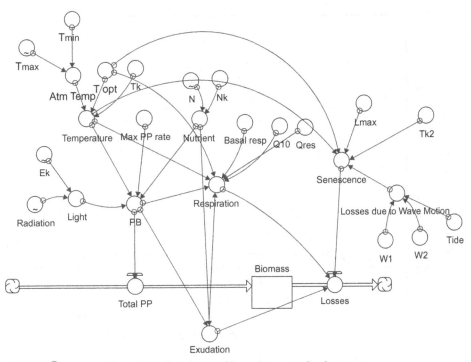

Figure 6.7 STELLA® conceptual model of *Spartina maritima* primary production.

nutrient resources, temperature of the system, and phonological processes that lead to senescence (Couto et al., 2014). If correctly calibrated the models can translate modeled data into real and accurate data, providing reliable predictions of the studied phenomenon according to the target conditions. Taking one of the most abundant halophytes in the Mediterranean systems into account, *S. maritima* and its primary production model, it is possible to clarify the effect of temperature increase on its primary productivity and thus carbon harvesting (Figure 6.8). The increase in temperature apparently had a positive effect in the model, as the plant biomass was promoted by the rising temperature by harvesting more carbon from the atmosphere. This can be combined by a series of secondary drivers, like increased nutrient availability under higher temperatures, higher daylight periods as well as higher water use efficiency (Duarte et al., 2013a). However, temperature increase will certainly affect the system in other ways. For example, it will cause sea level to increase due to oceanic thermal expansion, and also due to the melting of Arctic and Antarctic ice (Titus, 1991). This could be detrimental to salt marshes since a sea level increase would

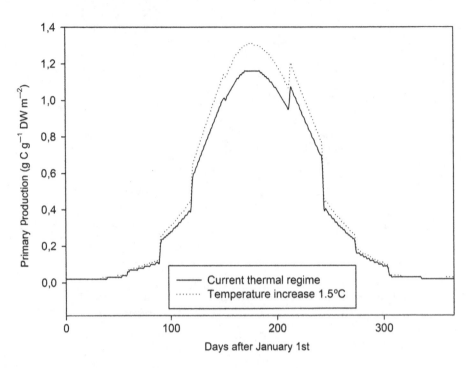

Figure 6.8 *Spartina maritima* model output for an annual primary production cycle under normal and future conditions.

place increased stress on certain marshes which might ultimately be submerged (Silliman et al., 2009).

Rising temperature seems to lead to a higher carbon standing stock, but the sediment carbon sink must also be addressed under this scenario. Currently, several studies are focusing on the effects of global warming on respiration in several ecosystems (e.g., Florides and Christodoulides, 2009; Shakun et al., 2012). These studies point to an important role of increasing temperature as a major contributor enhancing respiration and therefore the CO_2 fluxes to the atmosphere (Kirschbaum, 1995; Cox et al., 2000; Bond-Lamberty and Thomson, 2010). Again, ecosystem models can be key helpers for understanding and predicting these processes. Using a simpler approach, like generalized linear models, the simulations produced according to the IPCC predictions (IPCC, 2007), temperature seems to have a decreasing effect on the CO_2 effluxes due to sediment respiration (Figure 6.9). It is interesting to note that in other types of ecosystem such as semiarid grasslands, or arid regions, the effect of water content becomes dominant over temperature, in regulating soil respiration (Liu et al., 2009; Inglima et al., 2009). This points toward a differential effect of temperature in each

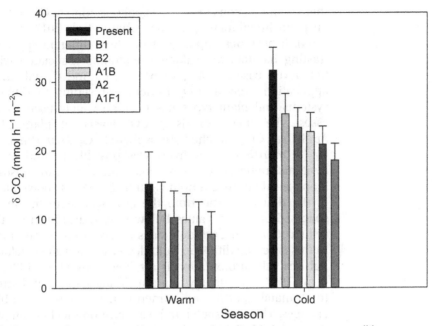

Figure 6.9 Sediment respiratory rates under present and predicted air temperature conditions.

ecosystem, as suggested by Singh et al. (2010). The effect of temperature may also vary seasonally, since below an optimal point, the effect of increasing temperature is quite the opposite of its effect over optimum values. Nevertheless, in the simulations produced according to the predicted IPCC scenarios, salt marshes tend to decrease their CO_2 emissions with increasing temperature, reinforcing their role as important carbon sinks (Caçador et al., 2004; Sousa et al., 2010a,b; Couto et al., 2013).

Both halophytes and their sediments present similar behaviors towards a common trend under rising temperature scenarios. Both system components act in the same direction: storing higher amounts of carbon. This can be interpreted as an ecosystem counteractive measure toward a reduction of the increasing temperature by decreasing the amounts of greenhouse gas emissions at the same time that the standing carbon stocks are increased.

6.7 CO_2 Rising in Salt Marshes: Improvement or Constraint?

In recent years several studies have been developed to evaluate the effects of CO_2 enrichment in different plant species, although there is still no consensus on a generalization of CO_2 effects based on plant functional type (Newingham et al., 2013; Nowak et al., 2004). It becomes important to address different species in contrasting habitats to evaluate constraints associated with rising CO_2 levels case by case, such as in C_3 and C_4 halophytes. The application Free-Air CO_2 Enrichment setups to different ecosystems and plant types has become widely used as a way to assess the effects of rising CO_2 levels on plant physiology. Applying a CO_2-enriched atmosphere to C_3 (*H. portulacoides*) and C_4 (*S. maritima*) halophytes was possible to understand their dynamics under these conditions combining both an isotopic and biophysical approach. Both C_3 and C_4 species showed a decrease in the diffusion ratio from the atmosphere to the intracellular spaces dropped to values below one, indicating that the intracellular CO_2 concentration stays below ambient one (Table 6.2). In this case, the diffusion ratio decreases (to values below one) as Rubisco full carboxylation is attained (Guy et al., 1986). Holtum et al. (1983) had already found a similar pattern and attributed it to stomatal aperture adjustments in response to ambient CO_2 changes. This was shown to be a common mechanism in both C_3 and C_4 species (Holtum et al., 1983). At a given environmental (CO_2) this diffusion coefficient is inversely proportional to the (CO_2) outside the plant, driving diffusion towards the leaf interior (Guy et al., 1986). This can be due to either an increase in photosynthetic capacity or to a decrease in stomatal conductance. A primary consequence of increased photosynthetic activity is a higher incorporation of atmospheric CO_2 in the plant tissues, thus increasing the $\delta^{13}C$ discrimination (Newingham et al., 2013). This was more evident in C_3 plants, in agreement with a higher increase in the cellular CO_2 concentration, supporting the hypothesis of Rubisco full carboxylation capacity and consequent photosynthetic enhancement. On the other hand, in C_4 species, intracellular (CO_2) concentration of the individuals exposed to CO_2 enrichment, showed only a small enhancement. Alongside, the $\delta^{13}C$ discrimination also didn't show any increased carbon incorporation, indicating that there was no photosynthetic capacity enhancement (Figure 6.10).

Observing the photochemical traits underlying these isotopic differences, the energetic mechanisms are elucidated (Figure 6.11). Analyzing the photochemical process through the perspective of PSII quantum efficiency, some differences arose. In *H. portulacoides* individuals exposed to high CO_2 levels there was a small increase in the maximum quantum efficiency, consistent with the findings from Cousins et al. (2002) and

Table 6.2 Diffusion Ratio (C_i/C_{atm}) Ratio and C_i of both *Halimione portulacoides* and *Spartina maritima* Exposed to 380 and 760 ppm of CO_2 (Average ± Standard Deviation)

	H. portulacoides		*S. maritima*	
	380 ppm	760 ppm	380 ppm	760 ppm
C_i/C_{atm}	1.15 ± 0.19^a	0.64 ± 0.19^a	0.51 ± 0.01^b	0.35 ± 0.04^b
C_i (ppm)	366.23 ± 11.55^a	548.14 ± 11.76^a	195.46 ± 3.62^a	265.10 ± 31.27^a

[a] $P < 0.01$.
[b] $P < 0.05$.

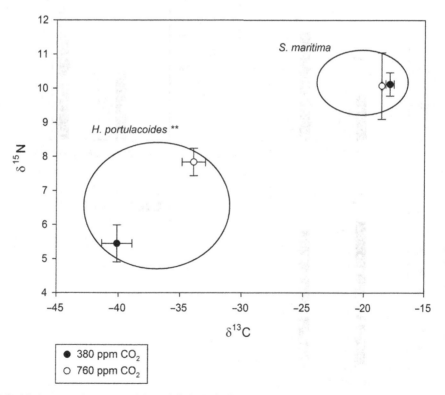

Figure 6.10 Stable isotope signatures of C_3 and C_4 halophytes under normal and CO_2-enriched atmosphere.

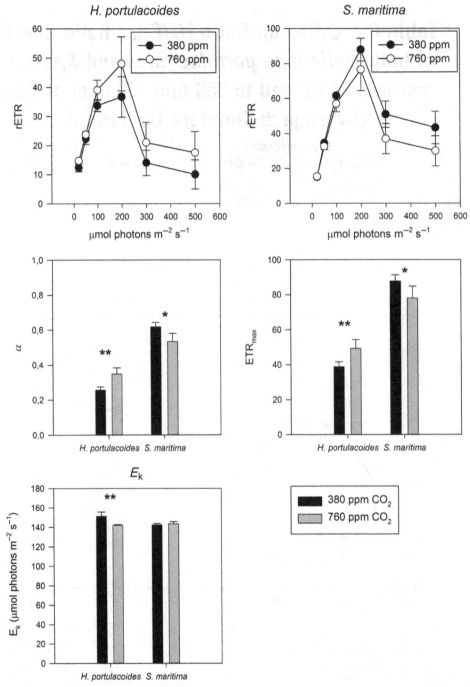

Figure 6.11 Electron transport rates (ETR), photosynthetic efficiency (α), and light saturation constants (E_k) of C_3 and C_4 halophytes under normal and elevated atmospheric CO_2 concentrations.

Wall et al. (2001). These authors suggested that this increase in PSII maximum efficiency is due to an increased ETR as a result of a higher efficiency in transporting absorbed quanta from the light-harvesting complexes (LHCs) to the PSII reaction centers, rather than a higher availability of electron acceptors downstream of PSII. This is supported by the higher values verified for maximum ETR rates inferred from the light curves. On the other hand, this species also showed low energy dissipation values, indicating a higher efficiency in the electronic transport under higher CO_2 levels. This fact has been attributed to an increasing demand for NADPH upon increasing CO_2 levels (Hymus et al., 1999). In this case, elevated CO_2 concentration under natural irradiance supports a higher electron transport rate (Long et al., 2004). As for *S. maritima* under CO_2 fertilization, it showed reduced PSII photosynthetic efficiencies and ETR, with a simultaneous increase in dissipated energy. As can be seen, changes were not mediated by alterations in electron fluxes, but by changes in redox-regulated non-photochemical quenching mechanisms, implying that, for a C_4 plant, photosynthesis is not substrate-limited (Wall et al., 2001). At midday conditions, high CO_2 availability increased linear electron transfer, dissipating the large variation in the stroma pH across thylakoid membranes as photosynthetic rates increased (Cousins et al., 2002). This accumulation of excessive reducing power needs to be dissipated or it will lead to damage in the thylakoid membrane system. Both these feedbacks to realistic future CO_2 concentrations are essential for future primary productivity models, as they are indicative of a possible abundance reduction of pioneer *S. maritima* and an increased biomass spreading of the sediment stabilizer, *H. portulacoides*. Thus, it becomes important to consider the inevitable changes in morphology and function of salt marshes, imposed by these atmospheric changes, both in terms of ecosystem functioning and loss of biodiversity (Duarte et al., 2014c).

If these changes are evident for plants, then the sediment and microflora will also be affected under CO_2 enrichment. Some studies suggest that marsh vegetation increases root-driven DOC release under CO_2 fertilization (Kim and Kang, 2008). DOC increase is expected to enhance organic matter decomposition through a priming effect (Allard et al., 2006). According to these authors, the priming effect would result in an increase in sediment organic matter, driven by plant rhizo-deposition of organic carbon forms, increasing the organic carbon sink in the sediments. These plant-driven carbon sources are not metabolized, but their increase stimulates sediment organic matter decomposition (Allard et al., 2006). This process is mostly mediated by an enhancement of the activity of several extracellular enzymatic activities (EEA) involved not only in the carbon cycle, but also in the mineralization of other nutrient forms (Shackle et al., 2006). Generally, rhizospheres accumulate higher labile carbon amounts than bulk sediments generated during photosynthesis exudates into the rhizosphere (Domanski et al., 2001; Jung et al., 2010). This way, DOC increase in the rhizosphere will enhance microbial activities, due to its C limitation under normal conditions (Jung et al., 2010). Another key factor, regulating the sediment carbon pool under elevated CO_2, is nitrogen concentration. In N-limited systems there is no observable activity increase under elevated CO_2 (Hungate et al., 2009). In salt marsh sediments colonized by halophytes there was an evident increase in microbial respiratory activity, pointing to a priming effect of elevated CO_2 in the rhizosphere community. Although these effects are markedly observable in the abundance and respiratory activity of the microorganisms inhabiting the halophyte rhizosphere, the ecosystem services provided were also suffered changes. One of the possible approaches undertaken for ecosystem function assessment is the evaluation of EEA of the sediments (Duarte et al., 2008, 2009, 2012, 2013b). Unlike some carbon and nitrogen enzymes, phosphatase showed a marked increase in the rhizosediments exposed to elevated CO_2 (Figure 6.12), pointing to an enhancement of the higher amounts of inorganic phosphate allowing higher growth rates (Kang et al., 2005). This was already reported for several other ecosystems like *Sphagnum*-dominated wetlands (Kang et al., 2005), tundra (Moorhead and Linkins, 1997), Mediterranean ecosystems (Dhillion et al., 1996), and grasslands (Ebersberger et al., 2003). This is in agreement with another mechanism of CO_2 interference in rhizosphere microbes. Increased concentrations of easily usable carbon sources, like monosaccharides, may inhibit some carbon-related enzymes (e.g., glucosidase), whilst other enzyme activities (e.g., phosphatase) may be increased to relieve limitation by other nutrients (Moorhead and Linkins, 1997; Barrett et al., 1998; Kang et al., 2001). Furthermore, actively growing vegetation under high atmospheric CO_2 can compete with microbes for organic nutrients, resulting in lower activities in some functional microbial groups (Freeman et al., 1998). The enzymatic activity of sulfatase also showed a marked increase under high CO_2. Sulfur is known to be an essential component of several amino acids, like cysteine and methionine. The high activity of this enzyme suggests a possible S-limitation as a result of CO_2 fertilization

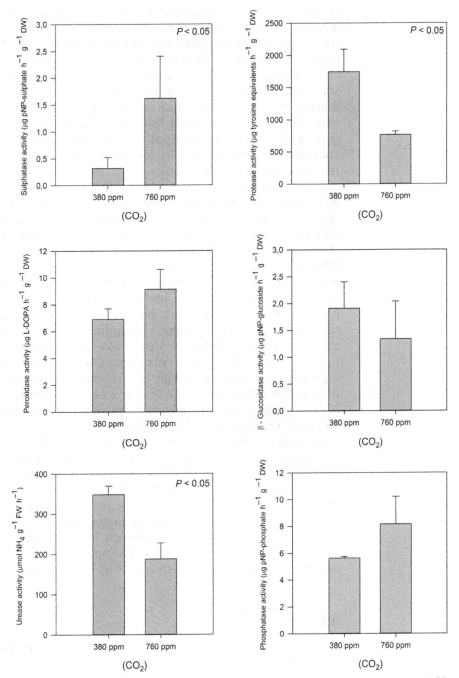

Figure 6.12 Rhizosediment extracellular enzymatic activities under normal and elevated atmospheric CO_2 concentrations.

(Kelley et al., 2011). The increase in microbial biomass and/or activity (as suggested by its proxy, dehydrogenase) increases the requirements in sulfur for protein synthesis and thus the acquisition rates of labile sulfur forms, like SO_4^{2-} (Kelley et al., 2011). Alongside with this need to acquire and synthesize amino acids for protein synthesis, there was also a marked decrease in protease activity. This is in accordance with the conclusions drawn from dehydrogenase activity, suggesting that there are no signs of N-limitation since N-linked enzymes all showed activity decreases under elevated CO_2 (Kelley et al., 2011; Kampichler et al., 1998). The same can be inferred if urease activity is observed. This enzyme produces NH_4 and CO_2 from urea hydrolysis (Askin and Kizilkaya, 2006). In this case, there was an inhibitory effect driven from excessive reaction products accumulation in the medium (Kampichler et al., 1998), mostly high NH_4 concentrations (Duarte et al., 2012), impairing urease activity. This is in agreement with the hypothesis provided by Langley and Megonigal (2010), pointing not to a local response from the plant or from the sediments, but to concerted feedback from the plant–sediment system. According to these authors, excessive phytoavailable N-forms, like NH_4, would lead to negative feedback from C_4 plants to CO_2 increase. Also, the exposure of sediments to high CO_2 concentrations contributed to a similar effect, since CO_2 is also a urease reaction product. In sum, high carbon supply from plants, resulting from CO_2 increase, changes microbial activities whether decreasing or enhancing it. Since these microorganisms are key players in the biogeochemical cycles occurring in salt marshes, these shifts will have impacts on the marsh and even on the estuarine system. The improvement of the recycling activity of organic N and P compounds and simultaneous depletion on the recycling capacity of C, has as major outcomes, an imbalance of the interconnections between biogeochemical cycles, affecting nutrient regeneration and supply to the primary producers, not only of the marsh but of all the estuarine system.

6.8 Final Remarks

Salt marshes have been considered as one of the most important ecosystems in relation to carbon harvesting. In fact, this is true although its halophytic composition and metabolic diversity are important shaping drivers of this ecosystem service. Nevertheless salt marsh vegetation constitutes alongside with its sediments a major carbon sink, with a high importance not only for the estuarine carbon balance but also for the global budgets. With high productivities also come high senescence rates, providing to the adjacent water bodies the necessary fuel for secondary production. In a planet under climate change, all these paradigms need to be re-evaluated. SLR, global warming, and rising atmospheric CO_2 levels will affect not only the halophytes, but the entire marsh ecosystem. Nevertheless these ecosystems seem to present an ecosystem feedback as counteractive measures to climate change, increasing their carbon sinks under climate change. This reinforces the marshes, and in particular the halophytes, as key players in the biogeochemical carbon cycle, both at the local and global scales.

References

Allard, V., Robin, C., Newton, P.C.D., Lieffering, M., Soussana, J.F., 2006. Short and long-term effects of elevated CO_2 on *Lolium perenne* rhizodeposition and its consequences on soil organic matter turnover and plant N yield. Soil Biol. Biochem. 38, 1178–1187.

Askin, T., Kizilkaya, R., 2006. Assessing spatial variability of soil enzyme activities in pasture topsoils using geostatistics. Eur. J. Soil Biol. 42, 230–237.

Baeta, A., Pinto, R., Valiela, I., Richard, P., Niquil, N., Marques, J.C., 2009. $\Delta^{15}N$ and $\delta^{13}C$ in the Mondego estuary food web: seasonal variation in producers and consumers. Mar. Environ. Res. 67, 109–116.

Barrett, D.J., Richardson, A.E., Gifford, R.M., 1998. Elevated atmospheric CO_2 concentrations increase wheat root phosphatase activity when growth is limited by phosphorus. Aust. J. Plant Physiol. 25, 87–93.

Beeftink, W.G., 1977. The coastal salt marshes of Western and Northern Europe: an ecological and phytosociological approach. In: Chapman, V.J. (Ed.), Wet Coastal Ecosystems. Elsevier, Amsterdam, pp. 109–155.

Bertness, M.D., Gough, L., Shumway, S.W., 1992. Salt tolerance and the distribution of fugitive salt marsh plants. Ecology. 72, 1832–1851.

Bluemle, J.P., Sabel, J.M., Karlén, W., 1999. Rate and magnitude of past global climate changes. Environ. Geosci. 6, 63–75.

Bond-Lamberty, B., Thomson, A., 2010. Temperature-associated increases in the global soil respiration record. Nature. 464, 579–583.

Borja, A., Dauer, D.M., 2008. Assessing the environmental quality status in estuarine and coastal systems: comparing methodologies and indices. Ecol. Indic. 8, 331–337.

Brevik, E.C., Houmburg, J.A., 2004. A 5000 year record of carbon sequestration from a coastal lagoon and wetland complex, Southern California, USA. Catena. 57, 221–232.

Bridgham, S.D., Megonigal, J.P., Keller, J.K., Bliss, N.B., Trettin, C., 2006. The carbon balance of North American wetlands. Wetlands. 26, 889–916.

Brumme, R., Beese, F., 1992. Effects of liming and nitrogen fertilization on emissions of CO_2 and N_2O from a temperate forest. J. Geophys. Res. 97, 12851–12858.

Caçador, I., Costa, A.L., Vale, C., 2004. Carbon storage in Tagus saltmarsh sediments. Water Air Soil Pollut. Focus. 4, 701–714.

Caçador, I., Caetano, M., Duarte, B., Vale, C., 2009. Stock and losses of trace metals from salt marsh plants. Mar. Environ. Res. 67, 75–82.

Caçador, I., Neto, J.M., Duarte, B., Barroso, D.V., Pinto, M., Marques, J.C., 2013. Development of an Angiosperm Quality Assessment Index (AquA—Index) for ecological quality evaluation of Portuguese water bodies—A multi-metric approach. Ecol. Indic. 25, 141–148.

Caetano, M., Bernárdez, P., Santos-Echeandia, J., Prego, R., Vale, C., 2011. Tidally driven N, P, Fe and Mn exchanges in salt marsh sediments of Tagus estuary (SW Europe). Environ. Monit. Assess. 184, 6541–6552.

Castaing, P., Allen, G.P., 1981. Mechanisms controlling seaward escape of suspended sediment from the Gironde: a macrotidal estuary in France. Mar. Geol. 40, 101–118.

Costa, A.L., Paixão, S.M., Caçador, I., Carolino, M., 2007. CLPP and EEA profiles of microbial communities in salt marsh sediments. J. Soils Sediments. 7, 418–425.

Cousins, A., Adam, N., Wall, G., Kimball, B., Pinter, P., Ottman, M., et al., 2002. Photosystem II energy use, non-photochemical quenching and the xanthophyll cycle in *Sorghum bicolor* grown under drought and free-air CO_2 enrichment (FACE) conditions. Plant Cell Environ. 25, 1551–1559.

Couto, T., Duarte, B., Caçador, I., Baeta, A., Marques, J.C., 2013. Salt marsh plants carbon storage in a temperate Atlantic estuary illustrated by a stable isotopic analysis based approach. Ecol. Indic. 32, 305–311.

Couto, T., Duarte, B., Martins, I., Caçador, I., Marques, J.C., 2014. Modelling the effects of global temperature increase on the growth of salt marsh plants. Appl. Ecol. Environ. Res. 12, 753–764.

Cox, P.M., Betts, R.A., Jones, C.D., Spall, S.A., Totterdell, I.J., 2000. Acceleration of global warming due to carbon-cycle feedbacks in a coupled climate model. Nature. 408, 184–187.

Dagg, M.J., Ammerman, J.W., Amon, R.M.W., Gardner, W.S., Green, R.E., Lohrenz, S.E., 2007. A review of water column processes influencing hypoxia in the Northern Gulf of Mexico. Estuar. Coasts. 30, 735–752.

Deegan, L.A., Hughes, J.E., Rountree, R.A., 2000. Salt marsh ecosystem support of marine transient species. In: Weinstein, M.P., Kreegers, D.A. (Eds.), Concepts and Controversies in Tidal Marsh Ecology. Kluwer Academic Publishers, The Netherlands, pp. 331–363.

Dhillion, S., Roy, J., Abrams, M., 1996. Assessing the impact of elevated CO_2 on soil microbial activity in a Mediterranean model ecosystem. Plant Soil. 187, 333–342.

Dias, J.M., Sousa, M.C., 2009. Numerical modeling of Ria Formosa tidal dynamics. J. Coast. Res. 56, 1345–1349.

Domanski, G., Kuzyakov, Y., Siniakina, D., Stahr, K., 2001. Carbon flows in the rhizosphere of ryegrass (*Lolium perenne*). J. Plant Nutr. Soil Sci. 164, 381–387.

Duarte, B., Delgado, M., Caçador, I., 2007. The role of citric acid in cadmium and nickel uptake and translocation, in *Halimione portulacoides*. Chemosphere. 69, 836–840.

Duarte, B., Reboreda, R., Caçador, I., 2008. Seasonal variation of Extracellular Enzymatic Activity (EEA) and its influence on metal speciation in a polluted salt marsh. Chemosphere. 73, 1056–1063.

Duarte, B., Raposo, P., Caçador, I., 2009. *Spartina maritima* (cordgrass) rhizosediment extracellular enzymatic activity and its role on organic matter decomposition and metal speciation processes. Mar. Ecol. 30, 65–73.

Duarte, B., Caetano, M., Almeida, P., Vale, C., Caçador, I., 2010. Accumulation and biological cycling of heavy metal in the root-sediment system of four salt marsh species, from Tagus estuary (Portugal). Environ. Pollut. 158, 1661–1668.

Duarte, B., Freitas, J., Caçador, I., 2012. Sediment microbial activities and physic-chemistry as progress indicators of salt marsh restoration processes. Ecol. Indic. 19, 231–239.

Duarte, B., Couto, T., Freitas, J., Valentim, J., Silva, H., Marques, J.C., et al., 2013a. Abiotic modulation of *S. maritima* photosynthetic ecotypic variations in different latitudinal populations. Estuar. Coast. Shelf Sci. 130, 127–137.

Duarte, B., Freitas, J., Couto, T., Silva, H., Marques, J.C., Caçador, I., 2013b. New multi-metric Salt marsh Sediment Microbial Index (SSMI) application to salt marsh sediments ecological status assessment. Ecol. Indic. 29, 390–397.

Duarte, B., Santos, D., Silva, H., Marques, J.C., Caçador, I., 2014a. Photochemical and biophysical feedbacks of C_3 and C_4 Mediterranean halophytes to atmospheric CO_2 enrichment confirmed by their stable isotope signatures. Plant Physiol. Biochem. 80, 10–22.

Duarte, B., Valentim, J.M., Dias, J.M., Silva, H., Marques, J.C., Caçador, I., 2014c. Modelling Sea Level Rise (SLR) impacts on salt marsh detrital outwelling C and N exports from an estuarine coastal lagoon to the ocean (Ria de Aveiro, Portugal). Ecol. Model. 289, 36–44.

Ebersberger, D., Niklaus, P.A., Kandeler, K., 2003. Long term CO_2 enrichment stimulates N-mineralization and enzyme activities in calcareous grassland. Soil Biol. Biochem. 35, 965–972.

Florides, G.A., Christodoulides, P., 2009. Global warming and carbon dioxide through sciences. Environ. Int. 35, 390–401.

Fragoso Jr., C.R., Ferreira, T.F., Marques, D.M., 2009. Modelagem Ecológica em Ecossistemas Aquáticos. Oficina de Textos, São Paulo, Brasil.

Freeman, C., Baxter, R., Farrar, J.F., Jones, S.E., Plum, S., Ashendon, T.W., et al., 1998. Could competition between plants and microbes regulate plant nutrition and atmospheric CO_2 concentration? Sci. Total Environ. 220, 181–184.

Freitas, J., Duarte, B., Caçador, I., 2014. Biogeochemical drivers of phosphatase activity in salt marsh sediments. J. Sea Res. 93, 57–62.

Gallardo, A., Schlesinger, W.H., 1994. Factors limiting microbial biomass in the mineral soil and forest floor of a warm-temperate forest. Soil Biol. Biochem. 26, 1409–1415.

Guo, H., Pennings, S.C., 2012. Mechanisms mediating plant distributions across estuarine landscapes in a low-latitude tidal estuary. Ecology. 93, 90–100.

Guy, R.D., Reid, D.M., Krouse, H.R., 1986. Factors affecting $^{13}C/^{12}C$ ratios of inland halophytes. I. Controlled studies on growth and isotopic composition of *Puccinellia nuttalliana*. Can. J. Bot. 64, 2693–2699.

Han, B., Wang, X.K., Ouyang, Z.Y., 2005. Saturation levels and carbon sequestration potentials of soil carbon pools in farmland ecosystems of China. Rural Eco Environ. 21, 6–11.

Holtum, I.A.M., O'Leary, M.H., Osmond, C.B., 1983. Effect of varying CO_2 partial pressure on photosynthesis and on carbon isotope composition of carbon-4 of malate from the Crassulacean acid metabolism plant *Katanchoe daigremontiana* Hamet et Perr. Plant Physiol. 71, 602–609.

Houghton, J., 1999. Global Warming: The Complete Briefing. fourth ed. Cambridge University Press, New York, NY.

Hungate, B.A., van Groenigen, K.J., Six, J., Jastrow, J.D., Luo, Y., de Graaff, M.A., et al., 2009. Assessing the effects of elevated carbon dioxide on soil carbon: a comparison of four meta-analyses. Glob. Change Biol. 15, 2020–2034.

Hymus, G.J., Ellsworth, D.S., Baker, N.R., Long, S.P., 1999. Does free-air carbon dioxide enrichment affect photochemical energy use by evergreen trees in different seasons? A chlorophyll fluorescence study of mature loblolly pine. Plant Physiol. 120, 1183–1191.

Ibañez, C., Curco, A., Day Jr., J.W., Prat, N., 2000. Structure and productivity of microtidal Mediterranean coastal marshes. In: Weinstein, M.P., Kreeger, D.A. (Eds.), Concepts and Controversies in Tidal Marsh Ecology. Kluwer Academic Publishers, London, p. 875.

Inglima, I., Alberti, G., Bertolini, T., Vaccari, F.P., Gioli, B., Miglietta, F., et al., 2009. Precipitation pulses enhance respiration of Mediterranean ecosystems: the balance between organic and inorganic components of increased soil CO_2 efflux. Glob. Change Biol. 15, 1289–1301.

Intergovernmental Panel on Climate Change (IPCC), 2007. Climate Change 2007: The Physical Science Basis. Cambridge University PressIPCC), 2007, New York, NY.

Jørgensen, S.E., Fath, B.D., 2011. Fundamentals of Ecological Modelling. fourth ed. Elsevier, USA.

Jung, S., Lee, S.-H., Park, S.-S., Kang, H., 2010. Effects of elevated CO_2 on organic matter decomposition capacities and community structure of sulfate-reducing bacteria in salt marsh sediment. J. Ecol. Field Biol. 33, 261–270.

Kampichler, C., Kandeler, E., Bardgett, R.D., Jones, T.H., Thompson, J., 1998. Impact of elevated atmospheric CO_2 concentration on soil microbial biomass and activity in a complex, weedy field model ecosystem. Glob. Change Biol. 4, 335–346.

Kang, H., Kim, S.-Y., Fenner, N., Freeman, C., 2005. Shifts of soil enzyme activities in wetlands exposed to elevated CO_2. Sci. Total Environ. 337, 207–212.

Kang, H.J., Freeman, C., Ashendon, T.W., 2001. Effects of elevated CO_2 on fen peat biogeochemistry. Sci. Total Environ. 279, 45–50.

Kelley, A.M., Fay, P.A., Polley, H.W., Gill, R.A., Jackson, R.B., 2011. Atmospheric CO_2 and soil extracellular enzyme activity: a meta-analysis and CO_2 gradient experiment. Ecosphere. 2, 1–20.

Kim, S.-Y., Kang, H., 2008. Effects of elevated CO_2 on below-ground processes in temperate marsh microcosms. Hydrobiologia. 605, 123–130.

Kirschbaum, M.U.F., 1995. The temperature dependence of soil organic matter decomposition, and the effect of global warming on soil organic C storage. Soil Biol. Biochem. 27, 753–760.

Langley, J.A., Megonigal, J.P., 2010. Ecosystem response to elevated CO_2 levels limited by nitrogen-induced plant species shift. Nature. 466, 96–99.

Lee, K.-H., Jose, S., 2003. Soil respiration, fine root production, and microbial biomass in cottonwood and loblolly pine plantations along a nitrogen fertilization gradient. Forest Ecol. Manage. 185, 263–273.

Lefeuvre, J.C., Bertru, G., Burel, F., Brient, L., Créach, V., Gueuné, Y., et al., 1994. Comparative studies on salt marsh processes: Mont Saint-Michel bay, a multi-disciplinary study. In: Mitsch, W.J. (Ed.), Global Wetlands: Old World and New. Elsevier Science B.V, Amsterdam, pp. 215–234.

Lefeuvre, J.C., Laffaille, P., Feunteun, E., Bouchard, V., Radureau, A., 2003. Biodiversity in salt marshes: from patrimonial value to ecosystem functioning. The case study of the Mont-Saint-Michel bay. C. R. Biol. 326, 125–131.

Lillebø, A.I., Flindt, M.R., Pardal, M.A., Marques, J.C., 2006. The effect of *Zostera noltii*, *S. maritima* and *Scirpus maritimus* on sediment pore-water profiles in a temperate intertidal estuary. Hydrobiologia. 555, 175–183.

Liu, W., Zhang, Z., Wan, S., 2009. Predominant role of water in regulating soil and microbial respiration and their responses to climate change in a semiarid grassland. Glob. Change Biol. 15, 184–195.

Lloyd, J., Farquhar, G.D., 1994. [13]C discrimination during CO_2 assimilation by the terrestrial biosphere. Oecologia. 99, 201–215.

Long, S.P., Ainsworth, E.A., Rogers, A., Ort, D.R., 2004. Rising atmospheric carbon dioxide: plants FACE the future. Annu. Rev. Plant Biol. 55, 591–628.

Lovelock, C.E., 2008. Soil respiration and belowground carbon allocation in mangrove forests. Ecosystems. 11, 342–354.

Ludemann, H., Arth, I., Wiesack, W., 2000. Spatial changes in the bacterial community structure along a vertical oxygen gradient in flooded paddy soil cores. Appl. Environ. Microbiol. 66, 754–762.

Macintosh, D.J., 1981. The importance of mangrove swamps to coastal fisheries and aquaculture. In: Proceedings of the Seminar on Some Aspects of Inland Aquaculture. Mangalore, Karnataka, 14 and 15 July, 1980, pp. 27–33.

Madigan, M.T., Martinko, J.M., Dunlap, P.V., Clark, D.P., 2009. Brock Biology of Microorganisms. Pearson Benjamin Cummings, USA.

Magenheimer, J.F., Moore, T.R., Chmura, G.L., Daoust, R.J., 1996. Methane and carbon dioxide flux from a macrotidal salt marsh, Bay of Fundy, New Brunswick. Estuaries. 19, 139–145.

Marinucci, A.C., 1982. Trophic importance of *Spartina alterniflora* production and decomposition to the marsh estuarine ecosystem. Biol. Conserv. 22, 35–58.

Middelburg, J.J., Nieuwenhuize, J., Lubberts, R.K., van de Plassche, O., 1997. Organic carbon isotope systematics of coastal marshes. Estuar. Coast. Shelf Sci. 45, 681–687.

Moorhead, D.L., Linkins, A.E., 1997. Elevated CO_2 alters belowground exoenzyme activities in tussock tundra. Plant Soil. 189, 321–329.

Munari, C., Mistri, M., 2008. The performance of benthic indicators of ecological change in Adriatic coastal lagoons: throwing the baby with the water? Mar. Pollut. Bull. 56, 95–105.

Newingham, B., Vanier, C., Charlet, T., Ogle, K., Smith, S., Nowak, R., 2013. No cumulative effect of 10 years of elevated [CO_2] on perennial plant biomass components in the Mojave Desert. Glob. Change Biol. 19, 2168–2189.

Nowak, R.S., Ellsworth, D.S., Smith, S.D., 2004. Functional responses of plants to elevated atmospheric CO_2—do photosynthetic and productivity data from FACE experiments support early predictions? New Phytol. 162, 253–280.

Odum, E.P., 1968. A research challenge: evaluating the productivity of coastal and estuarine water. In: Proceedings of the Second Sea Grant Conference, University of Rhode Island, pp. 63–64.

Odum, E.P., 1980. The status of three ecosystem-level hypotheses regarding salt marsh estuaries: tidal subsidy, outwelling and detritus based food chains. In: Kennedy, V.S. (Ed.), Estuarine Perspectives. Academic Press, New York, NY, USA, pp. 485–495.

Orchard, V.A., Cook, F.J., 1983. Relationship between soil respiration and soil moisture. Soil Biol. Biochem. 15, 447–453.

Pennings, S.C., Bertness, M.D., 2001. Salt marsh communities. In: Bertness, M.D., Gaines, S.D., Hay, M.E. (Eds.), Marine Community Ecology. Sinauer Associates, Sunderland, MA, USA, p. 550.

Prilha, O., Smolander, A., 1999. Nitrogen transformations in soil under *Pinus sylvestris*, *Picea abies* and *Betula pendula* at two forest sites. Soil Biol. Biochem. 31, 965–977.

Ravit, B., Ehrenfeld, J.G., Haggblom, M.M., 2003. A comparison of sediment microbial communities associated with *Phragmites australis* and *Spartina alterniflora* in two brackish wetlands of New Jersey. Estuaries. 26 (2B), 465–474.

Reed, D.J., Spercer, T., Murray, A.L., French, J.R., Leonard, L., 1999. Marsh surface sediment deposition and the role of tidal creeks: implications for created and managed coastal marshes. J. Coast. Conserv. 5, 81–90.

Richert, M., Saarnio, S., Juutinen, S., Silvola, J., Augustin, J., Merbach, W., 2000. Distribution of assimilated carbon in the system *Phragmites australis*-waterlogged peat soil after carbon-14 pulse labeling. Biol. Fert. Soils. 32, 1–7.

Shackle, V., Freeman, C., Reynolds, B., 2006. Exogenous enzyme supplements to promote treatment efficiency in constructed wetlands. Sci. Total Environ. 361, 18–24.

Shakun, J.D., Clark, P.U., He, F., Marcott, S.A., Mix, A.C., Liu, Z., et al., 2012. Global warming preceded by increasing carbon dioxide concentrations during the last deglaciation. Nature. 484, 49–55.

Sharpe, P.J., Baldwin, A.H., 2012. Tidal marsh plant community response to sea-level rise: a mesocosm study. Aquat. Bot. 101, 34–40.

Silliman, B.R., Grosholz, E.D., Bertness, M.D., 2009. Human Impacts on Salt Marshes: A Global Perspective. University of California Press, USA.

Silva, J., Santos, R., Calleja, M.L., Duarte, C.M., 2005. Submerged versus air-exposed intertidal macrophyte productivity: from physiological to community-level assessments. J. Exp. Mar. Biol. Ecol. 317, 87–95.

Singh, B.K., Bardgett, R.D., Smith, P., Reay, D.S., 2010. Microorganisms and climate change: terrestrial feedbacks and mitigation options. Nat. Rev. 8, 779–790.

Smolander, A., Kitunen, V., 2002. Soil microbial activities and characteristics of dissolved organic C and N in relation to tree species. Soil Biol. Biochem. 34, 651–660.

Snedaker, S.C., 1978. Mangroves: their value and perpetuation. Nat. Resour. 14, 6–13.

Sousa, A.I., Lillebø, A.I., Pardal, M.A., Caçador, I., 2010a. The influence of *S. maritima* on carbon retention capacity in salt marshes from warm-temperate estuaries. Mar. Pollut. Bull. 61, 215–223.

Sousa, A.I., Lillebø, A.I., Pardal, M.A., Caçador, I., 2010b. Productivity and nutrient cycling in salt marshes: contribution to ecosystem health. Estuar. Coast. Shelf Sci. 87, 640–646.

Titus, J.G., 1991. Greenhouse effect and coastal wetland policy: how Americans could abandon an area the size of Massachusetts at minimum cost. Environ. Manage. 15, 39–58.

Townend, I., Fletcher, C., Knappen, M., Rossington, K., 2011. A review of salt marsh dynamics. Water Environ. J. 25, 477–488.

Valentim, J.M., Vaz, N., Silva, H., Duarte, B., Caçador, I., Dias, J.M., 2013. Tagus estuary and Ria de aveiro salt marsh dynamics and the impact of sea level rise. Estuar. Coast. Shelf Sci. 130, 138–151.

Vernberg, F.J., 1993. Salt-marsh processes: a review. Environ. Toxicol. Chem. 12, 2167–2195.

Wall, G.W., Brooks, T.J., Adam, N.R., Cousins, A.B., Kimball, B.A., Pinter, P.J., et al., 2001. Elevated atmospheric CO_2 improved sorghum plant water status by ameliorating the adverse effects of drought. New Phytol. 152, 231–248.

Zhou, J., Wu, Y., Zhang, J., Kang, Q., Liu, Z., 2006. Carbon and nitrogen composition and stable isotope as potential indicators of source and fate of organic matter in the salt marsh of the Changjiang Estuary, China. Chemosphere. 65, 310–317.

FOOD SECURITY IN THE FACE OF SALINITY, DROUGHT, CLIMATE CHANGE, AND POPULATION GROWTH

John Cheeseman

Department of Plant Biology, University of Illinois at Urbana-Champaign, Urbana, IL, USA

7.1   Introduction

At this point in human history, the problem of food security is larger than it has ever been, with more than 800 million people chronically hungry and millions more at risk (FAO, 2013). Despite progress in some parts of the world to reduce hunger, in other areas, particularly in Africa and the Middle East, the hungry population is growing.

The problem will not get better by itself. Global climate change, including rising average temperatures, more severe droughts and more extreme weather variability will present ever-increasing challenges in marginal, already-stressed agricultural ecosystems. Moreover, no efforts to reduce hunger will be made easier by rapid population growth. In the next 35 years, the population of Earth is projected to increase by more than the current populations of China and India combined (FAO, 2011b). Most of this growth will be in developing countries, especially concentrated in the poorest communities in urban areas.

The contributions included in this volume represent a diversity of topics associated with halophyte physiology, molecular biology, and potential agricultural uses, largely focusing on dry lands including Bangladesh, Pakistan, the Arabian peninsula and across northern Africa to Morocco. These are areas affected by salinity, either because they are coastal or because inappropriate irrigation schemes have degraded soil and depleted or salinized groundwater. The world's major crops are proving

M.A. Khan, M. Ozturk, B. Gul, & M.Z. Ahmed (Eds): Halophytes for Food Security in Dry Lands.
DOI: http://dx.doi.org/10.1016/B978-0-12-801854-5.00007-8

inadequate to supply the calories, proteins, fats, and nutrients people need in these areas. New crops are needed, specifically appropriate to the myriad of ecological conditions in the region. The central question around which this volume is organized is whether halophytes can be a useful or important tool for assuring food security to people living in this region and under these conditions.

In this chapter, I will concentrate on the food security issue itself, especially contrasting the problems in Pakistan, one of the more at-risk countries in the region and in the world, and the oil-rich countries of the Arabian peninsula, especially Saudi Arabia and Qatar. I will also consider the potential for modifying old crops or developing new ones, particularly for Pakistan, recognizing the limitations imposed by local climate, societal instability, and the global economic environment.

7.2 The Problem of Food Security

Even among plant scientists, there is a well-ingrained perception that there is enough food in the world, but that the problem is distribution. If one takes the total production of grain (as the major food source by weight) and divides it by the total world population, this is perhaps true, although it is difficult to maintain confidence even at this level if one also projects future production and future population. But even at this level, it is clear that "distribution" is complicated by the fact that most suppliers have little apparent interest in distributing food to the more than 800 million hungry people in the world (FAO, 2013). This includes, in particular, commodity speculators who, in the 2007–2008 food price crisis, demonstrated their primary interest in creating and maintaining high prices, thus greatly reducing accessibility to those in poorer countries who were already spending two-thirds of their income or more on food (Ghosh, 2009; Rapsomanikis, 2009; Wahl, 2009). The problem is exacerbated by the increasing competition for nonfood uses of crops, for example, biofuels (Pimentel et al., 2009), and by rising standards of living, especially increased meat consumption, that alter global market structure.

With this in mind, it is worthwhile to examine the problem of food security more wholly. The broadest and most widely accepted definition of food security is that provided by the FAO, that is, as a "situation that exists when all people, at all times, have physical, social, and economic access to sufficient, safe, and nutritious food that meets their dietary needs and food preferences for an active and healthy life" (FAO, 2013). Neither the status of food security nor its achievability is uniform throughout the world, within individual countries, or over time.

There are four key dimensions to the problem (FAO, 2013):

- Availability—there must be adequate quantities for everyone, and the food must provide adequate amounts of calories, protein, fat, and micronutrients
- Accessibility—people must have both sufficient resources to obtain the food and physical access to it. Infrastructural components, such as roads and safe water supplies, are important factors affecting accessibility
- Usability—this includes having the resources for proper handling, storage and preparation, including safe and adequate water and sanitation
- Stability—supplies must be consistently available over time. Global climate change, including such immediate variables as temperature and precipitation, will affect both land suitability (including salinization) and crop yields (Battisti and Naylor, 2009; Schmidhuber and Tubiello, 2007), especially in poorer, drier nations at low latitudes.

Availability is more complex than simply something to fill your stomach; grains alone do not constitute "food." In Bangladesh, for example, the percentage of children who are underweight is falling as incomes are rising, and malnutrition is widespread (Ahmed et al., 2012). However, weight alone is not health. Nearly 80% of the calories consumed are cereals. Partly as a result of the poor overall nutritional quality in these seeds, more than 40% of women are anemic (FAO, 2013; Hyder et al., 2007). With CO_2 elevation and the associated reduction of micronutrients in C_3 grains, especially zinc and iron (Myers et al., 2014), this fraction may increase.

In many regions, food security, if it exists, is fragile. All four dimensions of food security are negatively impacted by political instability (Jenkins and Scanlan, 2001). Poverty, poor health, and malnutrition, in turn, are (and always have been) drivers for increasing armed conflict (Paveliuc-Olariu, 2013; Pinstrup-Andersen and Shimokawa, 2008). Moreover, a positive correlation between the occurrence of civil war and temperature suggests that armed conflict incidences may increase by more than 50% by 2030 associated with global climate change in the already hotter, drier and under-nourished regions of the world (Burke et al., 2009). This will further increase food insecurity.

Shabaz et al. (2012a) noted that conflict, along with the massive monsoon flooding in 2010 in Khyber Pakhtunkhwa province in north-west Pakistan, compounded the food security crisis in an area that was already one of the poorest in the country. The situation, however, is hardly better in the southern Sindh province where 72% of households are food insecure, and half the population suffers from nutritionally related stunting (Fazal et al., 2013), again exacerbated by conflict and flooding.

The issue of food security is not limited to poor countries, however. This was brought into clear focus in oil-rich countries of the Arabian peninsula during the food price crisis of 2007–2008. These countries were able to weather the crisis only because of their financial ability to buffer their local markets. One result of that has been their accelerating acquisition of farmland in developing countries. This is both a means of circumventing the water-limited agricultural systems of the investing countries (essentially importing water already processed through plants into food), and an expression of their distrust of the function of the global market place (Von Braun and Meinzen-Dick, 2009). Saudi Arabia, for example, has been reported to have removed its subsidy on water use for irrigation, much of it derived from desalination of seawater or over-pumping of groundwater, substituting instead food grown in Pakistan specifically for their consumption (Allouche, 2011). Qatar has similarly been in negotiations with Ghana to grow crops specifically for Qatar (Laessing, 2010). Von Braun and Meinzen-Dick (2009) present a table of 57 instances of such "land grabbing" [sic] between 2006 and 2009, the target countries being poor and with serious hunger issues of their own. They note that while these projects can inject much needed funds into the poorest areas of poor countries, they also threaten local access to and control of land and food resources by people in those areas.

Because of their wealth, had the investor countries the land and water resources themselves, they would technically be able to be highly efficient, although the history of their forays into agriculture using groundwater or desalinization of seawater does not reflect this possibility (Brown, 2009; Postel, 1998). Their oil-based economies could well support highly efficient planting, irrigating, fertilizing, and harvesting of crops. Whether this technology and investment can or will be made available to the targets of their investment is unclear, but I have found little reason to be optimistic in the near term.

At the same time, it should be noted that the capacities of the oil producers for outsourcing their food production and purchasing food security are highly dependent upon the world maintaining its dependence on fossil fuels, and hence on continued CO_2 emissions, warming, and climate change. If the world as a whole were to curb emissions, or if the Gulf countries' oil reserves were to be depleted, the economies and sustainability of these countries would be in question. This is not the problem for today, however, as no serious efforts to curb emissions have been taken by any country. Alternately, if warming is not curbed, the associated political, economic and social instability (Ulrichsen, 2009, 2011) may also overwhelm their short-term successes at buying food. Again, population promises to make all aspects of these relationships more difficult.

The possible solutions to the food security problem depend in large part on where in the world the problem is. In the dry countries from Pakistan and Bangladesh to Morocco, the problem is different from those facing countries in East Asia or sub-Saharan Africa. As noted in the "Introduction", the present discussion is most concerned with the former group, where drought and salinity are critical limitations. These areas may be saline either because they are coastal and have little available freshwater, or because inappropriate irrigation practices and related causes have led to salinization of their groundwater and soils. To solve this, crops will need to be able to integrate their salt management with other aspects of metabolism, especially with producing a product edible by and nutritious to humans or their animals. Whereas the genetic manipulation of crops to improve their nutritional value may be a viable solution to address deficiencies in the nutritional qualities of the plants themselves (Ronald, 2011), no such relatively simple solution is possible in the case of salt tolerance (Cheeseman, 2013).

Even within this region, the food security problems and solutions are not uniform. In countries with financial wealth with soil, water, and agriculture poverty, the problems are likely to grow in the next 35 years. Since the dawn of the petrochemical era, both the indigenous and guest worker populations in the oil-rich countries of the Arabian peninsula have grown well beyond what was previously sustainable. Today, 80% of the people in Qatar are non-Qatari (Qatar Information eXchange, 2014). In the working population, 94% are migrants/guest workers. The total population rose 5.6% in 2013, down from 14% in 2000 (World Bank, 2014). Also in 2013, the population of Saudi Arabia increased by 1.9%, Oman by 9.2%, and Kuwait by 3.6%. The ramifications of this for the whole region are huge; they are at risk of becoming very poor if or when the oil economy fades. Clearly, they will not be able to fall back on their own agricultural resources for food security.

Today's poor countries, on the other hand, and especially the poorest regions within those countries, are already experiencing a worst-case scenario. In a sense, they are prepared or preparing for things to come but, like the rich countries, they also have nothing to fall back on. Moreover, the worst case could get worse. The population of Pakistan, for example, has been growing at 2.2% since the 1980s (Anonymous, 2013), whereas wheat and rice production have been increasing more slowly (2.0% and 1.1% respectively, Khan et al., 2010).

7.3 The Problem of Salinity in Agriculture

Although there has been significant progress in reducing hunger world-wide in the last 25 years and there are promising signs for continued improvements in relatively stable, relatively productive parts of the world (FAO, 2013; Godfray et al., 2010), the outlook is less optimistic in areas affected by both drought and salinity. That salinity is a challenge and a major limitation to agriculture has been recognized for about 6000 years (Jacobsen and Adams, 1958). It has been a significant and recognized limitation to agricultural productivity in countries with industrial agriculture for more than 100 years (Lawton and Weathers, 1989). The relationships between salt and plants (at least one step removed from agriculture itself) have been a major area of study for more than 60 years (e.g., Epstein, 1956). In this latter phase, the salt relations of plants have been approached at the physiological, biochemical, and molecular levels as each sub-discipline emerged and flourished, leading to increasingly reductionist approaches with increasingly tenuous ties to organisms or to agriculture.

In the last roughly 20 years, molecular approaches have come to dominate basic research associated with salt, transport and stress tolerance, and land salinization and agricultural effects have increasingly been invoked as justification for research projects. Unfortunately, this has generally been little more than an invocation, usually at the start of the "Introduction" or "Discussion" section, and frequently as a conclusion (e.g., "our results show gene x is central to increasing salt tolerance in y"). The processes targeted have been quite limited and most often observational (e.g., simple photosynthetic rates, proline concentrations, activities of a few enzymes, sodium "exclusion," or worse, growth of *Arabidopsis* seedlings for 2 weeks in Petri plates with added glucose as a carbon source). The genes targeted have also been limited; although more than a hundred genes have been considered at some level, a few putative transporter genes and transcription factors are most highly represented (Cheeseman, 2013). Efforts to use molecularly based approaches to improve the salt tolerance of existing crops (especially wheat and rice) have made only painfully slow progress; even their successes are not all that impressive (Ashraf, 2010; Colmer et al., 2006; James et al., 2012; Munns, 2005; Munns et al., 2012; Schubert et al., 2009).

Today, more than 34 MHa are salt-affected and the area is increasing rapidly, with Pakistan being one of the most affected countries (FAO, 2011b). Food security puts a different spin and different urgency on the problem; once it is invoked, it is no longer sufficient to say that salinity is a growing problem. At stake are the lives of millions of people, and they are not winning.

7.4 Fitting Crops to the Environment—A Place for Halophytes?

Having dismissed molecular- or laboratory-based approaches to improving the salt tolerance of plants, I will briefly discuss what I consider to be four more viable options for improving or developing crops to meet the food security needs of dry and saline regions.

7.4.1 Option 1

Develop local cultivars of existing crops for local use; for example, wheat and rice, the two most important of the world's grains, are already major crops in Pakistan. In the sense that a large fraction of the literature on salinity and agriculture focuses on this approach, it is the obvious alternative. Such efforts have been underway in Pakistan, for example, for more than 45 years (Mahar et al., 2003; Shah et al., 1969), and more than 100 wheat landraces from Nepal and Pakistan have shown unexpectedly high salt tolerance (Martin et al., 1994). More recently, Shahzad et al. (2012) analyzed 190 landraces of wheat, including 130 from Pakistan, correlating molecular and morphological markers. They identified 12 SSR markers associated with tolerance to NaCl up to 250 mM, five of which were identified with four or more selectable traits. However, if these have been deployed in Pakistan (or countries with similar climatic conditions), they have not resulted in the yield increments needed to grow food production at the rate of population growth, or reduce under-nutrition. Because of the freshwater limitations on the Arabian peninsula, the costs of desalinization and the water use inefficiencies of C_3 crops, this is not an option for those countries.

7.4.2 Option 2

Expand efforts to cross crop species (especially wheat) with naturally salt-resistant relatives. Even within the Triticeae, there are more than 30 species that could be used for this (Colmer et al., 2006). In Pakistan, studies to improve wheat in this way have included efforts to cross it with *Aegilops ovata* (Poaceae) (Shafqat, 2002). Such efforts, particularly in wheat and rice, have a long history but have had only moderate success (Colmer et al., 2006). Perhaps most important for the present discussion, they have also not led to cultivars that can be deployed successfully in the poorest and most saline areas of the world, nor does a concerted effort appear to be pointed in that direction.

7.4.3 Option 3

Expand the species used in agriculture beyond the major grain crops. The premise here is that there are significant potential food crops lurking in diets around the world that could be expanded to new regions and contribute to local food security. Identifying them, moving them, establishing their cultivation (especially under drought and saline conditions) and developing local acceptance would be important aspects of exercising this option.

In Pakistan, for example, Shahbaz et al. (2012b), noted the importance of vegetables, not just grains, in providing calories, vitamins, proteins, and nutrients, and their importance to the diets of poor people there and in other parts of the developing world. Therefore, they studied the salt tolerance of pea, okra, tomato, eggplant, pepper, carrot, broccoli, cauliflower, and potato, and the strategies that were being used to enhance their salt tolerance. Such crops would also be important contributors in reducing the micronutrient limitations already in place and destined to increase as C_3 grains are subjected to higher atmospheric CO_2 levels (Myers et al., 2014).

Indeed, the number of plants which have served and are still serving as foods throughout the world is quite large even though they are poorly represented in the biological and agronomic literature. For example, Harlan (1992) provides a list of 88 genera harvested for food by native Australians, most of which are also represented in other countries as domesticated species (including *Chenopodium*, *Glycine*, *Ipomoea*, *Musa*, *Oryza*, *Solanum*, *Vigna*, and *Vitis*). In addition, he documented more than 250 plants cultivated in other regions of the world.

Prestcott-Allen and Prestcott-Allen (1990) listed 103 crops contributing 90% of the world's food supply based on FAO data, several of which were agglomerations (e.g., pulses, or roots and tubers). They noted, however, that some countries with unique and rich agricultural histories (e.g., Ethiopia) were not represented in their tables, so the number of globally or locally important crops is actually significantly higher.

Among the "minor crops," one in particular stands out for its transformation since these lists were published: quinoa (*Chenopodium quinoa*). This was an indigenous crop at high elevations in Bolivia, but its acceptance has experienced rapid expansion because it is highly nutritious, it can be substituted for other grains in many uses, and its domestication has not proceeded to the point at which reduced genetic variation compromises the development of new cultivars (FAO, 2011a). It is also naturally salt-tolerant.

7.4.4 Option 4

Localized development of new crops for local consumption. This option may be the only alternative in countries or regions experiencing low political stability (including ethnic disparities), poor infrastructure, or few if any financial resources. Essentially, it involves implementing domestication processes similar to those which led to the world's major grain crops, but in a much shorter period of time and in both biologically and sociologically stressful environments. The efforts would build on local knowledge and traditional uses of plants, and be accomplished largely by relatively uneducated farmers with the assistance (if possible) of extension specialists, but without reliable contributions from governments, universities, or international breeders. Much as the concept of de novo domestication of new crops is daunting, with concerted effort, it could be done in as little as 20–30 plant generations (Hancock, 2012). If perennials were targeted with the idea that they would remain perennials, it is conceivable that appropriate genotypes could be selected even more rapidly, the case of *Aeluropus lagopoides* in Pakistan being an example (Ahmed et al., 2013).

Local salt-tolerant relatives of major grain crops might themselves be such crops of the future. In fact, the development of halophytic crops is not a particularly new idea (e.g., Glenn et al., 1999; Rozema and Flowers, 2008; Yensen, 2006), but progress in accomplishing it has been slow, and in few cases has the intent been to develop food crops. In large part, this reflects the fact that plant biologists, crop scientists, and politicians in the developed world are not hungry, nor are their families, and thus, that funding has been difficult to obtain.

I envision this option as best suited to countries with an agricultural tradition, and question whether the historically non-agricultural oil-producing countries of the Arabian peninsula (for example) could follow this path. They do, however, have an endemic halophytic flora (from Shahina Ghazanfar, unpublished data), although they have managed it poorly; it is often difficult to find, having been subjected to oil spills and overgrazing. Whether this could be developed along with seawater irrigation for food and fodder will depend on the foresight and will of those ruling the countries. In any case, it is unlikely to support the population that has accrued and continues to grow rapidly.

7.5   Concluding Remarks

We are at a point of decision: do we attempt to develop crops to assure food security for saline and dry countries over the next 35 years, or do we let the people in those countries continue to fend for themselves. How we make this decision, or even if we make it, remains to be seen. Whether new crops can be developed rapidly enough to bring food security to all and assure its continuity will depend on many factors, including the will of governments and international organizations to seriously fund the efforts, and the willingness of people, especially poor people in poor countries, to adopt new crops and cropping methods if they are developed by "outsiders."

In 1995, Flowers and Yeo wrote: "although salinity might be of profound local importance, it has not yet had sufficient impact on regional agricultural production to warrant the effort necessary to produce new salt-tolerant cultivars." This is a very telling statement in that "regional agricultural production" largely refers to "regions" in the most developed countries, and the "effort" that has not been expended is effort by large agro-businesses. Without their involvement, what can we expect for dry, saline countries in our lifetimes and those of our children? The rich, now able to buy food without limit, are at peril because of their inability to produce locally. In an era of increasing CO_2, increasing temperatures, more frequent droughts, and population growth that will put ever-increasing pressure on water and land resources, a prosperous future is far from certain. For the poor, and particularly (in this volume) in those countries stretching from Bangladesh across the Arabian Peninsula and North Africa to Morocco, the situation is already dire. It does not seem likely to get better, and there is a very real possibility that they will have to go it alone.

References

Ahmed, M.Z., Shimazaki, T., Gulzar, S., Kikuchi, A., Gul, B., Khan, M.A., et al., 2013. The influence of genes regulating transmembrane transport of Na^+ on the salt resistance of *Aeluropus lagopoides*. Funct. Plant Biol. 40 (9), 860−871.

Ahmed, T., Mahfuz, M., Ireen, S., Ahmed, A.M.S., Rahman, S., Islam, M.M., et al., 2012. Nutrition of children and women in Bangladesh: trends and directions for the future. J. Health Popul. Nutr. 30, 1−11.

Allouche, J., 2011. The sustainability and resilience of global water and food systems: political analysis of the interplay between security, resource scarcity, political systems and global trade. Food Policy. 36 (Suppl. 1), S3−S8.

Anonymous, 2013. Pakistan. UNICEF, New York, NY, http://www.unicef.org/infobycountry/pakistan_pakistan_statistics.html.

Ashraf, M., 2010. Inducing drought tolerance in plants: recent advances. Biotechnol. Adv. 28, 169−183.

Battisti, D.S., Naylor, R.L., 2009. Historical warnings of future food insecurity with unprecedented seasonal heat. Science. 323, 240−244.

Brown, L., 2009. Plan B 4.0. W.W. Norton & Co., New York, NY.

Burke, M.B., Miguel, E., Satyanath, S., Dykema, J.A., Lobell, D.B., 2009. Warming increases the risk of civil war in Africa. Proc. Natl. Acad. Sci. 106, 20670−20674.

Cheeseman, J.M., 2013. The integration of activity in saline environments: problems and perspectives. Funct. Plant Biol. 40, 759−774.

Colmer, T.D., Flowers, T.J., Munns, R., 2006. Use of wild relatives to improve salt tolerance in wheat. J. Exp. Bot. 57, 1059−1078.

Epstein, E., 1956. Mineral nutrition of plants: mechanisms of uptake and transport. Annu. Rev. Plant Physiol. 7, 1−24.

FAO, 2011a. Quinoa: An Ancient Crop to Contribute to Global Food Security. Food and Agricultural Organization of the United Nations, Rome, http://www.fao.org/docrep/017/aq287e/aq287e.pdf.

FAO, 2011b. The State of the World's Land and Water Resources for Food and Agriculture (SOLAW)—Managing Systems at Risk. Earthscan, Abingdon, UK, http://www.fao.org/docrep/017/i1688e/i1688e.pdf.

FAO, 2013. The State of Food Insecurity in the World 2013: The Multiple Dimensions of Food Security. Food and Agricultural Organization of the United Nations, Rome, http://www.fao.org/publications/sofi/2013/en/.

Fazal, S., Valdettaro, P.M., Friedman, J., Basquin, C., Pietzsch, S., 2013. Towards improved food and nutrition security in Sindh Province, Pakistan. IDS Bull. 44, 21−30.

Flowers, T.J., Yeo, A.R., 1995. Breeding for salinity resistance in crop plants: where next? Aust. J. Plant Physiol. 22, 875−884.

Ghosh, J., 2009. The unnatural coupling: food and global finance. Presented at IDEAs: Re-regulating global finance in the light of global crisis, Tsinghua University, Beijing China, April 9−11, 2009, 20 pp. <http://www.levyinstitute.org/pubs/GEMconf2009/presentations/JayatiGhosh-Panel_II_paper.pdf>.

Glenn, E.P., Brown, J.J., Blumwald, E., 1999. Salt tolerance and crop potential of halophytes. CRC Crit. Rev. Plant Sci. 18, 227−255.

Godfray, H.C.J., Beddington, J.R., Crute, I.R., Haddad, L., Lawrence, D., Muir, J.F., et al., 2010. Food security: the challenge of feeding 9 billion people. Science. 327, 812–818.

Hancock, J.F., 2012. Plant Evolution and the Origin of Crop Species. third ed. CAB International, Oxford, UK.

Harlan, J.R., 1992. Crops and Man. second ed. American Society of Agronomy, Madison, WI.

Hyder, S.Z., Persson, L.-Å., Chowdhury, A., Ekström, E.-C., 2007. Anaemia among non-pregnant women in rural Bangladesh. Public Health Nutr. 4, 79–83.

Jacobsen, T., Adams, R.M., 1958. Salt and silt in ancient Mesopotamian agriculture. Science. 128, 1251–1258.

James, R.A., Blake, C., Zwart, A.B., Hare, R.A., Rathjen, A.J., Munns, R., 2012. Impact of ancestral wheat sodium exclusion genes *Nax1* and *Nax2* on grain yield of durum wheat on saline soils. Funct. Plant Biol. 39, 609–618.

Jenkins, J.C., Scanlan, S.J., 2001. Food security in less developed countries, 1970 to 1990. Am. Sociol. Rev. 66, 718–744.

Khan, S., Ahmad, F., Sadaf, S., Kashif, R., 2010. Crops Area and Production (by Districts) (1981–82 to 2008–09). Volume I: Food and Cash Crops. Government of Pakistan, Statistics Division, Federal Bureau of Statistics (Economic Wing), Islamabad, Pakistan, http://www.pbs.gov.pk/sites/default/files/agriculture_statistics/publications/area_and_production_by_districts_1981-82to2008-09/Area-and-Production-By-Districts-for-28-Years.PDF.

Laessing, U., 2010. Ghana in Talks with Qatar to Set JV Agriculture Firm. Reuters, http://www.reuters.com/article/2010/12/04/ghana-qatar-agriculture-idAFLDE6B305E20101204.

Lawton, H., Weathers, L., 1989. The origins of citrus research in California. In: Reuther, W., Calavan, E., Carman, G. (Eds.), The Citrus Industry: Crop Protection, Postharvest Technology, and Early History of Citrus Research in California. The Regents of the University of California, Oakland, CA, pp. 281–335.

Mahar, A.R., Memon, J.A., Abro, S.A., Hollington, P.A., 2003. Response of few newly developed salt-tolerant wheat landrace selections under natural environmental conditions. Pak. J. Bot. 35, 865–869.

Martin, P.K., Ambrose, M.J., Koebner, R.M.D., 1994. A wheat germplasm survey uncovers salt tolerance in genotypes not exposed to salt stress in the course of their selection. Aspects Appl. Biol.215–222.

Munns, R., 2005. Genes and salt tolerance: bringing them together. N. Phytol. 167, 645–663.

Munns, R., James, R.A., Xu, B., Athman, A., Conn, S.J., Jordans, C., et al., 2012. Wheat grain yield on saline soils is improved by an ancestral Na^+ transporter gene. Nat. Biotechnol.1–7.

Myers, S.S., Zanobetti, A., Kloog, I., Huybers, P., Leakey, A.D.B., Bloom, A.J., et al., 2014. Increasing CO_2 threatens human nutrition. Nature. 510, 139–142.

Paveliuc-Olariu, C., 2013. Food scarcity as a trigger for civil unrest. AAB Bioflux. 5, 174–178.

Pimentel, D., Marklein, A., Toth, M.A., Karpoff, M.N., Paul, G.S., McCormack, R., et al., 2009. Food versus biofuels: environmental and economic costs. Hum. Ecol. 37, 1–12.

Pinstrup-Andersen, P., Shimokawa, S., 2008. Do poverty and poor health and nutrition increase the risk of armed conflict onset? Food Policy. 33, 513–520.

Postel, S.L., 1998. Water for food production: will there be enough in 2025? Bioscience. 48, 629–637.

Prescott-Allen, R., Prescott-Allen, C., 1990. How many plants feed the world. Conserv. Biol. 4, 365–374.

Qatar Information eXchange, 2014. Census 2010—Demographic Characteristics. Ministry of Development Planning and Statistics, Doha, Qatar, http://www.qsa.gov.qa/eng/index.htm.

Rapsomanikis, G., 2009. The 2007—2008 Food Price Episode: Impact and Policies in Eastern and Southern Africa. FAO, Rome.

Ronald, P., 2011. Plant genetics, sustainable agriculture and global food security. Genetics. 188, 11—20.

Rozema, J., Flowers, T., 2008. Crops for a salinized world. Science. 322, 1478—1480.

Schmidhuber, J., Tubiello, F.N., 2007. Global food security under climate change. Proc. Natl. Acad. Sci. 104, 19703—19708.

Schubert, S., Neubert, A., Schierholt, A., Sumer, A., Zorb, C., 2009. Development of salt-resistant maize hybrids: the combination of physiological strategies using conventional breeding methods. Plant Sci. 177, 196—202.

Shafqat, F., 2002. *Aegilops ovata*: a potential gene source for improvement of salt tolerance of wheat. In: Ahmad, R., Malik, K.A. (Eds.), Prospects for Saline Agriculture. Kluwer Academic Publishers, pp. 123—130.

Shah, S.G.A., Aslam, M., Shah, S.U.D., 1969. Salt tolerance studies of ten wheat varieties in artificially salinized soil. West Pak. J. Agric. Res. 7, 6—15.

Shahbaz, B., Shah, Q.A., Suleri, A.Q., Commins, S., Malik, A.A., 2012a. Livelihoods, Basic Services and Social Protection in North-Western Pakistan. Sustainable Policy Development Institute, Working Papers, Secure Livelihoods Research Consortium, London, 70 pp. http://www.sdpi.org/publications/files/Livelihoods-basic-services-and-social-protection-in-North-Western-Pakistan.pdf.

Shahbaz, M., Ashraf, M., Al-Qurainy, F., Harris, P.J.C., 2012b. Salt tolerance in selected vegetable crops. CRC Crit. Rev. Plant Sci. 31, 303—320.

Shahzad, A., Ahmad, M., Iqbal, M., Ahmed, I., Ali, G.M., 2012. Evaluation of wheat landrace genotypes for salinity tolerance at vegetative stage by using morphological and molecular markers. Genet. Mol. Res. 11, 679—692.

Ulrichsen, K.C., 2009. Internal and external security in the Arab Gulf States. Middle East Policy. 16, 39—58.

Ulrichsen, K.C., 2011. The geopolitics of insecurity in the horn of Africa and the Arabian Peninsula. Middle East Policy. 18, 120—135.

Von Braun, J., Meinzen-Dick, R., 2009. "Land Grabbing" by Foreign Investors in Developing Countries: Risks and Opportunities. International Food Policy Research Institute, Washington, DC, 9 pp. http://www.ifpri.org/sites/default/files/publications/bp013all.pdf.

Wahl, P., 2009. Food Speculation: The Main Factor of the Price Bubble in 2008 (*Briefing paper*). World Economy, Ecology and Development, Berlin, 16 pp. http://www2.weed-online.org/uploads/weed_food_speculation.pdf.

World Bank, 2014. Population Growth (% Annual). World Bank Group, Washington, DC, http://data.worldbank.org/indicator/SP.POP.GROW.

Yensen, N., 2006. Halophyte uses for the twenty-first century. In: Khan, M., Weber, D. (Eds.), Ecophysiology of High Salinity Tolerant Plants. Springer Science + Business, Berlin, pp. 367—396.

8

THE IMPORTANCE OF MANGROVE ECOSYSTEMS FOR NATURE PROTECTION AND FOOD PRODUCTIVITY: ACTIONS OF UNESCO'S MAN AND THE BIOSPHERE PROGRAMME

Miguel Clüsener-Godt and María Rosa Cárdenas Tomažič
UNESCO Man and the Biosphere Programme, Division of Ecological and Earth Sciences, Paris, France

8.1 UNESCO Normative Tools to Ensure the Protection of the Environment and Its Wise Use

In 1968, the intergovernmental expert "Biosphere Conference," held at the United Nations Educational, Scientific and Cultural Organization (UNESCO) headquarters in Paris, highlighted for the first time the growing importance of environmental problems and made a series of recommendations concerning these issues, which were adopted at intergovernmental level. Increasing recognition of the downward trends in environmental health and the state of natural resources led to agreement to promote the rational use and conservation of natural resources through interdisciplinary approaches, hand-in-hand with policy and management issues for conservation (German Commission for UNESCO, 2007).

This conference led to a new emphasis on protected areas within UNESCO and two normative international instruments were created. In 1971, UNESCO launched the Man and the Biosphere Programme (MAB), which certifies biosphere reserves.

M.A. Khan, M. Ozturk, B. Gul, & M.Z. Ahmed (Eds): Halophytes for Food Security in Dry Lands.
DOI: http://dx.doi.org/10.1016/B978-0-12-801854-5.00008-X

In 1972, UNESCO's Member States adopted the *International Convention for the Protection of the World Cultural and Natural Heritage*, which establishes natural and cultural sites (Salem and Ghabbour, 2013).

Biosphere reserves and World Heritage sites provide two different visions for reconciling the use of natural resources and conservation. Both visions recognize the interdependent relationship between human cultures and the natural environments upon which their survival depends.

Another tool was adopted in 1971 in Ramsar, Iran. The *Convention on Wetlands of International Importance*, or the Ramsar Convention, is an intergovernmental treaty whose member countries agree to maintain the ecological character of their wetlands and to work to ensure the sustainable use of their resources.[1] UNESCO serves as the depositary for the Ramsar Convention.

In addition to these normative instruments, UNESCO also ensures the protection of the environment and its wise use through the development of projects on Integrated Coastal and Island Conservation and Development, and through the Monitoring of Conservation of Forests and Coastal Zones, including mangrove areas.

8.2   The MAB and Its World Network of Biosphere Reserves

The MAB was launched with the aim of establishing a scientific basis for the improvement of relationships between people and their environments. This intergovernmental scientific programme engages with the international development agenda to address challenges linked to scientific, environmental, societal, and development issues in diverse ecosystems ranging from mountain regions to marine, coastal, and island areas; and from tropical forests to drylands and urban areas.

Biosphere reserves are areas comprising terrestrial, marine and coastal ecosystems, nominated by national governments and recognized by the MAB Programme. They promote solutions to integrate the conservation of biodiversity and biological resources with their sustainable use.[2]

---

[1] *Convention on Wetlands of International Importance especially as Waterfowl Habitat*, Ramsar (Iran), February 2, 1971, UN Treaty Series No 14583, as amended by the Paris Protocol December 1982, and Regina Amendments, May 1987.
[2] Statutory Framework of the World Network of Biosphere Reserves.

As of 2014, the World Network of Biosphere Reserves comprises 631 biosphere reserves in 119 countries worldwide. The network is a key component in efforts to attain the goals of conserving biological diversity, promoting economic development, and maintaining associated cultural values.

Each biosphere reserve must fulfill three complementary and mutually reinforcing functions. The *conservation function* encompasses the preservation of genetic resources, species, ecosystems, and landscapes. The *development function* comprises efforts to foster sustainable economic and human development. The *logistic support function* includes support for demonstration projects, environmental education and training, and research and monitoring related to local, national and global issues of conservation and sustainable development.[3]

Biosphere reserves are organized into three interrelated zones (Figure 8.1):

- The *core area* comprises a strictly protected zone that contributes to the conservation of biodiversity, landscapes, ecosystems, species, and genetic variation.
- The *buffer zone* surrounds or adjoins the core area, and is used for activities compatible with sound ecological practices that can reinforce scientific research, monitoring, training, and education.

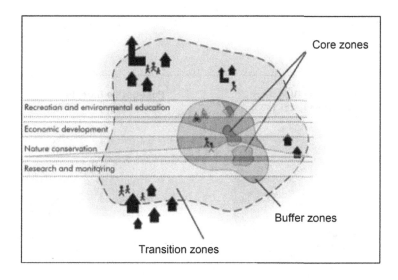

Figure 8.1 Biosphere reserve zonation. Source: http://intranet. floodmaster.de/.

---

[3]The Seville Strategy for Biosphere Reserves.

- The *transition zone* is where the greatest activity is allowed, and fosters economic and human development including environmental education, recreation, ecotourism, and applied and basic research. It includes a flexible transition area that incorporates different agricultural activities, settlements, and other uses, and in which different stakeholders, such as communities, management agencies, scientists, non-governmental organizations, cultural groups, and economic interests can work together to manage and develop the resources of the area in a sustainable manner.

In 1995, the *Seville Strategy for Biosphere Reserves* and the *Statutory Framework for the World Network of Biosphere Reserves* recommended a series of actions to ensure sustainable development for the twenty-first century.

One of the key directions was to "Ensure that all zones of biosphere reserves contribute appropriately to conservation, sustainable development and scientific understanding."

Until 1995, the core area represented 55% of the total area covered by biosphere reserves worldwide, while the buffer zone accounted for 29% and the transition zone 16%. Following the Seville Strategy, this trend changed and an increasing number of biosphere reserves began to incorporate a buffer zone and a transition area. Today, the core area represents 13% of the total area of biosphere reserves worldwide, while the buffer zone accounts for 34% and the transition zone 53% (Figure 8.2).

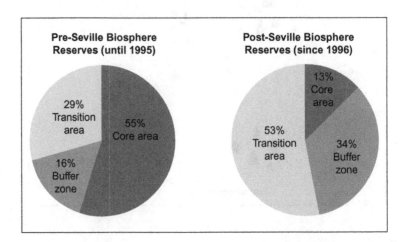

Figure 8.2 Evolution of biosphere reserves from 1995 to the present.

8.3 Distribution and Socio-Economic and Environmental Importance of Mangrove Ecosystems

Mangroves are found in 123 tropical and sub-tropical nations and cover 152,000 km^2. This equates to less than 1% of all tropical forests around the world, and less than 0.4% of the total global forest (Van Lavieren et al., 2012).

They are abundant and diverse along deltas, estuaries, and wetter coastlines. Over two-thirds of the world's mangroves are found in 12 countries. The largest areas are found on the wetter coastlines of South and Central America and West and Central Africa, and from northeast India through Southeast Asia to northern Australia (Figure 8.3) (ITTO, 2012).

Although mangrove ecosystems are rare globally they have a very high economic and ecological value because of the large range of ecosystem goods and services they offer. According to the *World Atlas of Mangroves* (Spalding et al., 2010), in ideal conditions, mangroves are one of the most productive ecosystems on earth and their economic value ranges from US$2000 to US$9000 per hectare per year. This value is significantly greater than potential values for any other alternative use, such as agriculture, aquaculture, or urban space, especially over the long term.

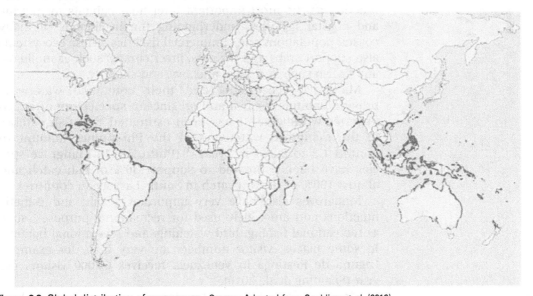

Figure 8.3 Global distribution of mangroves. Source: Adapted from Spalding et al. (2010).

Mangrove forests have higher levels of primary biomass productivity than most tropical or temperate forests because of their high effectiveness as global carbon stores and sinks (ITTO, 2012). It has been estimated that the total above-ground biomass is over 3700 Tg of carbon with a sequestration rate for carbon in the range of 14^{-17} Tg per year (Spalding et al., 2010).

Mangroves provide rot-resistant, high-value timber and the durability of the wood makes it a popular source of poles for buildings, boats, and fish-traps. Mangrove wood is also used in construction largely because of its resistance to termites. It also makes excellent fuelwood; however, in many cases it has been harvested unsustainably leading to major losses of mangrove cover.

The nutritional quality of mangroves is relatively poor and only a few communities are known to harvest and eat its fruits and propagules. Among the commonly used products is the sweet sap of *Nypa* palms, which can be consumed directly or fermented into an alcoholic drink or vinegar. Some mangrove species also produce a very good honey, and commercial apiculture is performed in different countries around the world, including Bangladesh, Cuba, Tanzania, the United States (Florida), and Viet Nam. Traditional communities in tropical coastal areas also use mangrove fruit, leaves, and bark for traditional medicine.

Mangroves also serve as nursery areas for fishery species and function as the most important inter-tidal habitats for marine and coastal fisheries, underpinning the livelihoods of many coastal populations and commercial fisheries. These ecosystems also support crabs and shrimps, invertebrates such as mollusks, and finfish such as mullet, anchovy, and snapper.

Mangrove ecosystems and their contained waterways provide habitats for fish and crustacean species important to fisheries worldwide. It has been estimated that fish catches in the mangrove waterways of the Philippines amount to around 1.3 to 8.8 kg h^{-1} year^{-1} (Pinto, 1987). Mangrove species have been estimated to support 30% of fish catch and almost 100% of shrimp catch in South-East Asian countries.

Mangroves also have very important scenic and esthetic functions and are widely used for recreational purposes, such as recreational fishing, bird watching, and recreational boating. In some places, visitor numbers are very high, for example, Laguna de Restinga in Venezuela receives 60,000 visitors per year (Spalding et al., 2010).

Mangroves are also important for wave attenuation and help to moderate coastal dynamics. In addition, their roots reduce erosion by holding soft sediments, and affect water movements by reducing wave height.

It has been widely reported that extensive areas of mangroves can decrease the loss of life and destruction caused by tsunamis. They shield coastlines by taking the first brunt of the impact and dispersing the energy of the wave as it passes through the mangrove forest. Studies have shown that wide mangrove areas sheltered some coastal areas which were close to the epicenter of the 2004 earthquake and tsunami. Simulations show that a coastal forest 200-m wide can reduce the hydraulic force of a 3-m tsunami by at least 80%, and the flow velocity by 70% (Forbes and Broadhead, 2007).

Desert and semi-desert countries have a high potential for the development of mangrove ecosystems. The whole Middle East region has some of the most arid coastlines in the world. Very few rivers reach the sea and mangroves are limited to lagoons and tidal creeks. However, mangroves have a high ecological value and in some areas constitute the only occurring trees. They offer unique foraging opportunities for livestock and nesting places for birds. Moreover, mangrove ecosystems function as the nursery for many pelagic organisms, which form the beginning of the food chain in coastal waters. Many crab and fish species have their origin in these areas. Mangroves are also used as a feeding ground for goats and camels, and in some areas as nesting grounds for sea turtles and other large animals. Fishing and collection of shells and firewood are also frequently reported.

However, in spite of the significant value of their ecosystem services, mangroves are still often seen as worthless wastelands available for other uses.

Commercial and high-intensity uses have altered the nature of the human–mangrove relationship, leading to considerable and continuing deforestation in almost all countries. According to FAO, between 1980 and 2005 some $35,600 \text{ km}^2$ were lost. Even if there are no accurate estimates of the original size of mangrove cover, the consensus places it at $200,000 \text{ km}^2$ and further gauges that over one-quarter of this cover has disappeared due to human actions (Figure 8.4).

The loss of the world's mangrove cover is due mainly to the direct conversion of mangrove areas to other uses, including aquaculture, agriculture, and urban and industrial uses.

Aquaculture, especially shrimp aquaculture, has been the strongest driver of mangrove conversion in recent decades. Mangrove areas throughout Central and South America and Southeast Asia have been transformed due to extensive

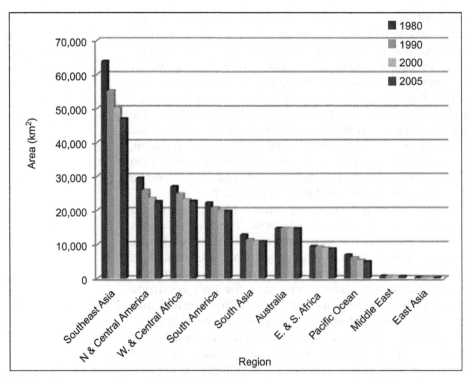

Figure 8.4 Decline in mangroves by region, 1980–2005. Source: Estimates based on 2007 FAO data.

aquaculture development. In many countries, this has led to a loss of 50% or more of mangrove cover (Saenger et al., 2013).

Extensive mangrove areas have been converted to arable or grazing land, even if the conversion cost is high and income from agriculture is comparatively low. This has occurred mainly in areas where population pressures are high and space is limited.

In Latin America, mangrove forests have been cleared and replaced with sugarcane and palm plantations, while in West Africa they have been replaced by rice paddies. Most of these conversions failed economically, ecologically, and socially. Yields are low and many mangrove soils become highly acidic and impossible to cultivate, leading to abandonment of the field (Cormier-Salem, 2006).

Mangroves are also often cleared and in-filled with dredged material to form low-lying areas for urban or industrial purposes. The remaining mangroves next to urban areas then quickly become degraded when they are used for fuel or building material.

It is critically important for the future of both mangroves and human societies that the roles and values of mangroves are properly evaluated and understood in order to preserve and/or use them in a sustainable way.

8.4 Actions of UNESCO's MAB

It is in this context that UNESCO's MAB has established several actions to contribute to the protection and sustainable use of mangroves.

In 2005, MAB established a 5-year cooperation programme with the International Tropical Timber Organization (ITTO), the International Society for Mangrove Ecosystems (ISME), the United Nations Food and Agricultural Organization (FAO), the United Nations Environmental Programme and its World Conservation Monitoring Centre (UNEP/WCMC), the United Nations University International Network for Water, Environment and Health (UNU/INWEH) and The Nature Conservancy (TNS), to gather worldwide data about the distribution and current status of mangrove ecosystems. The result was the *World Atlas of Mangroves* (Spalding et al., 2010). Available in English, French, and Spanish, the *Atlas* constitutes the first real global assessment of the state of mangroves, offering up-to-date and reliable coverage of nearly 99% of the world's mangroves, as well as statistics on biodiversity, habitat area, loss, and economic value. A total number of 73 species is detailed and range maps for all species are presented, including new maps derived from recent satellite imagery showing locations of the entire world's mangroves.

In 2012, the UNESCO/MAB Programme and its partners summarized the most relevant information from the *World Atlas of Mangroves* relating to management and policy development to produce the policy brief "Securing the Future of Mangroves" (Van Lavieren et al., 2012). The brief targets decision-makers and the broad public with a view to increasing awareness of mangroves and driving their effective protection and management at the policy level.

"Securing the Future of Mangroves" provides straightforward options for robust management and policy responses, and up-to-date information on the current status of mangrove ecosystems and their most important threats. It also describes the tools and measures that are currently available to assist the conservation and management of these ecosystems. It presents and analyses different lessons learned from around the world regarding conservation, management, and policy measures for the protection of mangroves.

In 2012, UNESCO also launched the "Floating Mangroves" project in cooperation with Lusail City, Qatar. The project monitors mangrove capacity for carbon sequestration to reduce atmospheric carbon levels. Qatar has the highest per capita carbon footprint in the world at 40.31 metric tonnes per capita.[4] One hectare of mangroves can sequestrate only about 1.5 million tonnes of carbon.

In 2013, the UNESCO Quito Office in partnership with the Permanent Commission for the South Pacific (CPPS) launched the Initiative on Mangroves and Sustainable Development, managed by Conservation International. This initiative aims to promote the conservation and sustainable use of mangroves, with a focus on valuing these ecosystem goods and services, addressing economic issues, and emphasizing the cultural, social, and spiritual aspects of mangroves.

As of 2014, there are 193 island and coastal biosphere reserves. Taking into account the importance of these areas, in 2012, MAB established the World Network of Island and Coastal Biosphere Reserves to study, implement, and disseminate island and coastal strategies to preserve biodiversity and heritage, promote sustainable development, and adapt to and mitigate the effects of climate change.

Island and coastal biosphere reserves, especially small islands in the Caribbean and the Asia-Pacific region, are highly vulnerable to climate change, the impacts of which cause poverty, natural disasters, depopulation, loss of traditional culture, and the detrimental effect of invasive species. These alter the balance of marine and terrestrial island ecosystems and cause irreversible loss of biodiversity.

In the two years since its formation, the Network has organized four international meetings and three training courses for managers from island and coastal biosphere reserves. It has also produced a website, a monthly newsletter, and four publications, as a means to exchange information.

Seventy-four of the 631 biosphere reserves that currently form part of the World Network of Biosphere Reserves include mangrove ecosystems, representing 11.73% of the total network. Africa has nine biosphere reserves in six countries that include mangrove ecosystems, the Arab States have three biosphere reserves in three countries, Asia and the Pacific have 27

---

[4]World Bank. CO_2 emissions (metric tonnes per capita) http://data.worldbank.org/indicator/EN.ATM.CO2E.PC.

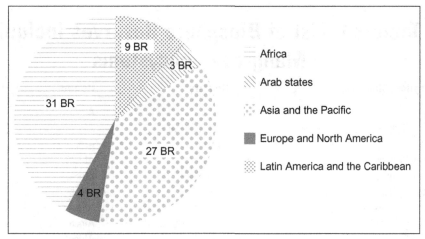

Figure 8.5 Regional distribution of biosphere reserves including mangrove ecosystems.

biosphere reserves in 12 countries, Europe and North America have three biosphere reserves in two countries and Latin America and the Caribbean have 31 biosphere reserves in 12 countries that include such ecosystems. With 13 biosphere reserves Mexico is the country with the most reserves with mangrove ecosystems. Conservation, reforestation, and sustainable management projects have taken place in some of these biosphere reserves (Figure 8.5 and Table 8.1).

8.5   Actions in Biosphere Reserves

The Palawan Biosphere Reserve in the Philippines was created in 1991. It covers around 14,000 km^2. The Province has a unique and diverse fauna and flora and is known as the country's "last ecological frontier." However, agriculture, fishing, mineral extraction, and offshore oil and natural gas, as well as tourism, have threatened the environment.

In 1992, a Strategic Environmental Plan for Palawan (SEP) was adopted, which led to the implementation of an Environmentally Critical Areas Network (ECAN). The SEP is a framework for the sustainable development of Palawan, taking an integrated approach to ecological viability and social acceptability. The ECAN is a graded system of protection and development control, which covers the whole of Palawan. It contains three components subdivided into management units graded from strictly protected to development areas. The terrestrial

Table 8.1 List of Biosphere Reserves Including Mangrove Ecosystems

Biosphere Reserve	Country	Region
Songor	Ghana	Africa
Boloma Bijagós	Guinea Bissau	Africa
Kiunga	Kenya	Africa
Malindi-Watamu	Kenya	Africa
Littoral de Toliara	Madagascar	Africa
Mananara Nord	Madagascar	Africa
Sahamalaza-Iles Radama	Madagascar	Africa
Delta du Fleuve Sénégal	Mauritania/Senegal	Africa
Delta de Saloum	Senegal	Africa
Al Reem	Qatar	Arab States
Marawah	United Arab Emirates	Arab States
Socotra Archipelago	Yemen	Arab States
Great Sandy	Australia	Asia and the Pacific
Mornington Peninsula and Western Port	Australia	Asia and the Pacific
Noosa	Australia	Asia and the Pacific
Prince Regent River	Australia	Asia and the Pacific
Wilson Promontory	Australia	Asia and the Pacific
Shankou Mangrove	China	Asia and the Pacific
Great Nicobar	India	Asia and the Pacific
Gulf of Mannar	India	Asia and the Pacific
Sunderban	India	Asia and the Pacific
Komodo	Indonesia	Asia and the Pacific
Siberut	Indonesia	Asia and the Pacific
Tanjung Puting	Indonesia	Asia and the Pacific
Wakatobi	Indonesia	Asia and the Pacific
Hara	Iran	Asia and the Pacific
Baa Atoll	Maldives	Asia and the Pacific
Utwe	Micronesia (Federal State of)	Asia and the Pacific
Ngaremeduu	Palau	Asia and the Pacific
Palawan	Philippines	Asia and the Pacific
Puerto Galera	Philippines	Asia and the Pacific
Bundala	Sri Lanka	Asia and the Pacific
Ranong	Thailand	Asia and the Pacific
Can Gio Mangrove	Viet Nam	Asia and the Pacific
Cat Ba	Viet Nam	Asia and the Pacific
Cu Lao Cham—Hoi An	Viet Nam	Asia and the Pacific
Kien Giang	Viet Nam	Asia and the Pacific

(Continued)

Table 8.1 (Continued)

Biosphere Reserve	Country	Region
Mui Ca Mau	Viet Nam	Asia and the Pacific
Red River Delta	Viet Nam	Asia and the Pacific
Archipel de la Guadeloupe	France	Europe and North America
Everglades and Dry Tortugas	United States of America	Europe and North America
Guanica	United States of America	Europe and North America
Virgin Islands	United States of America	Europe and North America
Mata Atlântica (including Sao Paolo Green Belt)	Brazil	Latin America and the Caribbean
Ciénaga Grande de Santa Marta	Colombia	Latin America and the Caribbean
Seaflower	Colombia	Latin America and the Caribbean
Agua y Paz	Costa Rica	Latin America and the Caribbean
La Amistad	Costa Rica/Panama	Latin America and the Caribbean
Baconao	Cuba	Latin America and the Caribbean
Buenavista	Cuba	Latin America and the Caribbean
Ciénaga de Zapata	Cuba	Latin America and the Caribbean
Cuchillas del Toa	Cuba	Latin America and the Caribbean
Península de Guanahacabibes	Cuba	Latin America and the Caribbean
Jaragua-Bahoruca-Enriquillo	Dominican Republic	Latin America and the Caribbean
Macizo del Cajas	Ecuador	Latin America and the Caribbean
Xiriualtique Jiquilizco	El Salvador	Latin America and the Caribbean
Maya	Guatemala	Latin America and the Caribbean
La Selle	Haiti	Latin America and the Caribbean
Rio Platano	Honduras	Latin America and the Caribbean
Arrecife Alacranes	Mexico	Latin America and the Caribbean
Banco Chinchorro	Mexico	Latin America and the Caribbean
Chamela Cuixmala	Mexico	Latin America and the Caribbean
El Vizcaino	Mexico	Latin America and the Caribbean
Huatulco	Mexico	Latin America and the Caribbean
Islas Marias	Mexico	Latin America and the Caribbean
La Encrucijada	Mexico	Latin America and the Caribbean
Los Tuxtlas	Mexico	Latin America and the Caribbean
Pantanos de Centla	Mexico	Latin America and the Caribbean
Ría Celestún	Mexico	Latin America and the Caribbean
Ría Lagartos	Mexico	Latin America and the Caribbean
Sian Ka'an	Mexico	Latin America and the Caribbean
Sistema Arrecifal Veracruzano	Mexico	Latin America and the Caribbean
Darien	Panama	Latin America and the Caribbean
Delta del Orinocco	Venezuela	Latin America and the Caribbean

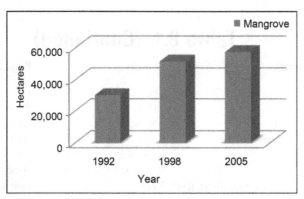

Figure 8.6 Mangroves on Palawan Island (PCSD, 2010).

component has a core zone, a buffer zone (subdivided into restricted, controlled and traditional use areas), and a multiple/manipulative use area.

In 1992, the total mangrove forest of Palawan was 50,602 ha. Since the establishment of the Strategic Environmental Plan for Palawan this area has increased to around 58,400 ha, and includes 31 species and 90% of the known mangrove species in the country. It accounts for 4% of the province's land area, and approximately 42% of the total remaining mangroves in the Philippines (PCSD, 2010; Regalo, 2013) (Figure 8.6).

The Can Gio Mangrove Biosphere Reserve, located in Vietnam, was designated as a biosphere reserve in 2000. Its 77 mangrove species (35 true mangroves and 42 associates) have contributed to the economic and environmental importance of the coastal zone. They provide timber, firewood, charcoal, tannin, food, and medicinals, but also act as the breeding ground for many species of marine organisms including shrimp, crab, fish, and water and migratory birds, as well as other economically terrestrial species of monkey, wild boars, and boas among others (Vietnam MAB National Committee, 2000).

The mangrove forest area was dramatically reduced during the two Indochina wars, due to over-exploitation for wood, fuelwood, and charcoal, and because of heavy herbicide spraying.

The local population with the support of national and international funds (Germany, Japan) has rehabilitated around 21,000 ha of mangroves. As a result, Can Gio has become one of the most beautiful and extensive sites of rehabilitated mangroves in the world with a high and varied biodiversity.

8.6 Conclusion

Mangrove ecosystems represent less than 1% of all tropical forests worldwide and less than 0.4% of the total global forest. However, their important economic and ecological role is now widely recognized. They provide a wide range of ecosystem goods and services ranging from food, fuelwood, and building materials to effective global carbon storage and sequestration, and protection of coastlines, among others. In many coastal territories the degree of dependence on mangroves is very high and there are no other alternative livelihoods. If mangroves disappear, many communities will lose their primary source of food and fuelwood and be at greater risk from storms and coastal erosion.

To ensure the effective management of mangrove ecosystems it is necessary to establish comprehensive legal and policy frameworks at the national level. To this end, it is vital to have access to sound long-term data on mangrove resources, their economic and ecological value, their functions and their responses to a range of pressures when taking rigorous policy and management decisions.

It is also important to reinforce technical, legal, and financial capacities for mangrove management, at various levels, through training courses and education programmes. These must take into account the priorities and need of the local community.

Finally, the success of any management intervention and legal and policy frameworks depends on the involvement of all relevant stakeholders and agencies, as well as the local communities, to ensure that they are well-adapted to local conditions.

References

Cormier-Salem, M.-C., 2006. Mangrove: changes and conflicts in claimed ownership, uses and purposes. In: Hoanh, C.T., Tuong, T.P., Gowing, J.W., Hardy, B. (Eds.), Environment and Livelihoods in Tropical Coastal Zones: Managing Agriculture–Fishery–Aquaculture Conflicts. CABI, Wallingford, UK and Cambridge, MA, US, pp. 163–176.

Forbes, K., Broadhead, J., 2007. The Role of Coastal Forests in the Mitigation of Tsunami Impacts. RAP Publication 2007/1. Food and Agriculture Organization of the United Nations (FAO), Regional Office for Asia and the Pacific, Bangkok.

German Commission for UNESCO, 2007. UNESCO Today: Magazine of the German Commission for UNESCO 2, 13.

ITTO (International Tropical Timber Organization), 2012. ITTO Tropical Forest Update 21, 2.

PCSD (Palawan Council for Sustainable Development), 2010. State of the Environment 2009 Updates. PCSD, Puerto Princesa City, Philippines.

Pinto, L., 1987. Environmental factors influencing the occurrence of juvenile fish in the mangroves of Pagbilao, Philippines. Hydrobiologia. 150, 283–301.

Regalo, A., 2013. Status monitoring of Palawan's mangrove resources. In: Presentation at the 1st Palawan Research and Policy Symposium, November 7–8, 2013, Palawan Island, Philippines.

Saenger, P., Gartside, D., Funge-Smith, S., 2013. A Review of Mangrove and Sea-Grass Ecosystems and Their Linkage to Fisheries and Fisheries Management. Food and Agriculture Organization of the United Nations (FAO), Regional Office for Asia and the Pacific, Bangkok.

Salem, B.B., Ghabbour, S.I., 2013. Joint biosphere reserves and world heritage sites. World Herit. Rev. 70, 26–27.

Spalding, M., Kainuma, M., Collins, L., 2010. World Atlas of Mangroves: A Collaborative Project of ITTO, ISME, FAO, UNEP-WCMC, UNESCO-MAB, UNU-INWEH and TNC. Earthscan, London.

Van Lavieren, H., Spalding, M., Alongi, D.M., Kainuma, M., Clüsener-Godt, M., Adeel, Z., 2012. Securing the Future of Mangroves. Policy Brief. UNESCO, Paris.

Vietnam MAB National Committee, 2000. Final report of the UNESCO/MAB Project on Valuation of the Mangrove Ecosystem in Can Gio Mangrove Biosphere Reserve, Viet Nam. Vietnam MAB National Committee, Hanoi.

9

THE POTENTIAL USE OF HALOPHYTES FOR THE DEVELOPMENT OF MARGINAL DRY AREAS IN MOROCCO

Salma Daoud[1], Khalid Elbrik[2], Naima Tachbibi[1], Laila Bouqbis[3], Meryem Brakez[1] and Moulay Chérif Harrouni[4]

[1]*Laboratory of Plant Biotechnologies, Faculty of Sciences, Ibn Zohr University, Agadir, Morocco* [2]*Faculty of Sciences, Ibn Zohr University, Agadir, Morocco* [3]*Polydisciplinary Faculty, Ibn Zohr University, Taroudant, Morocco* [4]*Hassan II Institute of Agronomy and Veterinary Medicine, Agadir, Morocco*

9.1 Introduction

Morocco is largely covered by semi-arid, arid, and desert climatical features due to its geographical location and owing to the high frequency of drought that has become structural (Debbarh and Badraoui, 2002). Food production is increasing in order to satisfy the demand of the increasing population and that depends heavily on irrigation and the efficient management of limited water resources. The expansion of irrigated agriculture and the intensive use of water resources combined with high evaporative rates, have inevitably given rise to the problems of salinity in the soil and underground water. Regions affected by salinity show a decline in crop production and hence in socio-economic development. However, areas which in the past presented an important diversity of natural resources are still inhabited by populations who are attached to their land and to their traditions. Most of these people have also inherited important local know-how from their ancestors who have managed to live in spite of the constraints since they have developed management techniques: the so-called "traditional" production systems that have enabled them to optimize agricultural production under extremely adverse conditions. This paper presents some considerations for the identification and the evaluation of the

M.A. Khan, M. Ozturk, B. Gul, & M.Z. Ahmed (Eds): Halophytes for Food Security in Dry Lands.
DOI: http://dx.doi.org/10.1016/B978-0-12-801854-5.00009-1

141

capacity of these marginalized areas to cope with climate change based on the capitalization of the local know-how and the young human resources. This can be achieved through valorization of traditional utilization of local halophytes and appropriate training for the introduction of a biosaline agricultural approach with the aim of restoring biodiversity and productivity.

9.2   Bio-Climate in Morocco

Due to its geographical location at the extreme northwest of Africa and its relief comprising four major mountainous chains, Morocco is characterized by a wide range of regional climatic conditions (oceanic plains, mountains, pre-Saharan, and Saharan regions). These climatic variations result from the interaction of several factors:

1. The latitudinal extension of Morocco from 21° to 36° in the northwest of the African continent, with a great opening, both on the Mediterranean sea in the north (500 km), and on the Atlantic ocean (3000 km) in the west (Anonymous, 2006a), as well as on the largest desert in the world, the Sahara in the south.

2. The topography that creates highly differentiated climatic zones: the Atlas mountains (3000 m average altitude) lie diagonally from the southwest to the northeast and constitute an obstacle to the prevailing northwesterly rain winds creating a dry area in the southeast on the edge of the Sahara. The Rif mountains (2000 m average altitude) in the north form a crescent-shaped barrier along the Mediterranean.

3. The location between two major centers of the atmospheric general circulation: the Azores anticyclone, an obstacle to the rainy disturbances originating from the polar front, and the Saharan depression (Anonymous, 2006b).

In view of these considerations, four bioclimatic zones are broadly identified for Morocco: (i) a humid and sub-humid zone (over 500 mm per year) in the extreme northwest and at high altitude (7% of the total area). It is characterized by short periods of drought and offers a high vegetation potential (with important biodiversity) enabling profitable rainfed agriculture; (ii) a semi-arid zone (250–500 mm per year) that mainly covers the coastal plains and continental plateaus in the north and west of the mountains of the High and Middle Atlas (15% of the total area). In this zone the dry summer period usually lasts 5–6 months. Rainfed agriculture is possible, but the yield is low; (iii) an arid zone which occurs in areas where the dry period exceeds 6 months. Temperature and precipitation conditions are such that the only possible activity is grazing; (iv) a Saharan zone where the practice of agriculture is impossible except in a few small areas (oases and wadi bottoms) with a special microclimate. The arid and Saharan zones receive less than 250 mm per year and represent 78% of the total area (Badraoui et al., 2000; Anonymous, 2006c, 2014). The climatic observations made during recent decades indicate that regions that were classified as humid or sub-humid are regressing towards a semiarid to arid climate (Anonymous, 2014; Driouech, 2010).

9.3 Biodiversity in Morocco

The geographical location and the climatic diversity of Morocco endow it with a wide diversity of ecosystems (inland, wetland, coastal, and marine areas) characterized by their flora and fauna richness with a Mediterranean specificity. Indeed, in the Mediterranean region, Moroccan biodiversity is second after Turkey, with an overall rate of 20% endemism. The flora of Morocco adds up to 7000 species identified so far and the fauna is rich, with 24,661 species (Anonymous, 2014). This diversity represents the resources underlying several economic activities (forestry, fishing, grazing, etc.) (Anonymous, 2006a). However, Morocco is a country predominantly arid since 93% of its total surface is semiarid, arid, or Saharan (Debbarh and Badraoui, 2002) and is strongly affected by climate change which is increasing its vulnerability. The sustainability of the country's socioeconomic development depends on the rational use of its resources in a sustainable development perspective (Anonymous, 2006a).

9.4 Vulnerability of Morocco to Climate Variations

Over the past three decades, Morocco has undergone five periods of severe drought combined with increasing demand for water. The global average warming throughout the country was estimated for 50 years to be about 1°C (Anonymous, 2006c, 2014). Moreover, spatial and temporal variability (random succession of dry and wet years) is accompanied by a significant overall decrease in the amounts of rain, ranging from 3% to 30% depending on the regions. Climatic projections made by the National Meteorology Direction predict an increase in average summer temperatures of 2−6°C and a 20% regression in average rainfall by the end of this century (Anonymous, 2014). Therefore, the water resources of the country, both surface and underground, are expected to further decline in the range of 15–20% by 2030 (Anonymous, 2006c). The evaporation rate is very high, especially in the continental parts, exacerbating the water deficit and bringing about salinization even in places where it was not previously prevalent. These areas are threatened by desertification to varying degrees due to water scarcity, loss of soil fertility, and pressure exerted by the growing population on natural resources to meet its needs (overgrazing and extension of cultivated lands). Natural resources are thus overexploited and their sustainability is threatened (Debbarh and Badraoui, 2002; Anonymous, 2009). This situation will likely have the strongest effect on Morocco since agriculture is of high importance for the country's economy and particularly for the poor in rural areas (Schilling et al., 2012). Agriculture contributes about 19% of the gross domestic product and supports an important part of the population since it provides employment for about 80% of the active population in rural areas (Anonymous, 2014).

With the frequency of drought years, the intensification of food production depends heavily on irrigation and the efficient management of limited water resources (Choukr-Allah and Harrouni, 1996). The rapidly increasing use of groundwater for irrigation in Morocco has allowed considerable agricultural growth, but in many areas such development becomes unsustainable due to overexploitation of aquifers and/or water and soil salinization (Anonymous, 2011).

9.5 Problems of Salinity in Morocco

At the present time, Morocco has more than 1 million ha of irrigated land. The expansion of irrigated agriculture and the intensive use of water resources combined with high evaporative rates, constitute a serious hazard to the sustainability of this land use system, especially in semiarid and arid areas where it often leads to degradation of soil quality (Badraoui et al., 2000). It has inevitably given rise to salinity problems in the soil and in underground water which have been observed in several regions of the country and are likely to spread as irrigation is intensified and irrigated areas are extended (Choukr-Allah and Harrouni, 1996; Anonymous, 2006a). Most irrigated land (500,000 ha) is affected by salinization to different degrees (Anonymous, 2006a), most of which is located in the Tafilalet, Ouarzazate, downstream Moulouya, Doukkala and Gharb perimeters (Figure 9.1) (Harrouni, 2000).

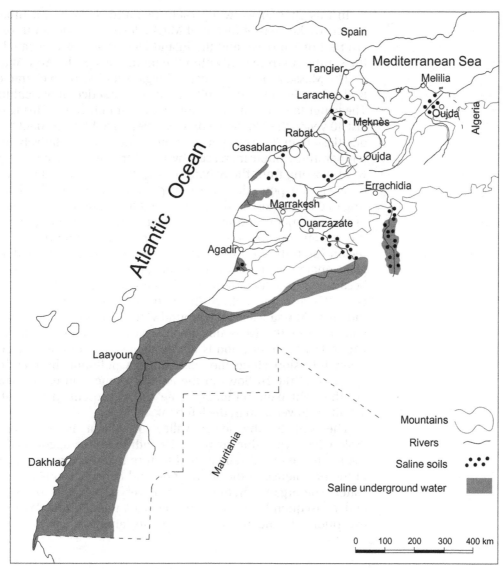

Figure 9.1 Salinity map of Morocco (Harrouni, 2000).

Salinization of soil and groundwater in irrigated areas is caused mainly by: poor drainage; rise of the saline water table and high evapotranspiration; irrigation with water with high risk of salinization and sodification (Badraoui et al., 2000). In coastal areas groundwater alteration is due to seawater intrusion because of overpumping and to pollution by the irrational use of fertilizers and pesticides in agriculture (Hssaisoune et al., 2012).

In this chapter we will present two arid regions of southern Morocco, the coastal region of Massa, located in the southwestern part of Morocco, and the inland Drâa river basin, located in the southeastern region within the mountains of the Anti-Atlas.

The Massa is one of the coastal regions of Morocco characterized by a climate that is suitable for the practice of agriculture. The water table is fed by the infiltration of rainwater. During the last 40 years, the Massa region has experienced five drought periods (less than 200 mm per year on average) and relatively high temperatures often increased by hot, turbulent, drying easterly wind blowing from the Sahara (*Chergui*). This makes the evapotranspiration potential exceed 1900 mm. The intensification of irrigated agriculture in the area has led to an annual water deficit of 155 mm^3 per year drawn from the aquifer. The hydrography of this region comprises the Massa river and its tributaries. In this region full utilization of surface water resources is reached and good-quality water available for agriculture is diminishing because of its salt content, which can reach 12.27 g L^{-1} in some areas (Hssaisoune et al., 2012). On the left bank of Massa river in the lower Massa valley, the groundwater is particularly salinized and most of the wells dug in this part are saline (Table 9.1). Groundwater salinization is due to the geology of the area, seawater intrusion along the coast and its pollution by recycling agricultural return flows in the aquifer (Bouchaou et al., 2008). On the right bank although some salt is present in the water, salinity is lower than in the left bank.

The soil in the lower valley of Massa is very saline (Solonchak type) due to natural conditions but aggravated by the existence of old salinas. Soil texture is rather heavy because of the prevalence of the silt fraction and hence it tends to retain water. The organic matter content is very low in the soil profile and consequently the total nitrogen content is low. With the exception of zinc the other elements analyzed are in rather

Table 9.1 Electrical Conductivity of the Water from Different Wells in the Lower Massa Valley

Location	EC (dS m^{-1})	
Left bank	5.95 ± 1.84	Min = 4.0, max = 10.1
Right bank	2.70 ± 0.62	Min = 1.5, max = 3.6

Values are means of 15 data ± SE.

good concentrations. The EC is very high and in consequence the sodium content also is very high. This will impede the sustainable development of any conventional crop.

The Drâa river basin, which is listed among the driest river basins of the world (Revenga et al., 1998) is situated in the transition zone to the Sahara (Ouhajou, 1996). The Drâa valley is located south of the Anti-Atlas Mountains (Heidecke, 2009). In the Drâa river basin agricultural production is highly dependent on irrigation as rainfall is not sufficient to cover crop water requirements. Precipitation decreases from north to south with the highest amounts coming from the High Atlas Mountains (snow). Melted snow constitutes the greatest part of water in the Mansour Addahbi reservoir built for water storage (Heidecke, 2009). Over a 200-km distance at midstream six palm tree oases are fed by the dam (Anonymous, 1995). Water supply for agriculture comes during long-term releases from the dam or from groundwater pumping from the shallow aquifers during dry years, when less surface water is available (Ouhajou, 1996). However, groundwater in this area is brackish to highly saline. The overall salinity increases in shallow groundwater from upstream to downstream reaching salt levels that constrain agricultural production, especially of salt-sensitive plants (Ouhajou, 1996; Anonymous, 1996). Indeed, investigation undertaken by Warner et al. (2013) revealed that groundwater salinity is lower ($0-4$ mS cm^{-1}) upstream compared to downstream where it can reach values of $4-17$ mS cm^{-1}.

The situation of the two aforementioned regions is similar to many other zones in Morocco and shows real salinity problems which inevitably affect agricultural production and the living conditions of the local population.

9.6 Agriculture in Massa and Drâa Valleys

Irrigated agriculture is the main activity of the local population in the Massa and Drâa regions. It is traditional and is practiced in small plots. Irrigation water is drawn from wells in the Massa valley and by traditional irrigation systems, especialy *Seguias* (open air channels) and *Khettaras* (underground aqueducts) in the palm oases of the Drâa valley. Agriculture is less intensive and is characterized by subsistence farming and small cultivated plots (0.5–1 ha). The combination of crop and livestock production provides living conditions for a large part of the rural population. In the region of Massa, cattle and sheep breeding is the main activity of villagers and local alfalfa varieties and maize are the only fodder crops cultivated but the yield hardly reaches 50% because of the arid climate and high salinity. The cultivation of these two species predominates because of their relative tolerance to salinity. Pulse crops are grown by some farmers for their own consumption and for sale in local markets. The average cultivated area does not exceed 1 ha per farm (Anonymous, 2013).

In the Drâa oases, the crop production system is characterized by storey or layer production. Palm trees constitute the upper layer providing some shade to less-tolerant trees such as olives, pomegranates, or figs and annual crops (cereals, fodder, vegetables) cover the rest of the plot (Anonymous, 2008).

Despite the constraints of arid lands and in particular oases, agricultural activity has always been marked by solidarity between people, the rational use of local resources (breeds and varieties) adapted to the conditions and the use of traditional skills in water management (*Seguias* and *Khettaras*). The result of these efforts is an agricultural production certainly with modest yields but with excellent nutritional and organoleptic qualities (Anonymous, 2008). However, increasing salinity levels in these regions constrain agricultural production, especially of salt-sensitive plants, and management models applied so far have been shown to be limited.

The alternative approach to the problem is the development of biosaline agriculture in these regions by making reasonable use of local plants that have high tolerance for salinity, that is, halophytes. For that, the farmers who are using traditional agricultural techniques require specific technical assistance and advice adapted to biosaline agriculture. This will improve their production and hence their income and their standard of living.

9.7 Potential Use of Halophytes in Areas Affected by Salinity

The southern regions of Morocco, both coastal and inland, are characterized by great biological diversity. In 2000, the oases of southern Morocco were declared a Biosphere Reserve by UNESCO (Anonymous, 2008). In the coastline region of Massa, several plant species, including halophytes, are protected within the Souss-Massa National Park. However, the resources of these areas are currently threatened by multiple environmental factors that influence species adaptation. Indeed, in parallel with edaphic changes, major changes are occurring in the most sensitive vegetation. Some groups disappear to give place to other, more adapted, that is, salt-tolerant populations (Bendaanoun, 1981). Table 9.2 shows a list of salt-tolerant species of southern Morocco. The list is established from the compilation of bibliographical data and from surveys carried out in the Drâa valley and the Massa coastline. Several studies on the Moroccan flora have been published but few are related specifically to halophytic flora of saline environments. Bendaanoun (1991) has conducted an ecological study of halophytes in coastal areas and estuaries. The other salt-tolerant species in Table 9.2 are compiled from ecological studies of the vegetation in arid environments of southern Morocco. Table 9.2 also gives a description of the traditional uses for the salt-tolerant species by local populations. This will enable the potential of their utilization and valorization for the sustainable rehabilitation of salt-affected areas to be determined.

The data presented in Table 9.2 show that in salt-affected areas, the local population has inherited an important local know-how for halophyte utilization, as they use them especially in traditional medicine, as fodder during dry years, as wood, and for fiber production. Indeed, in these areas, halophytes represent an important potential for valuation of the resources, land reclamation, dune stabilization, and landscaping. Many plants are of economic interest and should be promoted to alleviate the poverty of local communities.

9.8 Youth Potential in Arid Areas in Morocco

Morocco has a very young population with children under 15 years old representing 31.2% of the population and the active population (15–59 years old) constitutes the largest component with 62.2% (65.9% in cities and 57.2% in rural areas) increasing by 1.4% annually (Anonymous, 2004).

Since the youth and the active population represent an immaterial wealth and the driving force for social and economic development of a country, there are basic needs to satisfy in terms of education in Morocco. In the years to come, the number of people reaching working age will continue to grow. With the progress made in the schooling of children in rural areas, the majority of the population of working age will be able to easily acquire the necessary knowledge for a professional qualification. The effects on economic growth should be largely positive if a sufficient number of jobs is provided for this human capital (Anonymous, 2005). The Moroccan educational system

Table 9.2 Some Salt-Tolerant Plants of Southern Morocco and Their Local Traditional Uses

Family and Species	Traditional Uses
Aizoaceae	
Aizoon canariense[e]	Treament of intoxications; as soap[a]
Sesuvium portulacastrum[b,e]	As food[f] as well as fodder[f,g]
Mesembryanthemum nodiflorum[c]	Treatment of poisoning; as food (grains)[a]
Asclepiadaceae	
Calotropis procera[e]	Treatment of stomach upset, food poisoning, coughs, asthma, parasites, and helminthiases[a]
Pergularia tomentosa[e]	Tanning (animal hides); treatment of snake bites[a]
Asteraceae	
Launaea arborescens[e]	Treatment of tonsillitis; anthelmintic[d]
Senecio anteuphorbium[e]	Sedative for all abdominal and back pains (juice); rheumatism; burns; hemostatic in wound care and injury[a]
Andryala pinnatifida[e]	Gum (El-alk): breath purification and teeth cleaning[a]
Brassicaceae	
Cakile maritima[e]	Food and potential oilseed plant[k]
Caryophyllaceae	
Herniaria fontanesii[e]	As soap and fodder[a]
Spergularia marginata[e]	Treatment of cold and female infertility[a] and cystitis[o]
Chenopodiaceae	
Anabasis articulata[e]	Fodder; soap preparation; camel dermatitis care[a]
Arthrocnemum macrostachyum[c]	Grazing; soil stabilizer; firewood[h]
Atriplex halimus[c]	Food and fodder[a], leaves against kidney pain[o]
Atriplex portulacoïdes[c]	Fodder, ornamental and food (leaves)[i]
Atriplex parvifolia[e]	Effective against herpes simplex viral infection[j]
Chenopodium ambrosioides[e]	Treatment of gastrointestinal diseases, typhoid, child dysentery[a]
Chenopodium murale[e]	Against cold[a]
Centaurea aspera[e]	Treatment of sore throat and conjunctivitis treatment; digestive problems[a]
Halocnemum strobilaceum[c]	Occasionally grazed by camels and sheep; potassium source[a]
Obione portulacoides[c]	Fodder and food (leaves)[i]
Salsola baryosma[b]	Treatment of animal diseases[a]
Salsola foetida[b]	Treatment of intestinal worms, microbial infections, hypertension, arrhythmia, and gastritis; hair care[a]

(Continued)

Table 9.2 (Continued)

Family and Species	Traditional Uses
Salsola tetragona[e]	Treatment of intestinal worms, microbial infections, hypertension, arrhythmia, and gastritis[a]
Salsola vermiculata[e]	As soap[a]
Suaeda fruticosa[c]	Treatment of pruritus[a]
Suaeda maritima[b]	Treatment of pimples and pruritus[a]
Suaeda vera[b]	Treatment of pimples and pruritus[a]
Traganum nudatum[f]	Treatment of diarrhea, wounds, rheumatism, and dermatosis[i]
Convolvulaceae	
Convolvulus trabitianus[c]	As a mild purgative; cough treatment[a]
Cressa cretica[e]	Hair care; treatment of snake and scorpion bites
Cyperaceae	
Cyperus laevigatus[b]	Plaited mat making
Scirpus compactus[c]	Paper, mat, hat, and basket making[i]
Scirpus lacustris[c]	Astringent, diuretic and antibacterial (treatment of *Nocardia asteroides* and *Nocardia brasiliensis*)[m]
Scirpus maritimus[e]	Treatment of chest pain; skin care; natural deodorant.
Euphorbiaceae	
Euphorbia granulata[e]	Hair care; treatment of poisonous bites[d]
Euphorbia paralias[e]	Latex antiscabies[d]
Fabaceae	
Retama monosperma[e]	Purgative, vermifuge, and abortive[b]
Medicago sativa[e]	Fodder
Trifolium alexandrinum[e]	Fodder; fruits used in cough treatment[a]
Frankeniaceae	
Frankenia corymbosa[b]	Treatment of cystitis[o]
Frankenia pulverulenta[b,c]	Treatment of catarrh and nose and genito-urinary mucous discharges[i]
Juncaceae	
Juncus acutus[e]	Treatment of urogenital infections; fever; tremor mat manufacturing[a]
Juncus bufonius[b]	Diuretic (fruits)[a]
Juncus maritimus[b,c]	Treatment of urogenital infections, fever, tremor, cold[a,o]
Juncus rigidus[b,e]	Diuretic (fruits)[a]
Juncus subulatus[c]	Hepatoprotective activity[n]

(Continued)

Table 9.2 (Continued)

Family and Species	Traditional Uses
Lilliaceae	
Urginea maritima[e]	Treatment of bronchitis, cough, flu, and jaundice; diuretic; abortive[a]
Plumbaginaceae	
Limonium pruinosum[e]	Diuretic, food and fodder[a]
Limonium sinuatum[e]	Treatment of dysmenorrhea; hair care[a]
Portulacaceae	
Portulaca oleraceae[e]	Treatment of fever; abscess; burns; diabetes[a]

[a]Bellakhdar (1997).
[b]Hamada (2007).
[c]Bendaanoun (1991).
[d]Hmamouchi (1999).
[e]Authors' survey (2014).
[f]Mathieu and Meissa (2007).
[g]Lokhande et al. (2009).
[h]Heneidy and Bidak (2004).
[i]Al-Oudat and Qadir (2011).
[j]Ben Sassi et al (2008).
[k]Facciola (1990).
[l]OuldElhadj et al. (2003).
[m]Eshraghi et al. (2009).
[n]Abdel-Razik et al. (2009).
[o]Ghourri et al. (2014).
[p]Maghrani et al. (2005).

has known several reforms with the adoption of curricula more adapted to the local socioeconomic environment.

Since 2008, Morocco has adopted a new agricultural policy, the so-called Green Morocco Plan, to promote agriculture which is considered the main lever of the national economy for 2010–2020. This plan has distinguished between modern and traditional agriculture. In addition to its economic objective to develop an intensive agriculture in irrigated zones and favorable rainfed lands, the Green Morocco Plan also has a social objective through the promotion of traditional agriculture characterized by farm sizes smaller than 5 ha (70%) occupying 80% of the "useful agricultural surface". This support can be achieved by the creation of "territorial projects" and "technological progress," such as infrastructure development, technical assistance, and improved knowledge. This would include research progress (improvement of crop species and varieties adapted to the constraints to which traditional agriculture is confronted, including drought and salinity), a combination of traditional models such as rainfed and irrigated agriculture, validation of local know-how, education, and professionalization of agriculture and institutional innovations (Requier-Desjardins, 2010).

To contribute effectively to the implementation of the new agricultural policies, scientific research should adapt to the new needs of traditional agriculture. For that purpose, universities should propose new curricula that can meet these needs. Ideas should be developed to face ecosystem degradation (including salinization) and to raise awareness among the youth as to the importance of traditional know-how and its potential for wealth creation (Daoud, 2011). Training rural youth on the sustainable management of local resources, improvement of agricultural productivity in arid and salt-affected areas, and valuation of human capacities could create territorial projects that will eventually contribute to the stabilization of the population in rural zones while improving their income.

9.9 Conclusion

Areas affected by salinity in Morocco have great potential for the development of a biosaline agriculture based on the sustainable utilization of halophytic plants adapted to the local conditions. Nevertheless, it is important to make the best use of local know-how as inherited by the local populations. Therefore, in addition to the production of food for the local communities and their animals by cultivating salt-tolerant crops using sustainable agro-ecological technologies, biosaline agriculture could play a major role in the restoration of biodiversity and productivity of these regions.

The role of youth in these areas is essential to trigger a dynamic. To convince them to choose agriculture as an attractive profession, they need to see it as a good career option and this requires the creation of a "supportive environment." There is an urgent need for agricultural universities and research institutes in regions facing similar climatic constraints to act together, exchange their experiences and expertise and develop joint education, research, development and extension (ERD & E) programs which can meet the needs of their regions. This will ultimately lead to better agricultural productivity, sustainable local development, and bridging the gap between food requirements and food supply within the region.

References

Abdel-Razik, A.F., Elshamy, A.I., Nassar, M.I., El-Kousy, S.M., Hamdy, H., 2009. Chemical constituents and hepatoprotective activity of *Juncus subulatus*. Rev. Latinoam. Quím. 37, 70–84.

Al-Oudat, M., Qadir, M., 2011. The halophytic flora of Syria. International Center for Agricultural Research in the Dry Areas, Aleppo, Syria, viii + 186 pp.

Anonymous, 1995. Etude d'amélioration de l'exploitation des systèmes d'irrigation et de drainage de l'ORMVAO- Phase 1: Diagnostic de la situation actuelle. Rapport détaille du diagnostic de l'ORMVAO. vol 1. Ouarzazate. Morocco.

Anonymous, 1996. Etude de la salinite de sols dans la vallée du Drâa—Province de Ouarzazate. Mission II. Rapport definitif de l'ORMVAO Ouarzazate. Morocco.

Anonymous, 2004. RGPH: Recensement Général de la Population et de l'Habitat. Direction de la Statistique, Morocco.

Anonymous, 2005. 50 ans de développement humain & perspectives 2025. Centre des études et des recherches démographiques, Haut commissariat au plan. Rapport thématique: Démographie Marocaine: tendances passées et perspectives d'avenir. Morocco.

Anonymous, 2006a. Statistiques environnementales au Maroc, Haut commissariat au plan, Rabat—Morocco. Rapport du projet MEDSTAT-Environnement (MED-Env II), mis en oeuvre par le Plan Bleu sous la supervision technique d'Eurostat, pp. 1–108.

Anonymous, 2006b. Secrétariat d'Etat auprès du Ministère de l'Energie, des Mines, de l'Eau et de l'Environnement, Chargé de l'Eau et de l'Environnement. Département de l'Environnement. Etude vulnérabilité et adaptation du Maroc face aux changements climatiques Section I: Rapport de synthèse. Morocco.

Anonymous, 2006c. Secrétariat d'Etat auprès du Ministère de l'Energie, des Mines, de l'Eau et de l'Environnement, Chargé de l'Eau et de l'Environnement. Département de l'Environnement. Etude vulnérabilité et adaptation du Maroc face aux changements climatiques; Section III: Agriculture. Morocco.

Anonymous, 2008. Plan cadre de gestion de la Réserve de Biosphère des Oasis du Sud Marocain (RBOSM). Ministère de l'Agriculture, du Développement Rural et des Pêches Maritimes, Rabat, Morocco.

Anonymous, 2009. Plan d'Action National de Lutte Contre la Désertification (PAN LCD) Agadir: La lutte contre l'avancée du désert dans cinq préfectures de la Région Souss Massa Drâa, Comment s'y prendre? Eco consulting group, GTZ et DREF SO, Morocco.

Anonymous, 2011. Banque africaine de développement (BAD). Usage agricole des eaux souterraines et initiatives de gestion au Maghreb: Défis et opportunités pour un usage durable des aquifères, pp. 1–24.

Anonymous, 2013. Office Régionale de Mise en Valeur Agricole du Souss-Massa (ORMVA/SM). Monographie de la région du Souss Massa, 2012/2013. Morocco.

Anonymous, 2014. Politique du changement climatique au Maroc. Ministère délégué auprès du Ministère de l'Energie, des Mines, de l'Eau et de l'Environnement, chargé de l'Environnement, Morocco.

Badraoui, M., Agbani, M., Soudi, B., 2000. Evolution de la qualité des sols sous mise en valeur intensive au Maroc. Séminaire "Intensification agricole et qualité des sols et des eaux," Rabat, Morocco, Novembre 2–3, 2000.

Bellakhdar, J., 1997. La pharmacopée marocaine traditionnelle. Ibis Press, SL.

Ben Sassi, A., Harzallah-Skhiri, F., Bourgougnon, N., Aouni, M., 2008. Antiviral activity of some Tunisian medicinal plants against Herpes simplex virus type 1. Nat. Prod. Res. A. 22, 53–65.

Bendaanoun, M., 1981. Etude synécologique et syndynamique de la végétation halophile et halohygrophile de l'estuaire du Bou Regreg (littoral atlantique du Maroc). Applications et perspectives d'aménagement. Thèse de Docteur-Ingénieur, Université d'Aix-Marseille III. St-Jérôme, Fac. Sci. Tech. Spécialité: Es-Sciences naturelles, phytoécologie.

Bendaanoun, M., 1991. Contribution à l'étude écologique de la végétation halophile, halohygrophile et hygrophile des estuaires, lagunes, deltas et sebkhas du littoral atlantique et méditéranéen et du domaine continental du Maroc. Thèse de Doctorat de l'Université d'Aix-Marseille III. St-Jérôme, Fac. Sci. Tech. Spécialité: Es-Sciences naturelles, phytoécologie.

Bouchaou, L., Michelot, J.L., Vengosh, A., Hsissou, Y., Qurtobi, M., Gaye, C.B., et al., 2008. Application of multiple isotopic and geochemical tracers for investigation of recharge, salinization, and residence time of water in the Souss–Massa aquifer, southwestern Morocco. J. Hydrol. 352, 267–287.

Choukr-Allah, R., Harrouni, M.C., 1996. The potential use of halophytes under saline irrigation in Morocco. In: Symposium on the Conservation of Mangal Ecosystems, December 15–17, 1996, Al-Ain, United Arab Emirates.

Daoud, S., 2011. Climate Change, Agriculture and Youth: Supportive environment should be created for youth to choose agriculture as an attractive profession and good career option. Newsl. World Forum Clim. Change Agric. Food Secur., p. 4. ISSN: 2222-064X.

Debbarh, A., Badraoui, M., 2002. Irrigation et environnement au Maroc: Situation actuelle et perspectives. Serge Marlet et Pierre Ruelle (éditeurs scientifiques). Vers une maîtrise des impacts environnementaux de l'irrigation. Actes de l'atelier du PCSI, 28–29 mai 2002, Montpellier, France. CEMAGREF, CIRAD, IRD, Cédérom du CIRAD.

Driouech, F., 2010. Distribution des précipitations hivernales sur le Maroc dans le cadre d'un changement climatique: descente d'échelle et incertitudes. Thèse de Doctorat de l'Université de Toulouse. Spécialité: Sciences de l'Univers, de l'Environnement et de l'Espace.

Eshraghi, S., Amin, G., Othari, A., 2009. Evaluation of antibacterial properties and review of 10 medicinal herbs on preventing the growth of pathogenic Nocardia species. J. Med. Plants. 8, 60–78.

Facciola, S., 1990. Cornucopia—A Source Book of Edible Plants. Kampong Publications.

Ghourri, M., Zidane, L., Douira, A., 2014. La phytothérapie et les infections urinaires (La pyélonéphrite et la cystite) au Sahara Marocain (Tan-Tan). J. Animal Plant Sci. 20, 3171–3193.

Hammada, S., 2007. Etudes sur la végétation des zones humides du Maroc. Catalogue et analyse de la biodiversité floristique et identification des principaux groupements végétaux. Thèse de Doctorat d'Etat ès-Sciences. Spécialité: Ecologie végétale. Université Mohammed V, Faculté des Sciences, Rabat, Morocco.

Harrouni, NC., 2000. Halophytic Plants for Sustainable Development of Coastal and Salt Affected Areas in Morocco. Research report for UNESCO.

Heidecke, C., 2009. Economic Analysis of Water Use and Management in the Middle Drâa Valley in Morocco (Ph.D. thesis). Hohen Landwirtschaftlichen Fakultät der Rheinischen Friedrich-Wilhelms-Universität zu Bonn.

Heneidy, S.Z., Bidak, L.M., 2004. Potential uses of plant species of the coastal Mediterranean region, Egypt. Pak. J. Biol. Sci. 7, 1010–1023.

Hmamouchi, M., 1999. Les plantes médicinales et aromatiques marocaines: utilisation, biologie, écologie, chimie, pharmacologie, toxicologie, lexique.

Hssaisoune, M., Boutaleb, S., Benssaou, M., Beraaouz, E., Tagma, T., El Faskaoui, M., et al., 2012. Geophysical and structural analysis of the Souss-Massa aquifer: synthesis and hydrogeological implications. Geo-Eco-Trop. 36, 63–82.

Lokhande, V.H., Nikam, T.D., Suprasanna, P., 2009. *Sesuvium portulacastrum* (L.) L. a promising halophyte: cultivation, utilization and distribution in India. Genet. Resour. Crop Evol. 56, 741–747.

Maghrani, M., Michel, J.B., Eddouks, M., 2005. Hypoglycaemic activity of *Retama raetam* in rats. Phytother. Res. 19, 125–128.

Mathieu, G., Meissa, D., 2007. Traditional leafy vegetables in Senegal: diversity and medicinal uses. Afr. J. Tradit. Complement. Altern. Med. 4, 469–475.

Ouhajou, L., 1996. Espace hydraulique et société au Maroc: Cas des Systèmes d'irrigation dans la vallée du Drâa. Thèses et Mémoires. Faculté des Lettres et des Sciences Humaines, Université Ibn Zohr. Agadir, Morocco.

OuldElhadj, M.D., Hadj-Mahammed, M., Zabeirou, H., Abeirou, H., 2003. Courrier du Savoir. 3, 47–51.

Requier-Desjardins, M., 2010. Impacts des changements climatiques sur l'agriculture au Maroc et en Tunisie et priorités d'adaptation. Les Notes d'analyse du CIHEAM. No. 56–Mars 2010.

Revenga, C., Murray, S., Abramovitz, J., Hammond, A., 1998. Watersheds of the World: Ecological Value and Vulnerability. World Resources Institute and Worldwatch Institute, Washington, DC.

Schilling, J., Korbinian, P.F., Hertig, E., Scheffran, J., 2012. Climate change, vulnerability and adaptation in North Africa with focus on Morocco. Agric. Ecosyst. Environ. 156, 12–26.

Warner, N., Lgourna, Z., Bouchaou, L., Boutaleb, S., Tagma, T., Hsaissoune, M., et al., 2013. Integration of geochemical and isotopic tracers for elucidating water sources and salinization of shallow aquifers in the sub-Saharan Drâa Basin, Morocco. Appl. Geochem. 34, 140–151.

10

HALOPHYTE TRANSCRIPTOMICS: UNDERSTANDING MECHANISMS OF SALINITY TOLERANCE

Joann Diray-Arce[1], Bilquees Gul[2],
M. Ajmal Khan[2,3] and Brent Nielsen[1]

[1]Department of Microbiology and Molecular Biology, Brigham Young University, Provo, UT, USA [2]Institute of Sustainable Halophyte Utilization, University of Karachi, Karachi, Pakistan [3]Centre for Sustainable Development, College of Arts and Sciences, Qatar University, Doha, Qatar

10.1   Introduction

Studies of genomes and transcriptomes have rapidly advanced with next-generation sequencing (NGS) approaches. NGS technologies are utilized for single nucleotide polymorphism-based markers and draft sequencing of species without a reference genome (Wicker et al., 2006). These approaches have led to the discovery of markers that can be used to study genetic variations, population genetics, transcript profiling, mutations, and genetic associations for plant breeding (Qin et al., 2010). As NGS technologies have matured, RNA sequencing has become a preferred method for gene expression profiling (McGettigan, 2013), as it has the ability to identify transcripts and their expression over time and under different conditions. Transcriptome sequencing is less expensive than genome sequencing since only transcribed regions are investigated (Brautigam and Gowik, 2010).

Problems caused by high soil salinity for plants include the lowering of water potential leading to osmotic stress caused by cellular dehydration, toxicity of absorbed Na^+ and Cl^- ions which inhibits enzymatic activities and various cellular processes

M.A. Khan, M. Ozturk, B. Gul, & M.Z. Ahmed (Eds): Halophytes for Food Security in Dry Lands.
DOI: http://dx.doi.org/10.1016/B978-0-12-801854-5.00010-8

and the restriction of uptake of essential nutrients (Flowers and Colmer, 2008; Abideen et al., 2014). Plant salinity tolerance involves mechanisms at the physiological and molecular levels. Physiological response involves the adaptation of plants as the concentration of salt in the soil increases or the availability of water in the soil decreases (Hasegawa et al., 2000). Molecular mechanisms vary among halophyte species and involve a number of metabolites, genes, and pathways. In this chapter we discuss halophytes that have been characterized by NGS and implications for salt tolerance.

10.2 Transcriptome Sequencing Overview

Initial transcriptome studies relied on microarray analysis, qPCR, or real-time PCR techniques to measure gene expression. The development of NGS techniques provides high speed and throughput and projects can now be completed in weeks or days at lower costs. NGS technologies allow gene expression profiling, genome annotation, and discovery of non-coding RNA (Mutz et al., 2013). NGS technology obtains short sequence tags, 20−35 bases long, from each transcript in the sample. This allows detection of low-abundance RNAs, small RNAs, or other elements (Ansorge, 2009). The transcriptomics variant based on sequencing by synthesis is called short-read massively parallel sequencing or RNA-seq.

Methods for RNA sequencing are shown in Figure 10.1. Initial library preparation involves the isolation of RNA, which is converted to cDNA fragments with adaptors attached to one or both ends. The molecules are amplified, the libraries are quantified, analyzed for quality control and sequenced by high-throughput sequencers (Roche 454, Illumina, ABI SOLiD sequencing, PacBio, Ion Torrentor Helicos BioSciences) (Morozova and Marra, 2008). Bioinformatics is applied to the sequences generated. Pre-processing of data includes trimming of the sequencing adapters, error corrections, and elimination of poor-quality reads.

10.3 Applications of RNA Studies

Applications using RNA-seq data include mapping of short reads, detection of intron splicing junctions, isoform expression quantification, and differential expression analysis (Chen et al., 2011). For mapping-first methods, sequenced reads are mapped to the genome or transcriptome sequences for guided assembly. Low-quality reads are removed to prevent incorrect mapping.

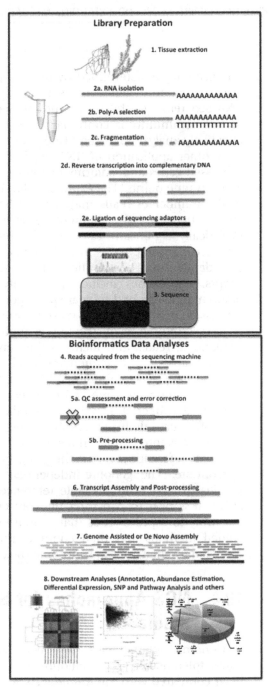

Figure 10.1 RNA sequencing flow chart. Total RNA is isolated and extracted from plant tissues of interest. RNA-Seq library preparation from Illumina involves poly-A RNA isolation, RNA fragmentation, reverse transcription to cDNA, adapter ligation and size selection. The product from the subsequent steps is then sequenced using a high-throughput RNA-sequencing machine. Raw RNA sequence data is then acquired and assessed for quality reads. Trimming, error correction and normalization allow to ensure good quality sequences. These reads will be assembled using a reference genome or de novo if it does not have a close relative genome available. Downstream analyses can be performed once metrics for a good transcriptome have been met. These may include differential expression studies, SNP and pathway analysis, estimation and annotation of transcripts.

The accuracy is determined by the mapping, therefore the best way to quantify genes or isoforms is to directly map the RNA-seq reads to the transcriptome sequences. The bioinformatics community is continually developing software to more effectively analyze RNA-seq (Trapnell et al., 2012).

Another application of RNA-seq is detection of differentially expressed genes and isoforms to compare conditions or samples at given time points. The expression level of transcripts is related to the number of reads mapped on them. Differences in read counts between two different experimental conditions at a statistically significant value can be regarded as differentially expressed. Several biases must be considered including sequencing depth, count distribution, library size, and length of transcripts. Approaches include probability distributions used by different pipelines and software packages for detecting differential expression between samples (Seyednasrollah et al., 2013).

Transcriptome reconstruction is another application of RNA-sequencing reads, and includes the genome-guided approach, which maps all the sequencing reads back to the reference genome, and genome-independent approach, which does not need a reference genome and directly assembles the reads into transcripts (Miller et al., 2010). Assembly using de novo techniques often uses de Bruijn graphs or the use of k-mers to assemble the reads into contigs. If a species already has a high-quality, complete reference genome, the genome-dependent approach is appropriate. The genome-independent approach is used for species that have no available reference genome (Miller et al., 2010). It is best to construct the transcriptome using de novo assembly to capture reads that cannot be obtained by genome-guided methods and then combine the results to produce a more comprehensive transcriptome (Box 10.1).

10.4 NGS Approaches for Salt-Tolerance Studies

Genomic technologies have been applied to study plant stress tolerance in some halophytes (Table 10.1), which have been compared with the *Arabidopsis* genome (Kant et al., 2008). *Thellungiella* spp. share many characteristics with *Arabidopsis* and are tolerant to salt and drought stresses (Griffith et al., 2007; Wong et al., 2006). Draft genomes for *Thellungiella parvula* and *Thellungiella salsuginea* were constructed with NGS to understand adaptation to abiotic stresses (Dassanayake et al., 2011; Wu et al., 2012).

Box 10.1 Transcriptome Assembly Features

Transcriptome (RNA-seq) sequencing: This technology analyzes RNA presence and measures the levels of transcripts and their isoforms using NGS technologies (Clarke et al., 2013).

De novo versus reference-based assembly: For species that do not have a reference genome, de novo reconstruction of transcriptomes using RNA-seq data is performed. Reference-based assembly uses the genome sequence to serve as a guide for transcriptome reconstruction (Clarke et al., 2013).

Assembly, Alignment, and Visualization

Overlap layout consensus: Assemblers developed for Sanger reads use an overlap layout consensus method which computes pairwise overlaps and captures the information in a graph. This method constructs a read graph and assigns reads as nodes and then creates a link between two nodes when the reads overlap is larger than a cutoff length. The computation of reads and consensus sequence of contigs is determined by the overlap graph (Kumar and Blaxter, 2010; Li et al., 2011; Miller et al., 2010).

De Bruijn graph approach: Reads are broken into smaller sequences or k-mers where k is the length in bases of the sequences. The k value is defined over a finite alphabet span, where k is a cyclic string where all words of length k appear exactly once in the sequence (Clarke et al., 2013; Compeau et al., 2011).

Sequence aligners: Alignments of transcriptome sequences reveal novel splice forms and sequence polymorphisms. Choosing an aligner is necessary to accurately detect transcripts expressed in a given cell or tissue type. Most aligners can increase accuracy by prioritizing alignments in which read pairs map consistently (Engstrom et al., 2013).

Gene annotation: Different approaches are used to predict biological information: structural annotation by identification of genomic elements (gene structure, coding regions, motifs, ORFs) and functional annotation (molecular function, biological processes, cellular component, regulations and interactions, and expression) (Garber et al., 2011; Stein, 2001).

Differential expression: RNA-seq measures the expression of specific gene products. Poorly replicated conditions, insufficient depths, or sequencing quality errors can lead to artifacts during differential analysis of the number of genes and transcripts showing significant fold changes in overall gene expression. A comparison between true replicates can reveal differences in gene transcripts from each condition while different tissues can show thousands of differentially expressed genes (Anders and Huber, 2010; Tarazona et al., 2011).

A number of genes are involved in the response to salinity, and have been grouped in the following categories (Xiong and Zhu, 2002): (i) genes that encode enzymes, transcription factors, hormones, detoxifiers, osmolytes, and those responsible for general metabolism; (ii) genes that function in water and ion uptake such as ABC transporters, ion transporters, aquaporins, ATP binding cassette transporters, antiporters, and those involved in the SOS pathway; (iii) those that are involved in regulation, such as protein kinases and phosphatases; and (iv) genes that

Table 10.1 Known Halophytes with Analyzed Transcriptomes or Genomes[1]

Species	Genome/ Transcriptome Information	Technology	Software Used	Purpose
Ceriops tagal (Liang et al., 2012)	432 DE transcripts 59 unigenes assembled	Microarray	LOWESS, SAM, BLASTX	Gene identification, differential expression, functional annotation
Eutrema salsugineum (Brassicaceae) (Yang et al., 2013)	241 Mb-genome 26,531 genes 137,652 bp exons	Sanger	Arachne, FGENESH, Genome Scan, BLAST	Phylogenetic analysis, genome assembly, synteny analysis, orthologue identification
Leymus chinensis (Gramineae) (Sun et al., 2013)	104,105 unigenes	454 FLX	LuCY, TagDust MIRA	Differential expression, annotation
Mesembryanthemum crystalinum (Kore-eda et al., 2004)	9733 expressed sequenced tags	cDNA library-dideoxy chain termination method	PHRED, CROSS-MATCH, PHRAP, BLASTX	EST assembly, functional categorization
Millettia pinnata (Huang et al., 2012)	54,596 unisequences, 65.8 Mb transcriptome	Illumina GA	SOAPdenovo	Gene annotation, differential expression
Populus euphratica (Zhang et al., 2013; Qiu et al., 2011)	86,777 unigenes	Illumina GA	SOAPdenovo, TGICL	De novo assembly, annotation, differential expression
Populus pruinosa (Zhang et al., 2013)	114,866 unique sequences	Illumina GA	SOAPdenovo, TGICL	De novo assembly, orthologue identification, annotation
Porteresia coarctata (Garg et al., 2014)	152,367 unique transcripts	Illumina GA II	Velvet, Oases, AbySS, Trinity, CLC Genomics, CDHIT	De novo assembly, gene ontology, pathway analysis
Reaumuria trigyna (Tamaricaceae) (Dang et al., 2013)	65,340 unigenes	IlluminaHi Seq 2000	SOAPdenovo, Blast2GO	De novo assembly, gene ontology, expression pattern analysis
Salicornia europaea (Fan et al., 2013)	57,151 unigenes	IlluminaHi Seq 2000	SOAPdenovo, ESTScan	De novo assembly, gene ontology, digital gene expression tag sequencing, differential expression

(Continued)

Table 10.1 (Continued)

Species	Genome/ Transcriptome Information	Technology	Software Used	Purpose
Salicornia europaea v.2 (Ma et al., 2013)	109,712 unigenes	IlluminaHi Seq 2000	Trinity, Blast2GO	De novo assembly, GO annotation, differential expression
Spartina maritima (Ferreira de Carvalho et al., 2013)	114,857 singletons	454 GS XLR70	GS Assembler v 2.3	De novo assemblies, GO annotation, polymorphism analysis
Spartina alterniflora (Ferreira de Carvalho et al., 2013)	58,298 singletons	454 GS XLR70	GS Assembler v 2.3	De novo assemblies, GO annotation, polymorphism analysis
Suaeda fruticosa Diray-Arce et al., 2015	54,526 unigenes	IlluminaHi seq 2000	Trinity, Oases, Velvet, CDHIT-EST, Blast2Go, Transdecoder	De novo assembly, GO annotation, differential expression
Suaeda maritima (Sahu and Shaw, 2009)	429 ESTs	PCR-based suppression subtractive hybridization	BLASTX, TIGR	SSH library construction, functional categorization
Schrenkiella parvula (*Thellungiela parvula*) (Oh et al., 2010)	21,619 contigs	454 GS FLX Titanium	Newbler, FGENESG, Repeat Masker, Pip maker	De novo assembly, annotation, synteny, comparative analyses of transcription, repeat identification
Thellungiella parvula (Dassanayake et al., 2011)	140 Mb genome	454 GS FLX Titanium, Illumina GA II	Newbler, ABySS, FGENESH, GENSCAN, BLAST, Blast2GO	Genome assembly, macrosynteny, ORF prediction and annotation
Thellungiella salsuginea (Lee et al., 2013)	42,810 unigenes	454 GS FLX Titanium	SFF Tools, MIRA, BLAST, MUSCLE, UGENE	De novo assembly, functional annotation, microRNA prediction, gene identification
Thellungiella salsuginea (Wu et al., 2012)	233.7 Mb genome	Illumina	ABySS, SOAPdenovo, Minimus2, MAUVE	Genome assembly, repetitive sequences identification, phylogenetic analyses, pathway analyses

[1]As of March 2015.

function to protect the cells against abiotic stress, such as late embryogenesis abundant (LEA) proteins, heat shock proteins, and osmoprotectants, such as dehydrin and osmotins.

10.5   Genes Involved in General Metabolism

This group includes genes that encode proteins for biosynthesis of osmolytes, hormones, and detoxification (Aslam et al., 2011). Some genes are responsible for abscisic acid signaling, which regulates plant germination, dormancy, and seed development. Others include antioxidants and enzymes that maintain the level of reactive oxygen species (ROS) to protect the cells from oxidative damage. Salt tolerance involves osmotic adjustment to maintain turgor, cell expansion, adjustments in photosynthesis and stomatal mechanisms, and plant growth. The sequestration of salt ions in the vacuole minimizes toxicity. Osmotic adjustment requires the accumulation of enzymes or osmolytes in the cytoplasm. The chemical nature of osmolytes varies from carbohydrates, polyols, and amino acids, and they are synthesized by halophytes and glycophytes in response to stress (Flowers and Colmer, 2008; Grigore et al., 2011).

10.5.1   Genes for Cell Maintenance

Genes responsible for transcription, translation, and post-translational modifications play a role in salt tolerance. Transcription factors in *Suaeda maritima* include ethylene-responsive element-binding protein, ethylene-responsive element, jasmonate and ethylene response factor (Sahu and Shaw, 2009). HDZip genes are involved in abscisic acid-related responses, such as water deficiency in *Arabidopsis* (Ariel et al., 2007). HDZip genes ATH -7, -12, -6, -21, -40, and -53 are overexpressed upon salt treatment (Söderman et al., 1996). Genes encoding pectin methyl-esterase inhibitor protein, glutathione S-transferases, and RNA transcription factors are up-regulated after NaCl treatment in *Salicornia*. Enzymes for cell wall metabolism and peroxidase are decreased at early stages of the treatment. There is an up-regulation after salt treatment of pectin methyl-esterase inhibitor family protein, aminotransferase, and unspecific anion channel. Down-regulated genes in the roots are involved in cell wall precursor synthesis and cellulose synthesis reducing plant lignifications (Fan et al., 2013).

A comparison study of salt-tolerant species of *Festuca rubra* ssp. *litoralis* and rice identified a differentially regulated WRKY-type transcription factor and a SUI homologous translation initiation factor in response to salinity (Diédhiou et al., 2009). WRKY transcription factor was also differentially regulated in *Suaeda fruticosa* in response to abiotic stress (Diray-Arce et al., 2015). Phosphorylation and O-linked β-N-acetylglucosamine (O-GlcNAc) modification of proteins are found in *S. maritima* under salt stress (Sahu and Shaw, 2009).

Schrenkiella parvula expresses genes encoding tetratrico peptide repeat protein 1 involved in flowering, glycine-rich protein for cell wall structure, phosphoenol pyruvate carboxykinase, carbonic anhydrase for C_4 assimilation, acyl coA-binding protein for fatty acid metabolism, and other genes that are involved in cell organization and plant growth (Jarvis et al., 2014). In *S. fruticosa* we have found f-box kelch protein for actin filament interaction, ribosomal proteins for translation, DNA-binding protein escarola-like for late flowering and leaf development, catepsin b-like cysteine protease for disease resistance, and glutathione S-transferase tau for increased protection against toxins to be up-regulated (Diray-Arce et al., 2015). Xyloglucan endotrans glycosylase/hydrolase (XTH) and expansin-3 are over-expressed in *S. maritima* (Sahu and Shaw, 2009) and *Ceriops tagal* (Liang et al., 2012) upon salt treatment. XTH catalyzes molecular grafting to maintain cell wall thickness and promote cell wall formation and elongation (Jan et al., 2004).

10.5.2 Stress Genes

High concentrations of ions are toxic to plants because of their effect on cell homeostasis, cytosolic enzyme activities, and photosynthetic and cellular metabolism. Salt stress leads to the closure of stomata, reducing carbon fixation and photosynthesis, loss of cell turgor due to hyperosmotic shock, inhibition of cell division and expansion, toxicity, and plant yield reduction (Aslam et al., 2011).

Millettia pinnata, a halophytic mangrove, has 21.9% of its genes differentially expressed. In roots, most of these genes are involved in gene expression, sulfur metabolic processes, redox, and secondary metabolic processes. In leaves, induced genes are involved in redox, cellular amino acid derivative metabolism, and cellular aromatic compound metabolic processes. Stress response genes are also activated, which might serve as protection from salt-induced deleterious effects (Huang et al., 2012). In *S. parvula*, differentially expressed genes include ABA insensitive-5,

D1-pyrroline-5-carboxylate synthase1, repressor of silencing 1, calcineurin B-like10, and are responsible for signaling under salt stress (Jarvis et al., 2014). In the root transcriptome of *S. maritima*, zeaxanthine eposidase, a precursor of ABA, and chaperone protein DNA J genes are up-regulated (Sahu and Shaw, 2009).

10.5.3 Photosynthetic Genes

Photosystem II family protein-coding genes (protein Z, d2 protein, cp43 chlorophyll protein) are up-regulated in salt-treated *S. fruticosa* (Diray-Arce et al., 2015). In *Salicornia europaea* photosynthetic genes, PSI and PSII pigment-binding proteins, b6f complex, and ATPase synthase CF1 were significantly induced (Fan et al., 2013). *Populus euphratica* expression of psbA proteins, D2 protein, and Rubisco large unit were decreased after 12 h of salt shock. Genes for plastidic and nuclear protein synthesis, genes with undefined functions, genes pointing to glycolysis and stress (a putative glutathione S-transferase and COBRA protein precursor) suggest the relationship of salinity with decreased photosystem II activity. Restored water potential after salinity shock causes an increase in calcineurin-like protein CLB activity, 1 aminocyclopropane-1-carboxylic acid oxidase, root organelle-specific genes psbA, and mitochondrial ATPase (Brinker et al., 2010).

10.5.4 Mitochondrial and ROS Related Genes

Salinity stress increases ROS that cause oxidative damage to cellular components (Dang et al., 2013). Thioredoxin, glutaredoxin, glutathione S-transferase family genes were found in *Reaumuria trigyna* (Dang et al., 2013), *S. maritima* (Sahu and Shaw, 2009) and *S. fruticosa* (Diray-Arce et al., 2015). The thioredoxin gene is involved in redox regulation in the apoplast, which regulates cell division, cell differentiation, pollen germination, and stress responses (Zhang et al., 2011). Superoxide dismutase is highly induced in halophytes, which rapidly dismutates superoxide radicals into oxygen and hydrogen peroxide. In *R. trigyna*, there is increased transcription of glutathione disulfide-reductase and glutathione S-transferases, enzymes for resisting oxidative stress and maintaining the reducing environment of the cell (Dang et al., 2013).

10.5.5 Proline and Other Amino Acids

Proline is concentrated in the cytosol, chloroplast, and vacuoles for osmotic adjustment in many species and also contributes to detoxification of ROS (Ketchum et al., 1991; Khan et al., 2000; Sucre and Suárez, 2011). Amino acid permease and proline transporter (ProT) were both up-regulated in the absence of salt and down-regulated at 10–500 mM salt concentration in *S. europaea* (Ma et al., 2013).

Glycinebetaine (GB) is up-regulated in plants exposed to dehydration (Lokhande and Suprasanna, 2012). The synthesis and accumulation of GB protects the cytoplasm from ion toxicity, dehydration, and temperature stress. It functions by stabilizing macromolecule structures and protecting photosystem II, and has been reported in many species (Khan et al., 2000; Lokhande et al., 2010). In *Atriplex nummularia*, GB is accumulated under salt stress and the transcript levels of S-adenosyl-L-methionine co-regulate with that of phosphoethanolamine N-methyl transferase (PEAMT) in response to salinity (Nedjimi and Daoud, 2009). In *S. maritima* the most overexpressed gene encodes PEAMT that is responsible for synthesis and accumulation of GB (Sahu and Shaw, 2009).

10.5.6 Genes Encoding Plant Hormones

There is a significant increase in plant biomass in some halophytes while there is a decreasing biomass in others at different salt conditions. Gibberellic acid (GA) genes are involved in the synthesis of gibberellin hormone, which regulates many aspects of the growth and development of plants. In *S. europaea*, GA genes were regulated at 200 mM NaCl, similar to the homologues of gibberellin 3-oxidase and gibberellin 20-oxidase in *Populus trichocarpa*. Two DELLA domain GRAS family transcription factors, inhibitors of plant growth, were down-regulated in plants with 200 mM salt (Ma et al., 2013). In *Arabidopsis*, bioactive GA is reduced through an increase in gibberellin 2-oxidase 7 (GA2ox7) that accumulates DELLA, which inhibits plant growth (Magome et al., 2008). However, down-regulation of GA2ox at 300 mM salt treatment in *S. fruticosa* deactivates bioactive GA. A decrease in GA2ox and DELLA in *S. fruticosa* favors plant growth upon salt treatment (Diray-Arce et al., 2015).

10.5.7 Genes Encoding Ion Transporters

ABC Transporters: Ion homeostasis involves the transport of ions, cellular uptake, sequestration of salt, and ion export. Plant cells require high K^+ (100–200 mM) and lower Na^+ (1 mM) to maintain osmotic balance. A large influx of extracellular Na^+ occurs in halophytes (Lokhande and Suprasanna, 2012). Several ion transporters such as high-affinity potassium transporters (HKT), low-affinity cation transporters, nonselective cation channels, cyclic nucleotide-gated channels, and glutamate-activated channels have been identified in halophytes (Horie and Schroeder, 2004). In *R. trigyna*, five vacuolar H^+ pumping pyrophosphatases (PPases) were detected and may generate a proton electrochemical gradient to compartmentalize excess Na^+ ions. Genes associated with K^+ transport composed the largest proportion of genes suggesting their importance in Na^+/K^+ homeostasis. Seven HKT1 genes for Na^+ influx were also salt-responsive. Other genes encode plasma membrane H^+-ATPases, vacuolar H^+-ATPases, and H^+-pyrophosphatases (Dang et al., 2013; Ahmed et al., 2013).

Most abundant transcripts in *S. parvula* under salt stress encode 17 transport-related proteins, including sodium and potassium ion transmembrane transporters, chloride channels, and ABC transporters. This halophyte and its relative *Eutrema salsugineum* highlighted the HKT1 Na^+/K^+ transporter (Wu et al., 2012). Highly enhanced expression of genes for cation-efflux transporters was observed in *Arabidopsis thaliana* (Jarvis et al., 2014). Studies in *Thellungiella* showed genes encoding transporters such as chloride channels and P-type H^+-ATPase. Chloride channels are groups of voltage-gated Cl^- channels that function in stabilizing cell membrane potential, regulating cell volume, and transcellular chloride transport (Hechenberger et al., 1996).

10.5.7.1 Antiporters

Ionic and osmotic equilibrium are necessary for plant salinity tolerance. Genes providing ionic stress protection are more abundant in *T. salsuginea* than in *Arabidopsis*. Studies have associated high Salt Overly Sensitive 1 (SOS1) expression levels with increased salt tolerance (Jarvis et al., 2014; Maughan et al., 2009). SOS1 is required for salt tolerance in *Arabidopsis* and encodes a plasma membrane Na^+/H^+ antiporter (Shi et al., 2000).

Studies showed that a plasma membrane Na^+/H^+ antiporter (SOS1), vacuolar Na^+/H^+ antiporter (NHX1), and a plasma membrane Na^+ transporter (HKT1) are important for salt tolerance (Bassil et al., 2011; Vera-Estrella et al., 2005). NHK1 is responsible for Na^+ sequestration and is up-regulated. Four genes that have strong homology to *A. thaliana* NHX2, *Mesembryanthemum crystallinum* and *Tetragoniate tragoniodes* NHX1 were slightly down-regulated and suggest that they play a role in mitigating the deleterious effects of high Na^+ levels in the cytosol and regulate intravacuolar K^+ and pH (Bassil et al., 2011). Halophytes have the ability to sequester large quantities of Na^+ into vacuoles. Cation/H^+ antiporters mediate these processes by vacuolar H^+-ATPase and H^+-PPase (Gaxiola et al., 2007).

10.5.7.2 Aquaporin

Aquaporin are intrinsic membrane proteins that serve as water-selective channels, and are involved in compartmentalization of water molecules. They likely play a role in maintaining osmosis and turgor of halophyte cells under salt stress (Dibas et al., 1998). *S. parvula* contains differentially expressed aquaporin genes, NOD26-like intrinsic protein (NIP) 5,1, and NIP 6,1 (Jarvis et al., 2014; Martínez-Ballesta et al., 2008). In *Poplar*, suppression of these genes prevents water loss during salt stress (Brinker et al., 2010).

10.6 Regulatory Molecules

Osmotic stress induces transmembrane histidine protein kinases and stretch-activated channels. Mitogen-activated protein kinases and phosphatases transduce signals for compatible osmolyte synthesis and ROS detoxification by antioxidants and regulate stress response (Senadheera and Maathuis, 2009). Brassinosteroid insensitive-1-associated receptor kinase acts synergistically with auxins and gibberellins by promoting cell elongation while protein phosphatase 2C (PP2C) regulates signal transduction pathways (Senadheera and Maathuis, 2009). In *Thellungiella*, A-type PP2C phosphatases are generally up-regulated in response to abscisic acid (ABA). SOS2, a protein kinase that phosphorylates SOS1 in *Thellungiella*, interacts directly with V-ATPase as part of its salt-tolerance mechanism (Lee et al., 2013). Serine-threonine protein kinase HT1, responsible for a reduced response to ABA or light, is decreased in salt-treated *S. fruticosa* (Diray-Arce et al., 2015).

10.7 LEA Protein Coding Genes

Late embryogenesis abundant (LEA) protein coding genes have been found to have a protective effect against desiccation or osmotic stresses due to water loss. They may function as chaperones to prevent denaturation of important proteins (Vinocur and Altman, 2005). Most genes encoding LEA proteins have abscisic acid response and/or low-temperature response elements in their promoters (Aslam et al., 2011). In *T. salsuginea*, the RAV (Related to ABI3 and VP1) gene family responds to high-salt and cold stresses (Wu et al., 2012). Osmotins are required for homeostasis by maintaining cell functions at low osmotic potentials and high ionic stress. Genes encoding cold-circadian rhythm RNA binding like protein and two isoforms of carbonic anhydrase are overexpressed in *Suaeda maritima* after salt treatment (Sahu and Shaw, 2009).

10.8 Other Genomic Elements

Several stresses can activate transposable elements (TE). A dramatic expansion of pericentromeric heterochromatin in *E. salsugineum* is hypothesized to be a result of stress-induced activation of TEs (Hundertmark and Hincha, 2008). There is a prevalence of CT-rich regions and a pyrimidine-rich region close to ATG initiation codon in *Thellungiella* 5′ UTR sequences, cytosolic cyclophilin ROC3, and transcription factor B3. Cyclophilins are abundant proteins induced under abiotic stress and transcription factor B3 is induced in specific developmental stages (Yang et al., 2013).

10.9 Pathways

Gene targets in *E. salsugineum* include four copies reported to post-transcriptionally regulate transcription factor NAC required for an ABA-independent pathway (Oh et al., 2007). In *Thellungiella*, genes involved in hormone pathways which include ZEP, AAO, and CYP707A families are all involved in ABA biosynthesis pathway contributing to salt tolerance (Kim et al., 2009). Calcium serves as a messenger in developmental processes in plants and the main mechanism for Na^+ extrusion is through the plasma membrane H^+-ATPase and Ca^{2+}-ATPase, which pump H^+ and Na^+ into the cell. This action removes a single calcium ion in exchange for the import of three sodium ions (Wu, 2012).

In *P. euphratica*, 40 metabolic pathways were changed under salt stress including carbohydrate pathway, amino acid, energy, lipid, secondary metabolite, cofactor and vitamin, terpenoid, and polyketide metabolism. ABA signaling and synthesis pathways exhibited highly induced genes under salt stress. At ZEP homologue zeaxanthine epoxidase and 9-cis-epoxycarotenoid dioxygenase increases ABA to improve drought and salt tolerance (Sun et al., 2013). Sodium accumulation induced genes involved in stress and signal transduction pathway with the involvement of calcium, ethylene, ABA signaling regulation, and biosynthesis, which play a role in drought and salinity responses (Qiu et al., 2011).

The elevation of sodium content increases root osmotic potential due to dehydration (Brinker et al., 2010). Calcium-signaling pathways were triggered after salt treatment as calcium-binding and calmodulin-binding proteins were enriched. This indicates that salt promotes auxin-signaling pathways to facilitate growth of *S. europaea* (Fan et al., 2013). The auxin-signaling pathway was considered to be critical during salt treatment because most differentially expressed genes showed increased expression.

10.10 Conclusions and Future Directions

The study of halophyte transcriptomes is still in its infancy. Many differentially expressed genes have been identified, and the results show that different species of halophytes utilize a variety of genes and pathways to establish salinity tolerance. Additional work in this area is warranted to increase our understanding of halophyte responses to salinity stress.

Acknowledgment

Support for halophyte research in the Nielsen, Khan, and Gul laboratories has been supported in part by a grant from the Pakistan—US Science & Technology Cooperation Program.

References

Abideen, Z., Koyro, H.-W., Huchzermeyer, B., Ahmed, M.Z., Gul, B., Khan, M.A., 2014. Moderate salinity stimulates growth and photosynthesis of *Phragmites karka* by water relations and tissue specific ion regulation. Environ. Exp. Bot. 105 (2014), 70—76.

Ahmed, M.Z., Shimazaki, T., Gulzar, S., Kikuchi, A., Gul, B., Khan, M.A., et al., 2013. The influence of genes regulating transmembrane transport of Na$^+$ on the salt resistance of *Aeluropus lagopoides*. Funct. Plant Biol. 40 (9), 860—871.

Anders, S., Huber, W., 2010. Differential expression analysis for sequence count data. Genome Biol. 11, R106.

Ansorge, W.J., 2009. Next-generation DNA sequencing techniques. New Biotechnol. 25, 195—203.

Ariel, F., Manavella, P., Dezar, C., Chan, R., 2007. The true story of HD-Zip family. Trends Plant Sci. 12, 419—426.

Aslam, R., Bostan, N., Amen, N., Maria, M., Safdar, W., 2011. A critical review on halophytes: salt tolerant plants. J. Med. Plants Res. 5, 7108—7118.

Bassil, E., Tajima, H., Liang, Y., Ohto, M., Ushijima, K., Nakano, R., et al., 2011. The *Arabidopsis* Na$^+$/H$^+$ antiporters NHX1 and NHX2 control vacuolar pH and K^+ homeostasis to regulate growth, flower development, and reproduction. Plant Cell. 23, 3482–3497.

Brautigam, A., Gowik, U., 2010. What can next generation sequencing do for you? Next generation sequencing as a valuable tool in plant research. Plant Biol. 12, 831–841.

Brinker, M., Brosche, M., Vinocur, B., Abo-Ogiala, A., Fayyaz, P., Janz, D., et al., 2010. Linking the salt transcriptome with physiological responses of a salt-resistant *Populus* species as a strategy to identify genes important for stress acclimation. Plant Physiol. 154, 1697–1709.

Chen, G., Wang, C., Shi, T., 2011. Overview of available methods for diverse RNA-Seq data analyses. Sci. China Life Sci. 54, 1121–1128.

Clarke, K., Yang, Y., Marsh, R., Xie, L., Zhang, K.K., 2013. Comparative analysis of de novo transcriptome assembly. Sci. China Life Sci. 56, 156–162.

Compeau, P.E.C., Pevzner, P., Tesler, G., 2011. How to apply de Bruijn graphs to genome assembly. Nat. Biotechnol. 29, 987–991.

Dang, Z., Zheng, L., Wang, J., Gao, Z., Wu, S., Qi, Z., et al., 2013. Transcriptomic profiling of the salt-stress response in the wild recretohalophyte *Reaumuria trigyna*. BMC Genomics. 14, 29.

Dassanayake, M., Oh, D., Haas, J.S., Hernandez, A., Hong, H., Ali, S., et al., 2011. The genome of the extremophile crucifer *Thellungiella parvula*. Nat. Genet. 43, 913–918.

Dibas, A.I., Mia, A.J., Yorio, T., 1998. Aquaporins (water channels): role in vasopressin-activated water transport. Exp. Biol. Med. 219, 183–199.

Diédhiou, C.J., Popova, O.V., Golldack, D., 2009. Comparison of salt-responsive gene regulation in rice and in the salt-tolerant *Festuca rubra* ssp. *litoralis*. Plant Signal. Behav. 4, 533–535.

Diray-Arce, J., Clement, M., Gul, B., Khan, M.A., Nielsen, B.L., 2015. Transcriptome assembly, profiling and differential gene expression analysis of the halophyte Suaeda fruticosa provides insights into salt tolerance. BMC Genomics. 16, 353.

Engstrom, P.G., Steijger, T., Sipos, B., Grant, G.R., Kahles, A., The, R.C., et al., 2013. Systematic evaluation of spliced alignment programs for RNA-seq data. Nat. Methods. 10, 1185–1191.

Fan, P., Nie, L., Jiang, P., Feng, J., Lu, S., Chen, X., et al., 2013. Transcriptome analysis of *Salicornia europaea* under saline conditions revealed the adaptive primary metabolic pathways as early events to facilitate salt adaptation. PLoS One. 8, e80595.

Ferreira de Carvalho, J., Poulain, J., Da Silva, C., Wincker, P., Michon-Coudouel, S., Dheilly, A., et al., 2013. Transcriptome de novo assembly from next-generation sequencing and comparative analyses in the hexaploid salt marsh species *Spartina maritima* and *Spartina alterniflora* (Poaceae). Heredity. 110, 181–193.

Flowers, T.J., Colmer, T.D., 2008. Salinity tolerance in halophytes. New Phytol. 179, 945–963.

Garber, M., Grabherr, M.G., Guttman, M., Trapnell, C., 2011. Computational methods for transcriptome annotation and quantification using RNA-seq. Nat. Methods. 8, 469–477.

Garg, R., Verma, M., Agrawal, S., Shankar, R., Majee, M., Jain, M., 2014. Deep transcriptome sequencing of wild halophyte rice, *Porteresia coarctata*, provides novel insights into the salinity and submergence tolerance factors. DNA Res. 21, 69–84.

Gaxiola, R.A., Palmgren, M.G., Schumachner, K., 2007. Plant proton pumps. FEBS Lett. 581, 2204–2214.

Griffith, M., Timonin, M., Wong, A.C.E., Gray, G.R., Akhter, S.R., Saldanha, M., et al., 2007. *Thellungiella*: an *Arabidopsis*-related model plant adapted to cold temperatures. Plant Cell Environ. 30, 529–538.

Grigore, M.N., Boscaiu, M., Vicente, O., 2011. Assessment of the relevance of osmolyte biosynthesis for salt tolerance of halophytes under natural conditions. Eur. J. Plant Sci. Biotechnol. 5, 12–19.

Hasegawa, P.M., Bressan, R.A., Zhu, J., Bohnert, H.J., 2000. Plant cellular and molecular responses to high salinity. Annu. Rev. Plant Physiol. Plant Mol. Biol. 51, 463–499.

Hechenberger, M., Schwappach, B., Fischer, W.N., Frommer, W.B., Jentsch, T.J., Steinmeyer, K., 1996. A family of putative chloride channels from *Arabidopsis* and functional complementation of a yeast strain with a CLC gene disruption. J. Biol. Chem. 271, 33632–33638.

Horie, T., Schroeder, J.I., 2004. Sodium transporters in plants. Diverse genes and physiological functions. Plant Physiol. 136, 2457–2462.

Huang, J., Lu, X., Yan, H., Chen, S., Zhang, W., Huang, R., et al., 2012. Transcriptome characterization and sequencing-based identification of salt-responsive genes in *Millettia pinnata*, a semi-mangrove plant. DNA Res. 19, 195–207.

Hundertmark, M., Hincha, D.K., 2008. LEA (Late Embryogenesis Abundant) proteins and their encoding genes in *Arabidopsis thaliana*. BMC Genomics. 9, 118.

Jan, A., Yang, G., Nakamura, H., Ichikawa, H., Kitano, H., Matsuoka, M., et al., 2004. Characterization of a xyloglucan endotransglucosylase gene that is up-regulated by gibberellin in rice. Plant Physiol. 136, 3670–3681.

Jarvis, D.E., Ryu, C., Beilstein, M.A., Schumaker, K.S., 2014. Distinct roles for SOS1 in the convergent evolution of salt tolerance in *Eutrema salsugineum* and *Schrenkiella parvula*. Mol. Biol. Evol. 31, 2094–2107.

Kant, S., Bi, Y.M., Weretilnyk, E., Barak, S., Rothstein, S.J., 2008. The *Arabidopsis* halophytic relative *Thellungiella halophila* tolerates nitrogen-limiting conditions by maintaining growth, nitrogen uptake, and assimilation. Plant Physiol. 147, 1168–1180.

Ketchum, R.E.B., Warren, R.S., Klima, L.J., Lopez-Gutiérrez, F., Nabors, M.W., 1991. The mechanism and regulation of proline accumulation in suspension cell cultures of the halophytic grass *Distichlis spicata* L. J. Plant Physiol. 137, 368–374.

Khan, M.A., Ungar, I.A., Showalter, A.M., 2000. The effect of salinity on the growth, water status, and ion content of a leaf succulent perennial halophyte, *Suaeda fruticosa* (L.) Forssk. J. Arid Environ. 45, 73–84.

Kim, J.H., Woo, H.R., Kim, J., Lim, P.O., Lee, I.C., Choi, S.H., 2009. Trifurcate feed-forward regulation of age-dependent cell death involving miR164 in *Arabidopsis*. Science. 323, 1053–1057.

Kore-eda, S., Cushman, M.A., Akselrod, I., Bufford, D., Fredrickson, M., Clark, E., et al., 2004. Transcript profiling of salinity stress responses by large-scale expressed sequence tag analysis in *Mesembryanthemum crystallinum*. Gene. 341, 83–92.

Kumar, S., Blaxter, M., 2010. Comparing de novo assemblers for 454 transcriptome data. BMC Genomics. 11, 571.

Lee, Y., Giorgi, F., Lohse, M., Kvederaviciute, K., Klages, S., Usadel, B., et al., 2013. Transcriptome sequencing and microarray design for functional genomics in the extremophile *Arabidopsis* relative *Thellungiella salsuginea (Eutrema salsugineum)*. BMC Genomics. 14, 793.

Li, Z., Chen, Y., Mu, D., Yuan, J., Shi, Y., Hao, Z., et al., 2011. Comparison of the two major classes of assembly algorithms: overlap-layout-consensus and de-bruijn graphs. Briefings in Func. Gen.

Liang, S., Fang, L., Zhou, R., Tang, T., Deng, S., Dong, S., et al., 2012. Transcriptional homeostasis of a mangrove species, *Ceriops tagal*, in saline environments, as revealed by microarray analysis. PLoS One. 7, e36499.

Lokhande, V., Suprasanna, P., 2012. Prospects of halophytes in understanding and managing abiotic stress tolerance. In: Ahmad, P., Prasad, M. (Eds.), Environmental Adaptations and Stress Tolerance of Plants in the Era of Climate Change. Springer, Maharashtra, India, 29–56.

Lokhande, V., Nikam, T., Penna, S., 2010. Differential osmotic adjustment to iso-osmotic NaCl and PEG stress in the *in vitro* cultures of *Sesuvium portulacastrum* L. J. Crop Sci. Biotechnol. 13, 251–256.

Ma, J., Zhang, M., Xiao, X., You, J., Wang, J., Wang, T., et al., 2013. Global transcriptome profiling of *Salicornia europaea* L. Shoots under NaCl treatment. PLoS One. 8, e65877.

Magome, H., Yamaguchi, S., Hanada, A., Kamiya, Y., Oda, K., 2008. The DDF1 transcriptional activator upregulates expression of a gibberellin-deactivating gene, GA2ox7, under high-salinity stress in *Arabidopsis*. Plant J. 56, 613–626.

Martínez-Ballesta, M.d.C., Bastías, E., Carvajal, M., 2008. Combined effect of boron and salinity on water transport: the role of aquaporins. Plant Signal. Behav. 3, 844–845.

Maughan, P.J., Turner, T.B., Coleman, C.E., Elzinga, D.B., Jellen, E.N., Morales, J.A., et al., 2009. Characterization of Salt Overly Sensitive 1 (SOS1) gene homoelogs in quinoa (*Chenopodium quinoa* Willd.). Genome. 52, 647–657.

McGettigan, P.A., 2013. Transcriptomics in the RNA-seq era. Curr. Opin. Chem. Biol. 17, 4–11.

Miller, J.R., Koren, S., Sutton, G., 2010. Assembly algorithms for next-generation sequencing data. Genomics. 95, 315–327.

Morozova, O., Marra, M.A., 2008. Applications of next-generation sequencing technologies in functional genomics. Genomics. 92, 255–264.

Mutz, K., Heilkenbrinker, A., Lönne, M., Walter, J.-G., Stahl, F., 2013. Transcriptome analysis using next-generation sequencing. Curr. Opin. Biotechnol. 24, 22–30.

Nedjimi, B., Daoud, Y., 2009. Cadmium accumulation in *Atriplex halimus* subsp. *schweinfurthii* and its influence on growth, proline, root hydraulic conductivity and nutrient uptake. Flora Morphol. Distrib. Funct. Ecol. Plants. 204, 316–324.

Oh, D., Gong, Q., Ulanov, A., Zhang, Q., Li, Y., Ma, W., et al., 2007. Sodium stress in the halophyte *Thellungiella halophila* and transcriptional changes in a thsos1-RNA interference line. J. Integr. Plant Biol. 49, 1484–1496.

Oh, D., Dassanayake, M., Haas, J.S., Kropornika, A., Wright, C., d'Urzo, M.P., et al., 2010. Genome structures and halophyte-specific gene expression of the extremophile *Thellungiella parvula* in comparison with *Thellungiella salsuginea (Thellungiella halophila)* and *Arabidopsis*. Plant Physiol. 154, 1040–1052.

Qin, Q.P., Zhang, L.L., Li, N.Y., Cui, Y.Y., Xu, K., 2010. Optimizing of cDNA preparation for next generation sequencing. Yi Chuan. 32, 974–977.

Qiu, Q., Ma, T., Hu, Q., Liu, B., Wu, Y., Zhou, H., et al., 2011. Genome-scale transcriptome analysis of the desert poplar, *Populus euphratica*. Tree Physiol. 31, 452–461.

Sahu, B.B., Shaw, B., 2009. Isolation, identification and expression analysis of salt-induced genes in *Suaeda maritima*, a natural halophyte, using PCR-based suppression subtractive hybridization. BMC Plant Biol. 9, 69.

Senadheera, P., Maathuis, F.J.M., 2009. Differentially regulated kinases and phosphatases in roots may contribute to inter-cultivar difference in rice salinity tolerance. Plant Signal. Behav. 4, 1163–1165.

Seyednasrollah, F., Laiho, A., Elo, L.L., 2013. Comparison of software packages for detecting differential expression in RNA-seq studies. Brief. Bioinform. 16, 59–70.

Shi, H., Ishitani, M., Kim, C., Zhu, J.-K., 2000. The *Arabidopsis thaliana* salt tolerance gene SOS1 encodes a putative Na$^+$/H$^+$ antiporter. Proc. Natl. Acad. Sci. 97, 6896–6901.

Söderman, E., Mattsson, J., Engström, P., 1996. The *Arabidopsis* homeobox gene ATHB-7 is induced by water deficit and by abscisic acid. Plant J. 10, 375–381.

Stein, L., 2001. Genome annotation: from sequence to biology. Nat. Rev. Genet. 2, 493–503.

Sucre, B., Suárez, N., 2011. Effect of salinity and PEG-induced water stress on water status, gas exchange, solute accumulation, and leaf growth in *Ipomoea pescaprae*. Environ. Exp. Bot. 70, 192–203.

Sun, Y., Wang, F., Wang, N., Dong, Y., Liu, Q., Zhao, L., et al., 2013. Transcriptome exploration in *Leymus chinensis* under saline-alkaline treatment using 454 pyrosequencing. PLoS One. 8, e53632.

Tarazona, S., García-Alcalde, F., Dopazo, J., Ferrer, A., Conesa, A., 2011. Differential expression in RNA-seq: a matter of depth. Genome Res. 21, 2213–2223.

Trapnell, C., Roberts, A., Goff, L., Pertea, G., Kim, D., Kelley, D.R., et al., 2012. Differential gene and transcript expression analysis of RNA-seq experiments with TopHat and Cufflinks. Nat. Protocols. 7, 562–578.

Vera-Estrella, R., Barkla, B.J., Garcia-Ramirez, L., Pantoja, O., 2005. Salt stress in *Thellungiella halophila* activates Na$^+$ transport mechanisms required for salinity tolerance. Plant Physiol. 139, 1507–1517.

Vinocur, B., Altman, A., 2005. Recent advances in engineering plant tolerance to abiotic stress: achievements and limitations. Curr. Opin. Biotechnol. 16, 123–132.

Wicker, T., Schlagenhauf, E., Graner, A., Close, T., Keller, B., Stein, N., 2006. 454 sequencing put to the test using the complex genome of barley. BMC Genomics. 7, 275.

Wong, C., Li, Y., Labbe, A., Guevara, D., Nuin, P., Whitty, B., et al., 2006. Transcriptional profiling implicates novel interactions between abiotic stress and hormonal responses in *Thellungiella*, a close relative of *Arabidopsis*. Plant Physiol. 140, 1437–1450.

Wu, H., Zhang, Z., Wang, J., Oh, D., Dassanayake, M., Liu, B., et al., 2012. Insights into salt tolerance from the genome of *Thellungiella salsuginea*. Proc. Natl. Acad. Sci. 109, 12219–12224.

Wu, Y., 2012. Unwinding and rewinding: double faces of helicase? J. Nucleic Acids. 2012, 140601.

Xiong, L., Zhu, J.-K., 2002. Molecular and genetic aspects of plant responses to osmotic stress. Plant Cell Environ. 25, 131–139.

Yang, R., Jarvis, D.J., Chen, H., Beilstein, M., Grimwood, J., Jenkins, J., et al., 2013. The reference genome of the halophytic plant *Eutrema salsugineum*. Front. Plant Sci. 4, 46.

Zhang, C., Zhao, B., Ge, W., Zhang, Y., Song, Y., Sun, D., et al., 2011. An apoplastic H-type thioredoxin is involved in the stress response through regulation of the apoplastic reactive oxygen species in rice. Plant Physiol. 157, 1884–1899.

Zhang, J., Xie, P., Lascoux, M., Meagher, T.R., Liu, J., 2013. Rapidly evolving genes and stress adaptation of two desert poplars, *Populus euphratica* and *P. pruinosa*. PLoS One. 8, e66370.

11

SUSTAINABLE DIVERSITY OF SALT-TOLERANT FODDER CROP—LIVESTOCK PRODUCTION SYSTEM THROUGH UTILIZATION OF SALINE NATURAL RESOURCES: EGYPT CASE STUDY

Hassan M. El Shaer[1] and A.J. Al Dakheel[2]
[1]*Desert Research Center, Mataria, Cairo, Egypt* [2]*International Center for Biosaline, Dubai, UAE*

11.1 Introduction

Populations in Egypt are growing so quickly (approaching 90 million people) that the arable lands and the available fresh water are unable to sustain the population increments. Salinity that leads to desertification is a serious problem with crucial impacts on agriculture development, in particular in arid and semi-arid zones of Egypt (Anonymous, 2009). It is believed that cultivation of salt-tolerant crops, using marginal resources, such as saline soils and irrigation water, has significant social and economic potential to solve the problems of food supply for human and animal feed shortages and to decrease its costs (El Shaer, 2010). These plants can grow in saline to extremely saline habitats and have particular characteristics which enable them to evade and/or resist and tolerate salinity by various eco-physiological mechanisms (Wassif et al., 1983). Such forage crops can constitute a major part of the yearly feeding program of animals since they provide a valuable reserve feed for animals, particularly under drought conditions (El Shaer, 2006). An

M.A. Khan, M. Ozturk, B. Gul, & M.Z. Ahmed (Eds): Halophytes for Food Security in Dry Lands.
DOI: http://dx.doi.org/10.1016/B978-0-12-801854-5.00011-X

177

agreement was signed between the International Center for Bio-saline Agriculture (ICBA), Dubai, UAE, and the Desert Research Center (DRC), Cairo, Egypt, for cooperation in a joint project to introduce the concept of a saline agriculture–livestock system to the local farmers in Sinai. The project was comprised of various activities to develop and supply sustainable integrated management practices to improve the farmers' livelihoods, resiliency to climate changes, and to increase the income of poor farmers relying on marginal resources in Sinai. This chapter aims to introduce and evaluate environmentally and economically feasible forage–livestock systems using marginal resources in Sinai.

11.2   Egypt's General Characteristics

Egypt is located in the north-eastern corner of Africa; with an area of slightly 1 million km^2. Around 95% of the population lives on about 4% of the total land in the Nile Valley, Nile Delta, and in coastal areas.

Egypt's climate is semi-desert, characterized by hot dry summers, moderate winters, and very little rainfall. The arid climate of Egypt is characterized by high evaporation rates (1500–2400 mm per year) and very low rainfall (5–200 mm per year) in winter, particularly in the delta and along the Mediterranean coast and west of the delta (EEAA, 1997). Environmental degradation that has resulted from current climate changes, including prolonged drought, land degradation, desertification, and loss of biodiversity, is presenting enormous challenges to achieving food security and eradication of poverty in the marginal regions (EEAA, 1997). It is worth mentioning that agriculture is a key sector in the Egyptian economy; it contributes about 40% to the gross domestic product (GDP) and 50% of overall employment (SADS, 2009). Because of the climate change and other human factors, Egypt is far from being self-sufficient in some food materials. However, climate change would make the situation even worse as a result of its expected adverse impact on the national production of many crops. Any decrease in the total supply of water, coupled with the expected increase in consumption due to the high population growth rates will have drastic impacts. Therefore, water management is one of the most important adaptation actions (El-Beltagy and Abo-Hadeed, 2008).

Livestock constitutes an important component of the agricultural sector, representing about 24.5% of the agricultural GDP with a value of around US$ 6.1 billion (SADS, 2009). The total numbers of animals were estimated as: 4.9 million cattle, 4.2 million buffalos, 5.2 million sheep, 4.3 million goats, and 141,537 camels, in addition to 1.4 million heads of other animals, such as horses, donkeys, etc. (Economic Affairs Sector, 2012). The amount of feed produced in 2012 decreased due partially to the impact of climate changes on feed ingredient cultivation (Economic Affairs Sector, 2012). The natural vegetation, as the principal feed resource in Egyptian deserts (El Shaer, 2006), is seasonally and drastically variable depending on rainfall. The range of vegetation in most of these regions is characterized by stands of shrubs and semi-shrubs with a cover of short-lived annual forbs and grasses (FAO, 2010). It is, generally, of low forage production and quality due to several environmental and management factors (El Shaer, 1981, 2004). The yield of this vegetation, as animal feed, does not cover the annual nutritional requirements for animals (El-Lakany, 1987; El Shaer, 2010). The main constraints that limit rangeland utilization and animal production in the Egyptian deserts could be briefly described as:

1. Inadequate feeding results from low pasture quality and productivity, especially in the rainfed areas, and inadequate diet formulation due to lack of farmers' knowledge of nutritional value and feed requirements.
2. Inadequate stock and poor-quality water in most range areas.
3. Climate change constraints, particularly high-frequency and long-term drought, soil salinization, and range degradation due to overgrazing and human factors.
4. Inadequate herd management practices leading to uncontrolled reproduction and health problems.

As reported by the FAO (2010), the indigenous rangelands provide only 5% of animal feed; it became necessary to find sustainable sources of feed resources, especially forage crops. Therefore, utilization of salt-affected soils and saline/brackish water resources for forage production could be a potential approach for alleviating the impact of climate changes in marginal areas of Egypt (Anonymous, 2009, 2013).

11.3 General Characteristics of Project Location in Sinai Region

11.3.1 North Sinai

El Tina plain (so-called Sahl El Tina) area, with an area of 50,000 acres, was selected for the project activities to represent the marginal ecosystem of North Sinai region. Poverty and inappropriate management practices are common among the local farmers; besides the marginal (saline) soil and water resources are the problems of agriculture development in this area. It was chosen to carry out the activities of the project at the farm level. It is located on the eastern side of the Suez Canal; characterized by semi-arid conditions (annual rainfall is about 160 mm per year). The soil texture, in brief, ranges from sandy loam to clay (Anonymous, 2013), the water table levels range from 5 to 70 cm and soil pH of the surface layer (0–30 cm) varies from 7.49 to 8.1. The soil is saline-alkaline, where soil salinity ranges from 5.46 to 33.3 dS m^{-1} and exchangeable sodium percentage ranges from 10.4% to 67.0%, indicating that the soil is moderately salty to very severely salt-affected. In addition, the soil is characterized by poor fertility, where organic matter content and nutrient levels (N, P, K) in the surface layer (0–30 cm) range between 0.08% to 0.38%, 0.08% to 0.22%, 0.21 to 1.4 mg per 100 g soil, and 8 to 20 ppm, respectively. It is clear that all of these nutrients lie in deficiency levels. The mixed Nile water (from El Salam Canal) is used for irrigation with salinity ranges between 1.6 and 2.3 dS m^{-1} (1024–1472 mg L^{-1}), respectively. The common irrigation system is flood irrigation but the participating farmers who contributed and benefited in the project activities were applying drip irrigation and/or sprinkler irrigation systems.

11.3.2 Ras Sudr Area

Ras Sudr Research Station (which belongs to Desert Research Center) was selected for the project activities at the research level. All studies related to seed production and feeding and nutritional evaluation of salt-tolerant fodder crops using sheep and goats were conducted at Ras Sudr Research Station.

The soil at the Research Station was characterized by a loamy sand texture where EC was approximately 10 ds m^{-1} depicting it as strongly saline. The $CaCO_3$, Cl, Na, Ca, Mg, and K concentrations were 40, 42, 11.6, 15.2, 10.8, and 7.5 mg per 100 mg, respectively. The main irrigation systems are drip irrigation and gated pipe irrigation systems using saline underground water. The chemical analysis of irrigation water was characterized by 7.22 pH, 14.3 EC (dS m^{-1}), 191 Na$^+$ (mg L^{-1}), 20.5 Ca^{++} (mg L^{-1}), 23.0 Mg^{++} (mg L^{-1}), 0.43 K$^+$ (mg L^{-1}), 95.9 Cl$^-$ (mg L^{-1}), 3.2 HCO$_3^-$ (mg L^{-1}), 0.40 CO$_3^-$ (mg L^{-1}), and 43.5 hardness (Anonymous, 2013).

11.4 Main Activities and Results

The main activities and achievements (as reported in Anonymous, 2013) could be briefly described as follows:

1. Development and transfer of farmer-based seed production technologies
2. Production and dissemination of a package of efficient forage production and utilization suitable for marginal environment
3. Fodder crop utilization and livestock production
4. Capacity building and economic assessment.

It is worth mentioning that all data and results that are presented herein (in this case study) were derived from the project annual reports (Anonymous, 2013) and some published papers of the project staff members through their Ph.D. and M.Sc. theses, as well as research papers.

11.4.1 Development and Transfer of Farmer-Based Seed Production Technologies

These activities were conducted in the fields of the participating farmers (in Sahl El Tina area) and at Ras Sudr Research Station (Anonymous, 2013).

It was important to identify progressive farmers in the selected benchmark sites and to train them on seed production. Therefore, four sites were chosen for seed production for sorghum and pearl millet; an improved management practice package was applied (the details were presented in Anonymous, 2013). The selected farmers were grown the grasses for seed production and green forage as well to feed their livestock.

Several genotypes of different summer and winter fodder crops were also evaluated at Ras Sudr Research Station in Sinai as shown in Table 11.1.

However, in order to develop a seed production technology package, particular practices were applied as summarized in Table 11.2.

Through applying these practices, the following yields for barley, safflower, and triticale were recorded:

1. 500 kg for four genotypes of barley CHK 6, CHK 37, CHK 38, and CHK 52
2. 200 kg of 54 genotypes of safflower and 150 kg of 36 genotypes of triticale
3. 300 kg of sorghum genotypes
4. 500 kg for pearl millet genotypes
5. About 4000 sesbania seedlings were distributed to farmers.

Table 11.1 The Most Salt-Tolerant Genotypes of Different Fodder Crops

Crop		The Tolerant Genotypes
Sorghum	3	S 35/csv 15/JJ 1041
Pearl millet	4	IP 13150/IP 19612/S pop I and II
Barley	4	L 1/L 10/L 11
Fodder beet	4	L 1/L 10/L 9/L 11
Triticale	4	L 5/L 6/L 9/L 7
Safflower	15	L 4/L 6/L 11/L 13/L 18/L 25/L 19/L 4/L 30/L 37/L 43/L 47/L 51

Table 11.2 Integrated Management Package for Annual Grasses Seed Production Applied at Ras Sudr Research Station

Practice	Particulars
Seed rate	Seed 120 kg/ha
Number of seedlings/hill	2 seedlings/hill for seed
Spacing	10 cm in barley, triticale and sunflower
Fertilization	Spraying mixture of amino Acids, twice during vegetative phase based on morphological characters and twice during and after flowering
	Organic manure, elemental sulphur and effective microorganisms at the rate of 50 tonnes/ha, 150–200 kg/ha and 50 liter/ha, respectively

11.4.2 Production and Dissemination of a Package of Efficient Forage Production and Utilization Suitable for Marginal Environment

Improved integrated water management through improved water use efficiency and transfer of an integrated management package (IMP) for grain and fodder production to a minimum of 120 farmers were achieved. Both drip irrigation systems and sprinkler irrigation systems were used at the farms of the participating farmers, who were well trained by the project staff on the improved management package (IPM). It is worth mentioning that the farmers applied such IPM on one-half of their lands, while the other half was traditionally cultivated (using a flood irrigation system) using their own methods, in order to compare between the results obtained by traditional management and those obtained by applying IPM. The results were collected from 50 and four directly participating farmers in the summer and winter seasons, respectively. Those farmers were considered as Pilot Farmers for the Farmers' Field Schools programs. The applied IPM using drip or/and sprinkler irrigation systems consisted of:

- Selection of forage crops introduced from ICBA
- Leveling soil surface using laser technique
- Addition of organic manure, elemental sulfur and effective microorganisms at the rate of 50 tonnes ha^{-1}, 250–300 kg ha^{-1}, and 50 L ha^{-1}, respectively, and EM solution 50 L ha^{-1}
- Special planting procedure to lower the salinity in the soil around the germinating seeds
- Application of mineral fertilizers at the rate varied according to crop type
- Application of leaching requirements with irrigation water to keep the salts in solution and flush them below the root zone. The drainage system was developed at each pilot farm by adding plant residues inside old surface drainage channels and installing new surface drainage channels
- Irrigation using a special time schedule for each crop to maintain a relatively high level of soil moisture to achieve periodic leaching of the soil. In this respect, developing an irrigation system using spill pipes was applied at five pilot farms (2 ha each).

11.4.2.1 Winter Season

Fodder beet, Egyptian clover, and barley were cultivated on 8, 30, and 12 farmers' fields, respectively. Although 120 farmers participated in these activities, the data were collected only from 50 of the directly participating farmers (Pilot Farmers), who are well trained and offered some production materials, such as irrigation equipment and seeds from the project fund. Soil preparation practices were carried out for all crops. The fertilization program of fodder beet plant included 75 kg P_2O_5 ha^{-1} as mono calcium phosphate (super phosphate) and 300 kg S ha^{-1} added during soil preparation and 120 kg N ha^{-1} as ammonium sulfate added in two equal doses on the sowing date and after 60 days from the sowing date. Moreover, some agricultural practices and irrigation methods were modified as previously mentioned. In each field, the crop yields resulting from traditional and improved management practices were compared. The main results obtained can be summarized (Anonymous, 2013) as follows:

- In the 2011/2012 growing season, the percentages of increments as a result of applying IPM of fodder beet yielded 92.11 tonnes ha^{-1} and reached 21.86% as compared to those obtained by traditional management. The increment percentages varied according to yield part and management type. Meanwhile, in the 2012/2013 growing season, the percentages of increments reached 7.90% as compared to traditional management. The reduction in the increment was due to raising farmers' skill in application of the package components.
- Planting Egyptian clover on the 1st October showed the highest fresh and dry yield (90 tonnes ha^{-1}) for all cuts at both seeding rates (75 and 100 kg ha^{-1}), either with traditional or improved management practices. Increasing seeding rate to 100 kg ha^{-1} led to an increase in fresh and dry yields of clover cuts, but the values associated with improved management were greater than those of traditional management at both seeding rate and all planting date. Application of 20−25% of leaching requirements increased the yield of crops under consideration.

11.4.2.2 Summer Season

Pearl millet was cultivated in 8.0 ha at four farms in the El Tina plain area. Farmers preferred to grow pearl millet to sorghum because it produced more yield than the sorghum crop in recent years in this region (Anonymous, 2009). On an overall average basis, the fresh yield of 130.21 tonnes ha^{-1} was recorded when the IPM were applied. The results showed that the average total fresh yield of pearl millet was increased by 50.7%, 25.5%, and 4.74%, in the 2011, 2012, and 2013 growing seasons, respectively, regardless of the sites of farms, when IMP were applied as compared to the application of traditional practices.

Moreover, using low-quality water for irrigation could contribute to the process of secondary salinization, which is accelerating due to the application of poor management practices (Wassif et al., 1983). Surface soil samples (0−30 cm) were collected from fields of improved (IMP) and traditional management practices at the end of summer season 2012 to evaluate the effect of the IPM on soil salinity properties. The results of soil properties under improved management conditions were compared with those obtained under traditional conditions. The main results indicated, briefly, that soil properties under IPM practices were improved due mainly to: soil pH values were decreased by 2–6%; EC_e value was decreased by about 30%; soil organic matter content reached 0.56%, the reduction percent of exchangeable Na% ranged from 16% to 68% and the availability of N, P, and K increased by different percentages.

The impacts of irrigation and drainage management modification were evaluated. It is briefly, concluded, that the spill pipes irrigation system with a long steady sloped field led to a reduction in the amount of irrigation water reaching 14%, 15%, 9%, and 11% for barley, berseem, pearl millet, and fodder beet crops, respectively. Moreover, the highest average value of water use efficiency ($27 \, \text{kg m}^{-3}$) was obtained with the application of improved management practices for pearl millet crop. In general, the average values of water use efficiency were 16, 20, and $21 \, \text{kg m}^{-3}$ for fodder beet, Egyptian clover, and barley crops, respectively.

11.4.3 Fodder Crop Utilization and Livestock Production

Evaluation of the feeding values of the introduced forage crops was conducted at the research level (Ras Sudr Research Station), while the impact of feeding livestock on such fodder crops were evaluated at both research and farm levels. These forage crops were fed to animals as an individual fodder material or mixed or incorporated with other feed materials (Fahmy et al., 2010; Anonymous, 2013; Helal et al., 2013). The potential salt-tolerant fodder plants were: (i) pearl millet (*Pennisetum americanum*), (ii) sorghum, (iii) fodder beet (*Beta vulgaris*), (iv) *Panicum turgedum*, (v) alfalfa (*Medicago sativa*), (vi) saltbush (*Atriplex nummularia*), (vii) *Kochia indica*, and (viii) *Leucaena leucocephala* and *Sesbania sesban*.

Intensive results were derived from a series of experiments carried out during the project (Fayed et al., 2010; El-Essawy et al., 2011; Helal et al., 2013). Some data as examples will be reported in the following subsections.

11.4.3.1   Chemical Composition of Fodder Crop Species

Data on the chemical composition of the best potential forage species as feed materials (Table 11.3, Anonymous, 2013), generally concluded that:

- *Atriplex nummularia* contained low organic matter content due to a high concentration of ash content.
- All the forage feed materials contained variable concentrations of crude protein (CP) ranging from 7.18% to 18.82%. Leucaena and alfalfa showed the highest CP content (16.64% and 18.82%, respectively). They could potentially be protein-rich plant sources for supplementation. Quansah and Makkar (2012) recommended that forages containing CP ≥ 16% are very good, >8 and <16% are good, and ≤ 8% are fair based on a minimum CP of 15–18% for adequate growth in ruminants. Therefore, the CP content of the studied forage species is categorized into three groups, namely "very good," "good," and "fair." *Leucaena leucocephala* was categorized as a very good source of CP (16.64%), while *A. nummularia* was categorized as good (12.16%). *Panicum turgidum* and pearl millet (*P. americanum*), as grasses, were considered as fair (being 7.26% and 7.18%, respectively).

Table 11.3 Overall Means of the Chemical Composition of Salt-Tolerant Fodder Crops (as % on DM Basis) as Cited from Anonymous (2013)

Items	Pearl Millet	Sorghum	*Kochia indica*	*Atriplex nummularia*	*Leucaena*	*Panicum turgidum*	Alfalfa	Fodder Beet
ASH	11.49	14.5	17.2	25.9	9.30	11.96	7.55	12.75
CP	7.18	13.2	11.1	12.16	16.64	7.26	18.82	8.50
CF	28.23	24.9	27.8	17.97	26.1	27.93	20.63	7.75
EE	2.76	2.86	11.82	1.28	1.42	1.35	2.51	2.43
NFE	50.34	44.56	43.08	42.69	46.54	51.5	50.49	68.57
OM	88.51	85.5	82.8	74.1	90.7	88.04	92.45	87.25
CHO	40.07	23.4	40.0	60.25	43.75	32.5	38.51	67.61
NDF	59.93	76.6	59.5	42.8	56.25	67.5	61.49	32.39
ADF	37.45	28.3	39.3	17.45	25.45	38.32	47.72	19.47
Cellulose	27.35	25.1	20.2	11.70	21.13	20. 22	20.78	11.22
Hemicellulose	22.48	48.3	28.4	25.35	30.8	29.18	13.77	12.92

- All introduced fodder crops attained a greater content of organic matter (OM) except *A. nummularia* (saltbush) which was characterized by high concentrations of ash and Na (25.9% and 1.92%, respectively). *Atriplex nummularia*, as a halophytic shrub, showed the lowest OM, crude fiber (CF), ether extracts (EE), nitrogen free extracts (NFE), neutral detergent fiber (NDF), acid detergent fiber (ADF), potassium and metabolizable energy values compared to other forages. These findings are in agreement with those reported for the same plant species by Ramirez et al. (2000).
- Fodder beet and all grasses species such as pearl millet, sorghum, and panicum are considered as good sources of energy fodder crops. They should be mixed with alfalfa or Leucaena to formulate balanced rations under saline conditions. It is also concluded that pearl millet (*P. americanum*), *A. nummularia*, *L. leucocephala*, and *P. turgidum* are characterized by high vegetative yields and could have great potential as sources of livestock fodders (Anonymous, 2009; Fahmy et al., 2010). These authors concluded that such forage species could be nutritious feed materials for livestock since they contain enough protein and energy to cover their nutritional requirements (El Shaer, 2006). Furthermore, feeding animals these nitrogen-rich fodder crops mixed with the tested grasses and fodder beet (as a source of energy) would positively support live weight maintenance and growth as reported by many investigators (El-Essawy et al., 2011; Abdou et al., 2011; Helal et al., 2013).

11.4.3.2 Evaluate the Nutritive Value of the Introduced Fodder Crop Species

This case study aimed at shedding some light, briefly, on the nutritive value of some introduced fodder crops fed to livestock. Therefore, few examples for research studies are presented herein. An experiment was conducted by El-Essawy et al. (2011) to study the effect of feeding different sources of nitrogen-rich fodders with dietary inclusion of fodder beet as an energy source on nutrient digestibility, intake, and nitrogen balance. These fodder crops were: alfalfa (*M. sativa*) (R1), Atriplex (*A. nummularia*) (R2) and Leucaena (*L. leucocephala*) (R3). The experiment was performed on 15 adult sheep rams (46 + 0.47 kg average body weight) which were divided into three equal groups. The plant species were nutritionally evaluated through three digestibility trials. Considerable variations were observed among the chemical composition of the studied forage plants.

Leucaena leucocephala attained the highest content of CP. According to the recommendations of Quansah and Makkar (2012), *L. leucocephala* and alfalfa were categorized as very good sources of protein, while *A. nummularia* was categorized as good. However, many salt-tolerant plants contain high levels of non-protein nitrogen. Benjamin et al. (1992) reported that 42% of the nitrogen in *Atriplex barclayana* was nonprotein nitrogen. Similar trends were obtained for *A. nummularia* and *Atriplex halimus* grown under saline conditions in Egypt (El Shaer, 2010). Therefore, feeding animals on *Atriplex* spp. should be supplemented with any available sources of energy, such as barley and molasses (Masters et al., 2001; El Shaer, 2010).

It appeared that *L. leucocephala* is a protein and energy-rich fodder crop that could play an important role in providing a balanced diet to small ruminants in salt-affected regions. Such figures and trends were compatible with those reported by Youssef et al. (2009) and Abdou et al. (2011). The same authors also reported that significant ($P \leq 0.05$) differences in the digestibility coefficients of dry matter (DM), OM, CF, CP, EE, NDF, ADF, and acid detergent lignin (ADL) were shown among the animal groups. On the other hand, animals fed on alfalfa indicated the highest dry matter intake (DMI g kg^{-1} Bw) and CP intake (CPI g kg^{-1} Bw). Alfalfa and Atriplex contained the highest total digestible nutrients (TDN) and digestible crude protein (DCP) values. As a supplementary fodder generally, *Atriplex* spp. should not be more than 25–30% of the sheep's diet. Casson et al. (1996) suggested that the high-salt content of salt-land forage plants is likely to be the major determinant of palatability and that dilution of salt content through the availability of other feed resources would be necessary to improve intake and performance.

Another study was conducted to evaluate the nutritional value of four salt-tolerant plant species fed to goats (Helal et al., 2013). Four salt-tolerant plane species were separately cultivated in the salt-affected soil of the Research Station Farm and irrigated with underground saline water. These were: pearl millet (shandweel grass) (*P. americanum*), *A. nummularia*, *L. leucocephala*, and *P. turgidum*. Each forage species was harvested individually and total yield of nutrients was determined. During the digestibility trial, 28 Black Desert male goats were divided randomly into four equal groups and offered one of the following rations as follow: G1: pearl millet; G2: pearl millet + *A. nummularia*; G3: pearl millet + *L. leucocephala*; and G4: pearl millet + *P. turgidum*. Daily voluntary feed intake (VFI) of each forage and body weight changes were recorded. The results of Helal et al. (2013) showed that *L. leucocephala* was categorized as a very good source of protein (16.64%), energy (2.46 Mcal kg^{-1} DM), and organic matter

Table 11.4 Nutrient Yield of the Cultivated Salt-Tolerant Fodder Crops (on average basis, tonnes ha^{-1} year^{-1}) as Cited from Helal et al. (2013)

Items	Forage			
	Pennisetum americanum	*Atriplex nummularia*	*Leucaena leucocephala*	*Panicum turgidum*
Fresh yield	35	23	37	30
DM yield	12	8	12	11
CP yield	0.862	0.973	2	0.800
TDN yield	7	4.03	7.71	6.32
DCP yield	0.426	0.359	0.956	0.433
No. of goats[a]	51	45	123	56

[a]Number of goats that could be raised on 1 ha of fodder crop per year.

(90.8%). As shown in Table 11.4, all the tested forages showed comparable amounts of DM yield (around 12 tonnes ha^{-1} per year), except *A. nummularia*, which attained the lowest yield (8 tonnes ha^{-1} per year). Results of the digestibility trial pointed out that *L. leucocephala* was highly preferred by goats compared with the other forages and exhibited superiority of CP, TDN, and DCP production. The highest values of total forage consumption were recorded for animals fed pearl millet mixed with *L. leucocephala* in G3 (1428 g DM h^{-1} d^{-1}) followed by those fed pearl millet plus *P. turgidum* in G4 (1177 g DM h^{-1} d^{-1}). Goats in G3 were superior ($P \leq 0.05$) in total DMI, CPI, and OMI (Table 11.4). It is concluded that a combination of pearl millet with either *L. leucocephala* (G3) or with *P. turgidum* (G4) can be used as good-quality feed material.

11.4.3.3 Evaluate the Reproductive and Productive Performance of Sheep and Goats Fed the Tested Plant Species

A series of experiments have been carried out to determine the productive and reproductive performance of sheep and goats under saline conditions in Sinai (Shaker, 2009; El-Bassiony, 2013; El-Hawy, 2013; Shaker et al., 2014), in order to improve the reproductive and productive efficiency of small ruminants in such regions. Some research studies are given herein as examples for the obtained results that focused on the reproductive efficiency of Shami goats fed salt-tolerant fodder crops in Sinai.

A study was conducted to evaluate the effect of feeding salt-tolerant fodder plants on the reproductive performance of Shami female goats during different physiological stages (El-Hawy, 2013). In this study, 40 adult female Shami goats were used in four groups (ten each). Animals in groups 1 and 2 were offered berseem hay while groups 3 and 4 were offered alfalfa (as a salt-tolerant fodder grown at Ras Sudr Research Station). Groups 1 and 3 were offered fresh tap drinking water, while the other two groups (2 and 4) were offered saline groundwater (6000 ppm). Body weight changes, gain, and additional reproductive and productive traits as well as milk production were measured. The main obtained data are summarized in Table 11.5. However, the general trends of the study showed that:

1. The conception rate was insignificantly increased in alfalfa groups as compared to berseem hay groups. Animals in groups 3 and 4 recorded 90% and 80%, while those in groups 2 and 1 recorded 90% and 75%, respectively.
2. Group 4 recorded a higher percentage of kidding rate than groups 2 and 3 (83.3%), while those in group 1 recorded 75%.
3. Bodyweight of the control kids (group 1) was heavier than other experimental groups followed by kids weaned from does of groups 2 and 4.
4. Nevertheless, the animals fed salt-tolerant fodders and saline water had heavier birth and weaning weight than those of hay animals.

Table 11.5 The Reproductive Performance of Shami Goats as Affected by Feeding Salt-Tolerant Plants (Cited from El-Hawy, 2013)

Item	Experimental Groups			
	G1	G2	G3	G4
Conception rate, %	75	90	90	80
Kidding rate, %	75*	83.4*	91.7*	83.3*
Average litter size	1.6	1.1	1.3	1.5
% Kids weaned	76.9	72.7	75.0	77.8

G1: Animals fed Berseem hay and offered fresh drinking water.
G2: Animals fed Berseem hay and offered saline drinking water.
G3: Animals fed salt-tolerant fodders and offered fresh drinking water.
G4: Animals fed salt-tolerant fodders and offered saline drinking water.
*($P < 0.05$)

5. Weight gain of hay groups (1 and 2) was higher than those fed salt-tolerant plants (groups 3 and 4). On the other hand, drinking saline water resulted in an increase in birth weight and a decrease in weaning weight and weight again.

6. Animals that drank saline water had lower milk yields than those that drank tap water, while there were no significant differences between animals fed salt-tolerant plants and their counterparts fed control diets. Groups of tap water animal's recorded higher fat, protein, lactose, ash, and total solids percentages than those that drank saline water.

On the other hand, a study was carried out to evaluate and assess the impact of salinity on the reproductive performance of growing Shami males from weaning to sexual maturity throughout 1 year (Ashour et al., 2013). Twenty-eight growing Shami male kids (2.5–3.0 months old and 12.94 ± 0.64 kg average live body weight) were randomly assigned into four equal groups. The animals were fed and treated similarly to the treatments of El-Hawy (2013). Live body weight, testosterone level, and testicular measurements were measured monthly and age and weight at puberty were estimated. The results demonstrated that Shami bucks in groups 1 and 3 reached sexual maturity at 171 and 177 days old, respectively, followed by group 2 (186 days old), while group 4 was the most delayed in reaching the age of puberty (191 days old). Semen quality, testicular and reproductive organ measurements were better ($P < 0.05$) in group 1 (control) followed by group 3 bucks than those in groups 2 and 4.

Moreover, the impact of feeding salt-tolerant plants on physiological, behavioral, hematological, and biochemical parameters, as well as some physical, milk production, and chemical properties of coat fibers of sheep and goats under Sinai conditions, were also evaluated (Ashour et al., 2013; Younis et al., 2012; Hanan et al., 2014; Desouky and El-Shaer, 2014; El-Gendy et al, 2014; Shaker, 2014).

11.4.4 Capacity Building and Economic Assessment

Several activities have been implemented through the extension, information dissemination, and economic assessment components of the project. Improving capacity-building activities included field days, training courses, preparation of several brochures relevant to project activities, and regional farmer field school.

The number of all farmer attendees increased (from 45 to 65 farmers) during field days, which reflects the interest of farmers in participating in the project activities (Anonymous, 2013). The targets from these training courses were to improve the farmers' skills and solve their problems. The related topics in the training courses were: insect and pesticide control, irrigation systems, crop patterns, seed production, silage and feed block making, livestock management and production, and dairy products. The regional farmer field school was organized to help farmers to discover many aspects of crop management through regular field observation, sharing and learning from their collective experiences.

Assessment and quantification of profitability and impact of the introduced forage–livestock production packages on the livelihood of farmers were determined. A random sampling technique was used in selecting the sample (100 direct beneficiary farmers from Sahl El Tina). Data used were basically primary data collected based on socioeconomic variables, such as age, farm size, gender, educational status, income, costs, and returns. The analytical tools used for this evaluation included descriptive statistics and gross margin. The gross margin analysis was estimated from costs and returns in forage crop production. An economic study was carried out to compare the gross margin of different forage crops grown under both improved (IPM) and traditional management practices for winter and summer season crops. At the Sahl El Tina location, during winter season (2011–2012), the average net farm incomes of Egyptian clover, barley, and fodder beet cultivation using the IPM were 22.1%, 17.2%, and 41.3% higher than those cultivated using traditional practices, respectively. In summer season (2012), the average net farm incomes of pearl millet and sorghum cultivation under improved practices were 24.8% and 36.7% higher than cultivation under traditional practices, respectively. Economic evaluation of feeding animals on salt-tolerant fodders at the farmer levels, generally, showed an increase of about 60% in milk production; reduced feeding costs by about 40%.

11.5 Conclusions

Establishing irrigated forages, grown in salt-affected soils using poor-quality water is considered one of the best ways of overcoming shortages in feed, especially in the Egyptian desert areas. Barley, triticale, alfalfa (*M. sativa* L.), ryegrass (*Lolium perenne* L.), pearl millet (*Pennisetum glaucum* L.), Egyptian clover (*Trifolium alexandrinum* L.), *P. turgidum*, rhodes grass and fodder beet (*B. vulgaris* L.), *Acacia cyanophila*, *S. sesban*, *L. leucocephala*, and *Porsopis cheilanse* are among the most commonly recommended salt-tolerant forages under the saline conditions of Sinai due to its high biomass production, high nutritive values, and salt- or/and drought stress adaptability. Growing these forage species may contribute to the development of fragile resources in such marginal areas and enhance the living standards of local people.

Acknowledgments

The authors acknowledge financial support from the International Centre for Bio-saline Agriculture (ICBA) in Dubai, Desert Research Center (DRC), Islamic Development Bank (IDB), OPEC Fund for International Development (OFID), and International Fund for Agricultural Development (IFAD).

References

Abdou, A.R., Eid, E.Y., El-Essawy, A.M., Fayed, A.M., Helal, H.G., El Shaer, H.M., 2011. Effect of feeding different sources of energy on performance of goats fed Saltbush in Sinai. J. Am. Sci. 7, 1040–1050.

Anonymous, 2009. Final Report on Introduction of salt-tolerant forage production systems to salt-affected lands in Sinai Peninsula in Egypt: a pilot demonstration project, joint project between Desert Research Center (DRC), Egypt and the International Center Bio-saline Agriculture (ICBA), May 2009.

Anonymous, 2013. Regional project on: adaptation to climate change in WANA marginal environments through sustainable crop and livestock diversification, Annual Report, December 2013, ICBA (UAE)–DRC (Egypt).

Ashour, G., Badawy, M.T., Hafez, Y.M., El-Bassiony M.F., Ibrahim, N.H., 2013. Reproductive performance of growing Shami male kids under salinity conditions in South Sinai. In: The 23rd Annual Conference, February 3–7, Cairo/Ain El-Sokhna, Egypt. vol. 2, 195–214.

Benjamin, R.W., Oren, E., Katz, E., Becker, K., 1992. The apparent digestibility of *Atriplex barclayana* and its effect on nitrogen balance in sheep. Anim. Prod. 54, 259–264.

Casson, T., Warren, B.E., Schleuter, K., Parker, K., 1996. On farm sheep production from sheep pastures. Proc. Aust. Soc. Anim. Prod. 21, 173–176.

Desouky, M.M., El-Shaer, H.M., 2014. Quality of goats' milk soft cheese (Domiati-type) produced under salinity stress. Egypt. J. Dairy Sci. 42 (2), 175–188.

Economic Affairs Sector, 2012. Statistics of Livestock. Ministry of Agriculture and Land Reclamation (MALR), Egypt.

EEAA, 1997. A Report on OECP Climate Change Agriculture and Land Use Change Sectors Workshop. Cairo: Meridien Heliopolis Hotel, March 10, 1997.

El-Bassiony, M.F., 2013. Productive and Reproductive Responses of Growing Shami Goat Kids to Prolonged Saline Conditions in South Sinai (Ph.D. thesis). Faculty of Agriculture, Cairo University, Egypt.

El-Beltagy, A.T., Abo-Hadeed, A.F., 2008. The main pillars of the National Program for maximizing the water-use efficiency in the old land. The Research and Development Council. MOALR. (in Arabic). 30 page bulletin.

El-Essawy, A.M., Eid, E.Y., Fayed, A.M., Abdou, A.R., Helal, H.G., El Shaer, H.M., 2011. Influence of feeding fodder beet with different forages as nitrogen sources under saline conditions on Barki Rams performance in Southern Sinai. Egypt. J. Nutr. Feeds. 14, 191–205.

El-Gendy, M.H., Desouky, M.M., El-Shaer, H.M., Abdou, A.R., Helal, H.G., El Shaer, H.M., 2014. Physico-chemical properties and mineral profile of Shami goats'milk produced under desert condition. J. Food Dairy Sci. 6, 377–387.

El-Hawy, A., 2013. Reproductive Efficiency of Shami Goats in Salt Affected Lands in South Sinai (Ph.D. thesis). Faculty of Agriculture, Ain Shams University, Egypt.

El-Lakany, H.H., 1987. Protective and productive tree plantations for desert development. In: Proceeding of 2nd International Conference on Desert Development, January 25–31, 1987, Cairo, Egypt.

El Shaer, H.M., 1981. A Comparative Nutrition Study on Sheep and Goats Grazing Southern Sinai Desert Range with Supplements (Ph.D. thesis). Faculty of Agriculture, Ain Shams University, Cairo, Egypt.

El Shaer, H.M., 2004. Potentiality of halophytes as animal fodder under arid conditions of Egypt. Rangeland and pasture rehabilitation in mediterranean areas. Cahiers OPTIONS Mediterraneenes. 62, 369–374.

El Shaer, H.M., 2006. Halophytes as cash crops for animal feeds in arid and semi-arid regions. In: Ozturk, M., Waisel, Y., Khan, M.A., Gork, G. (Eds.), Bio-Saline Agriculture and High Salinity Tolerance in Plant. Birkhauser, Basel, pp. 117–128.

El Shaer, H.M., 2010. Halophytes and salt-tolerant plants as potential forage for ruminants in the Near East region. Small Rumin. Res. 91, 3–12.

FAO, 2010. Valuing Rangelands for the Ecosystem and Livelihood Services. In: Thirtieth FAO Regional Conference for the Near East. Khartoum, the Republic of the Sudan, December 4–8, 2010. Pub. NERC/10/INF/6 December 2010.

Fahmy, A.A., Youssef, K.M., El Shaer, H.M., 2010. Intake and nutritive value of some salt-tolerant fodder grasses for sheep under saline conditions of South Sinai, Egypt. Small Rumin. Res. 91, 3–12.

Fayed, A.M., El-Essawy, A.M., Eid, E.Y., Helal, H.G., Abdou, A.R., El Shaer, H.M., 2010. Utilization of Alfalfa and Atriplex for feeding sheep under saline conditions of South Sinai, Egypt. J. Am. Sci. 6, 1447–1460.

Hanan, Z., Amer-Ibrahim, N.H., Donia, G.R., Younis, F.E., Shaker, Y.M., 2014. Scrutinizing of trace elements and antioxidant enzymes changes in barki ewes fed salt-tolerant plants under south sinai conditions. J. Am. Sci. 10, 241–249.

Helal, H.G., Eid, E.Y., Nassar, M.S., El Shaer, H.M., 2013. Some nutritional studies of four salt-tolerant fodder crops fed to goats under saline conditions in Egypt. Egypt. J. Nutr. Feeds. 16, 65–77.

Masters, D.G., Norman, H.C., Dynes, R.A., 2001. Opportunities and limitations for animal production from saline land. Asian Aust. J. Anim. Sci. 14, 199–211 (Special issue).

Quansah, E.S., Makkar, H.P.S., 2012. Use of Lesser-known Plants and Plant Parts as Animal Feed Resources in Tropical Regions, FAO, Animal Production and Health Working Paper. No. 8. Rome.

Ramirez, R.G., Neira-Morales, R.R., Ledezma-Torres, R.A., Graribaldi-Gonzalez, C.A., 2000. Ruminal digestion characteristics and effective degradability of cell wall of browse species from northern Mexico. Small Rumin. Res. 36, 49–55.

SADS, 2009. Sustainable Agricultural Development Strategy towards 2030. Agricultural Research and Development Council. Arab Republic of Egypt, Ministry of Agriculture and Land Reclamation.

Shaker, Y.M., 2009. Ovarian response of Barki ewes fed *Atriplex nummularia* to estrus synchronization. Egypt. J. Basic Appl. Physiol. 8, 185–203.

Shaker, Y.M., 2014. Live body weight changes and physiological performance of Barki Sheep fed salt tolerant fodder crops under the arid conditions of Southern Sinai, Egypt. J. Am. Sci. 10, 78–88.

Shaker, Y.M., Ibrahim, N.H., Younis, F.E., El Shaer, H.M., 2014. Effect of feeding some salt tolerant fodder shrubs mixture on physiological performance of Shami goats in Southern Sinai, Egypt. J. Am. Sci. 10, 66–77.

Wassif, M.M., El-Bagouri, L.H., Sabet, S.A., Robishy, A.A., 1983. Effect of irrigation with saline water at different growth stages on yields and mineral composition of some cereal crops under conditions of highly calcareous soil. Desert Inst. Bull. A.R.E. 14, 41–56.

Younis, F.E., Abd, W.H., El Ghany, Helal, A., El Shaer, H.M., 2012. Study of haematological and biochemical parameters and some coat characteristics of sheep fed on salt tolerant plants. Egypt. J. Basic Appl. Physiol. 11, 71–371.

Youssef, K.M., Fahmy, A.A., El Essawy, A.M., El Shaer, H.M., 2009. Nutritional studies on *Pennisetum americanum* and *Kochia indica* fed to sheep under saline conditions of Sinai, Egypt. Am. Eurasian J. Agric. Environ. Sci. 5, 63–68.

12

INSIGHTS INTO THE ECOLOGY AND THE SALT TOLERANCE OF THE HALOPHYTE *CAKILE MARITIMA* USING MULTIDISCIPLINARY APPROACHES

Karim Ben Hamed[1], Ibtissem Ben Hamad[1,2], François Bouteau[2] and Chedly Abdelly[1]

[1]*Laboratoire des Plantes Extrêmophiles, Centre de Biotechnologie de Borj Cédria, Hammam Lif, Tunisia* [2]*Institut des Energies de Demain, Université Paris Diderot, Sorbonne Paris Cité, Paris, France*

12.1 Introduction

Worldwide, salt stress is one of the main environmental constraints that threaten global food security. It has been estimated that the world is losing about 1.5 Mha of arable land each year (at the rate of 3 ha min^{-1}) due to inappropriate irrigation practices and that 10–20 Mha of irrigated land deteriorate to zero productivity every year (Panta et al., 2014). Most agricultural plants are very sensitive to the presence of sodium in the soil. Interestingly, halophytes (salt-tolerant plants) have developed an original adaptation to cope with the presence of salt. Model plants are mainly glycophytes and although dedicated research concerning salt stress tolerance has produced a huge amount of knowledge, the need for halophytic models is growing. Recently, many papers have been published with the aim of presenting the advantages and limits of some alternative model plants like *Mesembryanthemum crystallinum* (Bohnert and Cushman, 2000), *Thellungiella* (Amtmann, 2009; Dassanayake et al., 2011), *Spartina alterniflora* (Subudhi and Baisakh, 2011), and

M.A. Khan, M. Ozturk, B. Gul, & M.Z. Ahmed (Eds): Halophytes for Food Security in Dry Lands.
DOI: http://dx.doi.org/10.1016/B978-0-12-801854-5.00012-1

Sesuvium portulacastrum (Lokhande et al., 2013) to try to define what a halophyte model species should be. Flowers and Colmer (2008) outlined some criteria that should be considered in the selection of model halophytes for future research. In fact, a model halophyte should belong to an order or a family with a significant number of halophyte species. It is well known that the two major groups of halophytes are the Chenopodiaceae with over 380 halophytic species and the Poaceae with over 140. Although with few halophyte species, Brassicaceae is also a family of well-known model halophytes like *Eutrema* or *Thellungiella* (*Thellungiella salsuginea*, *Thellungiella parvula*), *Cochlearia*, and *Lepidium*. A model halophyte, like the majority of halophytes should not use external bladders or salt glands for salt secretion. Other important features may be considered in the choice of these models, such as ploidy (diploids being preferred for genetic aspects), genome size, an annual habitat (to enable ease of experimentation), and amenability to genetic transformation and plant regeneration (Flowers and Colmer, 2008). We review what makes the coastal halophyte *Cakile maritima* special and discuss the possibility of proposing it as a new Brassicaceae model to examine salinity tolerance.

12.2 Latitudinal Distribution and Taxonomic Diversity

Cakile has diversified across a wide range of latitudes, naturally occurring as far north as northern Norway and as far south as the southern coasts of the Caribbean (Clausing et al., 2000). *Cakile maritima* is subdivided into a number of controversial subspecies based primarily on morphological characters, the shape of the distal segment, and leaf shape. The most recent subdivision of the species recognizes four subspecies: ssp. *baltica*, ssp. *maritima* (*integrifolia*), ssp. *Aegyptiaca*, and ssp. *euxina*. The ssp. *integrifolia* is found in northern Norway, on the shores of the Atlantic and the Mediterranean coast. It is largely replaced by ssp. *baltica* in most Baltic areas and completely replaced by ssp. *euxina* along the Black Sea. In northern Europe, it shares part of its distribution with ssp. *islandica*, and in the Mediterranean region and southern Portugal with ssp. *aegyptiaca*. *Cakile* has also been accidently introduced to western North America, Australia, and New Zealand (Cousens and Cousens, 2011). Replacement of *Cakile edentula* with *C. maritima*, in areas where both plants are native, has attracted attention for a number of reasons. First, it has been extremely well-documented (Barbour and Rodman, 1970; Heyligers, 1985), and there is no evidence that hybridization has occurred that might mask a true replacement event (Rodman, 1974, 1986). Secondly, the fact that replacement occurs simultaneously on both American and European continents argues against coincidence in favor of some underlying ecological mechanism (Rodman, 1986). Finally, beach communities appear to be relatively open to invasion due to low plant density and high levels of disturbance (Breckon and Barbour, 1974). Several explanations for the observed replacement have been offered. Interference from pollinators (Barbour and Rodman, 1970) is a possible mechanism—C. maritima has larger flowers—but C. edentula is capable of extensive self-fertilization. Boyd and Barbour (1993) concluded that replacement of C. edentula by C. maritima is due to the latter's ability to better tolerate fore-dune conditions and survive into second or third reproductive seasons. This latitudinal range makes Cakile an ideal system in which to study phenological adaptation to climate.

The genus Cakile has undergone a relatively recent diversification (Willis et al., 2014). Data on local phenology combined with genetic and geographic distribution data will help us understand how phenological adaptation has contributed to the diversification of the clade. Using the technique of amplified fragment length polymorphism (AFLP), Clausing et al. (2000) and Westberg and Kadereit (2009) found, in Europe, a remarkable differentiation between Atlantic and Mediterranean populations of C. maritima. Based on morphological and isoenzyme markers, a similar situation was reported in Tunisian populations of C. maritima (Gandour et al., 2008). This variability is most pronounced between the northern coasts and regions Cap Bon and Sahel. The geographic population structure of Cakile is explained by the movement of ocean currents that make it virtually impossible for seed exchange between populations of the northern coast and the remaining populations. Moreover, flow cytometry analysis showed that the Tunisian populations have a smaller genome than Atlantic populations (from Britanny) (Debez et al., 2013).

12.3 Dispersal and Environmental Adaptation

The genus Cakile has a unique fruit type, termed heteroarthrocarpy. Each fruit consists of two segments: a proximal segment, which remains attached to the maternal plant, and a distal segment, which detaches and has the potential to disperse long distances via wind or water (Figure 12.1). The potential for long-distance dispersal means that species can potentially experience a wide range of climatic conditions. The seeds can be dispersed by water and germination is not affected by salt-water (Westberg and Kadereit, 2009). Maun and Payne (1989) investigated the adaptive significance of fruit and seed dimorphism in *C. maritima* and two subspecies of *C. edentula*, *C. edentula* var. *edentula* and *C. edentula* var. *lacustris*. Dimorphic fruit segments were tested for differences in sizes of propagules, dispersal ability, germination behavior, and growth

Figure 12.1 *Cakile maritima* in its natural biotope (Location Raoued, NE Tunisia) (A). Aspect of the inflorescence (B) and of the fruit (C).

rate. With respect to dispersal, the upper fruit segments of all three taxa did not differ significantly in floating ability from lower fruit segments. Compared with the two groups of *C. edentula*, fruits of *C. maritima* remained afloat for a long time; 50% of the fruit segments of *C. maritima* remained floating after 100 days, compared with 21 days for *C. edentula* var. *lacustris* and 15 days for *C. edentula* var. *edentula*. Larger upper seeds of *C. edentula* var. *lacustris* emerged from significantly deeper depths than smaller lower seeds; the authors suggest that sand accretion may be a strong selective force in the evolution of *Cakile* seeds. Morphological dimorphism in the fruits of *C. edentula* var. *lacustris* is linked with physiological dimorphism in that lower fruit segments germinate better over a wider range of temperatures than do upper segments. Light did not alter the final germination percentage of upper or lower fruit segments, but inhibited the rate of germination. Recently, Debez et al. (2012) reported that the seeds of this halophyte appeared to be equipped with proteins encoded by constitutively expressed genes, a "survival kit" of high significance for this halophyte in its saline environment, since it allows to (i) deliver various nutritional protein and non-protein compounds both early during the germination process and for seedlings, (ii) readily accomplish germination by a machinery already present in the seed, and (iii) cope relatively successfully with salinity.

Pakeman and Lee (1991a,b) evaluated environmental factors contributing to differences in the performance of *C. maritima* in strandline and fore-dune habitats. These authors found a marked difference in growth, depending upon their position of establishment. Large, rapidly growing plants are associated with sand containing macro-algal litter, whereas small stunted plants are found on fore-dunes above the limit of tidal inundation. Edaphic factors and plant nutrient concentrations were measured throughout the 1988 growing season at a site in Wales to determine possible causes for these differences. Pakeman and Lee (1991a,b) found that the observed differences in performance were not associated with water or potassium availability, but that nitrogen availability was much higher in the strand habitat as compared with the fore-dunes, and the concentration of nitrogen (N) in the leaves of plants growing in the strand environment was higher than that measured for fore-dune plants. The authors conclude that nitrogen differences account for the observed growth differences, probably by controlling photosynthetic rates. Phosphorus may also be important (Pakeman and Lee, 1991a,b).

12.4 Basis of the Tolerance to Salinity

Despite the intensive work on unraveling salt-tolerance mechanisms in plants that has been done, much remains to be discovered, because salinity is a complex phenomenon, integrating responses at the whole plant, tissue, and cellular levels. Furthermore, the harmful impact of salinity arises from the combination of its osmotic, nutritional, and specific ion components. A potent genetic source for the improvement of salt tolerance in glycophytes might reside among wild populations of halophytes, which could be either used as a source of genes to be introduced into crop species by classical breeding or molecular methods, or domesticated into new, salt-resistant crops. *Cakile maritima* represents a promising species owing to its ecological and genetic attributes (Debez et al., 2013). I will describe in this section the research done in our laboratory towards understanding the basis of this Brassicaceae halophyte's tolerance to salt stress and establishing it as a model halophyte.

12.4.1 Physiological Mechanisms

Our investigations are multidisciplinary, bringing together eco-physiological, biochemical, and molecular approaches in order to identify the main determinants explaining the plant's performance during its whole life cycle. This includes (i) the characterization of the plant response toward increasing salinities at the early developmental stages (seed germination and seedling establishment), (ii) relating the changes in the phenotype of salt-challenged plants to the water status, the salt accumulation, the uptake, and the use-efficiency of the main nutrients (nitrogen, calcium, potassium, magnesium), the K^+ versus Na^+ selectivity, the photosynthetic performance, the antioxidant system (enzymes and molecules), and the metabolism of compatible osmolytes, and the efficiency of cellular sodium compartmentalization (by measuring the leaf V- and P-H^+-ATPase activity). Table 12.1 summarizes the response of *C. maritima* to salinity at the different stages of development. *Cakile maritima* displayed its halophytic behavior in the vegetative and fruiting stages under low salinity (Debez et al., 2004). Moreover, the plant accumulated increasing amounts of Na^+ and Cl^- in shoots. The strong correlation between the parameters of the water status of the leaves (succulence) and salt loading suggests the existence of mechanisms for compartmentalization of Na^+ and Cl^- in these organs. Two factors support

Table 12.1 Responses of *Cakile maritima* to Salinity at the Different Stages of Development

Stage of Development	Response to Salinity
Germination	NaCl inhibited germination only at concentrations higher than 200 mM, mainly by an osmotic effect (fully reversible after seed transfer to fresh water)
	Seeds can survive up to 4 months immersed in seawater and show up to 1 year floating capacity on seawater
Vegetative stage	Growth was optimal when treated with 50—100 mM NaCl (typical halophyte) for 6 weeks Threshold of salinity tolerance[a] amounts to 500 mM NaCl
Fruiting stage	Seed production was stimulated at 50—100 mM NaCl as compared to control treatment, and severely restricted at higher salt levels.

[a]Threshold of salinity tolerance refers to the substrate concentration leading to a growth decrease of 50% in comparison to plants without salinity.

this hypothesis: the significant increase in the activity of the vacuolar H^+-ATPase up to 300 mM NaCl and the absence of morphological structures responsible for the excretion of salt on the surface of leaves. Also, salt concentrations above 300 mM NaCl stimulate the activity of the proton pump located on the plasma membrane (Debez et al., 2006). Regarding photosynthesis, both the CO_2 uptake and the specific activity of Rubisco increased at the optimum salinity for the growth. The reduction in photosynthetic activity at supra-optimal concentrations was mainly due to stomatal closure (Debez et al., 2008). The proteomic data have shown that stimulation of the growth of the plant is accompanied by changes in leaf proteome, which result in a significant increase in the abundance of many proteins involved in cell growth (Debez et al., 2012). Regarding RuBisCO activase, for example, it plays a role in the reactivation of RuBisCO for CO_2 fixation and would act as chaperone. Similarly, the significant increase in the abundance of ATP synthase is related to the energy cost due to the increased activity of the vacuolar H^+ ATPase, which is involved in vacuolar compartmentalization of sodium (Debez et al., 2006). Finally, the significant increase in the abundance of catalase confirms the data acquired on antioxidant response in this species, which is not only constitutively expressed, but also stimulated under moderate salinities (Ben Amor et al., 2006). The combination of the emerging *omics* techniques with metabolic changes and eco-physiological investigations may allow a better understanding of integrated plant responses to salinity.

12.4.2 Early Responses to Salinity are Crucial to Distinguish Between Halophytes and Glycophytes

Research into the mechanisms of tolerance to salinity made until now in the laboratory was based on exposure of plants under controlled conditions to prolonged saline treatments. However, much work on the evaluation of salt tolerance in different species (glycophytes and halophytes) has shown that the time factor is crucial in determining the behavior of plants to salinity (Ellouzi et al., 2011). Thus, recent work is oriented towards the study of kinetic and close short time responses to salinity with a particular focus on the interactions between multiple responses and looking for pathways that are involved in the induction of adaptive mechanisms.

12.4.2.1 Early Osmotic and Ionic Effects of Salinity

Early osmotic and ionic responses to salt stress were compared in *C. maritima* and *Arabidopsis thaliana* (Ellouzi et al., 2011). The results showed that the growth of both Brassicaceae species decreases with increasing salinity. However, this effect is early in *C. maritima* and more severe in *A. thaliana*. The osmotic effect of salt resulted in a rapid and transient dehydration, signs of water disturbance, followed by rapid osmotic adjustment in *C. maritima* (16-h saline treatment). Complete rehydration took place at the end of treatment (after 72 h) in the halophyte. However, in *A. thaliana*, salinity caused extended and depressive effects on water supply. Mineral analysis showed differences both in the absorption and accumulation of sodium and potassium at the level of their distribution between the roots and aerial parts. *Cakile maritima* showed early uptake and transport of Na^+ from the roots to the aerial parts, according to the inclusive nature of this species (Figure 12.2). *Arabidopsis thaliana* has shown a rapid absorption of Na^+, which is retained in the roots and does not seem to be transported to the aerial parts, indicating a rather exclusive nature of this species associated with low ability to protect the photosynthetic organs against invasion by Na^+ (Ellouzi et al., 2011).

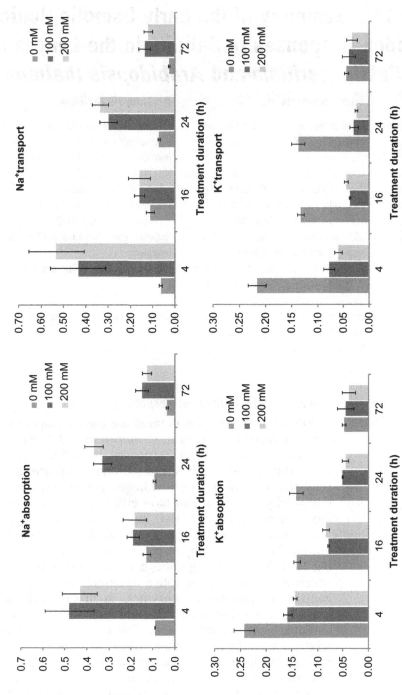

Figure 12.2 Rates of absorption and transport from root to shoot of Na$^+$ and K$^+$ in *Cakile maritima* at different NaCl concentrations. Absorption and transport rates are calculated according to Aleman et al. (2009) and expressed in μmol mg root DW^{-1} h^{-1}.

Table 12.2 Summary of the Early Osmotic, Ionic and Redox Responses to Salinity in the Leaves of *Cakile maritima* and *Arabidopsis thaliana*

	Cakile maritima	*Arabidopsis thaliana*
Water relations	Rapid dehydration/rehydration at early stages of the salt-stress response	The osmotic potential decreased linearly and in a sustained manner, both in roots and leaves, from the first hours of salt stress
Ion accumulation	Early accumulation of Na^+ and Cl^- in the leaves: indicating an inclusion strategy	Na^+ and Cl^- accumulate in the roots under salinity: excluder behavior
H_2O_2 and MDA	Reached a maximum within 4 h of salt stress and rapidly declined afterwards	Continued to rise during the entire experimentation period (72 h)
Antioxidant enzymes	SOD, CAT, and POX increased significantly after 4 h of stress application	Highest enzyme activities at 72 h of stress treatment
Tocopherols	The basal level was twofold higher than in *A. thaliana*. Nonsignificant changes in levels during salt stress	Decreased by 50% during salt stress
Ascorbate and glutathione	Reduced forms started rising after the rapid initial H_2O_2 and lipid peroxidation increase	Decreased significantly under salinity

12.4.2.2 *Early Antioxidant Responses*

Ellouzi et al. (2011) compared the early changes in oxidative status induced by salinity in *C. maritima* and *A. thaliana*. They showed distinctive early responses to salinity (Table 12.2). These differences clearly suggested that halophytes are quick to send stress signals through hydrogen peroxide (H_2O_2) and malondialdehyde (MDA), and have efficient enzymatic antioxidant mechanisms to scavenge H_2O_2 upon completion of signaling. In other words, it is not a detoxifying role for antioxidant enzymes but the fact that halophyte species are capable of rapidly inducing H_2O_2 levels, having enough antioxidant enzymes in stock, which gives them an adaptive advantage over glycophytes. In contrast to antioxidant enzymes, ascorbate and glutathione levels in *C. maritima* started rising after the rapid initial H_2O_2 and lipid peroxidation increase. This increase was not accompanied by an imbalance in the oxidation state of the ascorbate and glutathione pools. Successive increases in antioxidant enzymes and molecules (ascorbate and glutathione) in *C. maritima* indicated the coordinated action of both types of antioxidants in order to regulate the accumulation of H_2O_2. Comparing the tocopherol concentration pattern in *Cakile* and *Arabidopsis* suggested that *Cakile* may detoxify the salt-induced singlet oxygen production through direct quenching, whereas *Arabidopsis* does it through a chemical reaction (Bose et al., 2014). The above hypothesis should be confirmed by experimental evidence.

12.4.3 Cellular Mechanisms

Established protocols of tissue culture should enable large-scale production of halophytes for conservation, to cover barren lands for greenification and in saline areas for desalination purpose (Lokhande et al., 2010). For conservation of the species and taking into account the approach of large-scale multiplication, Ben Hamed et al. (2014) successfully developed an efficient protocol for the callogenesis from the aerial parts of *C. maritima*. The calluses are used for the establishment of cell suspension cultures (Figure 12.3), which could act as a source for the synthesis and production of valuable secondary metabolites.

In salt-stress studies, soil-less culture with nutrient solutions of known salt concentrations are preferred to field conditions where the salinity levels are heterogeneous and often combined with other environmental factors (Lokhande et al., 2013). In this regard, the use of cell and tissue culture facilitate study of the salt tolerance mechanism at the unorganized cellular or organized tissue level, and may provide information on the physiological, biochemical and growth responses to salt stress at different levels of tissue organization. Additionally, in vitro studies allow for relatively rapid responses, a short generation time and the use of controlled environments and the inferences obtained from in vitro cultures under salt stress may be directly applicable at the whole-plant level (Lokhande et al., 2013). Suspension cells of *C. maritima* were tested for their salt tolerance under similar saline conditions applied at the whole plant level. No significant difference was observed between untreated cells and the suspension cells growing in the presence of 100 mM NaCl. This result is different from what was already described for the seedlings; *C. maritima* is an obligate halophyte with an optimal growth at 100 mM NaCl (Debez et al., 2004). In addition, a strong reduction of cell suspension growth is observed for higher NaCl concentrations. The higher levels of tissue organization in the whole plant might efficiently distribute the effect of salt ions which was not found at the cellular levels. There are possibly other physiological factors operating at the whole plant level

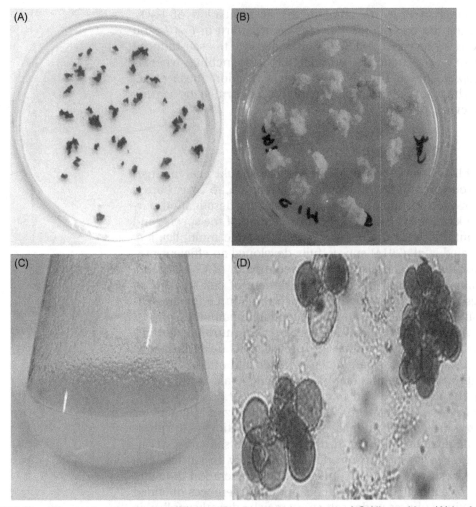

Figure 12.3 The different steps for the establishment of cell suspension culture of *Cakile maritima*. (A) Leaf slices in petri dishes filled with MS medium. (B) Calluses, 2–3 weeks after feeding. (C) Dense suspension cells in flasks, after 4 months feeding. (D) Six-day-old cell aggregates with rounded shape.

conferring salt tolerance which are lacking in the dedifferentiated and unorganized cell cultures. This could explain why suspension cells and whole plants respond differentially to salt stress.

NaCl induced the depolarization of the cell membrane in a dose-dependent manner. This depolarization was correlated with a large increase in whole cell ion current (unpublished data). This result suggests that nonselective cation channels are responsible for the influx of Na^+ through the plasma membrane. Interestingly, repolarization occurs in *Cakile* cells at lower NaCl concentrations, inducing a large decrease in whole cell ion current (unpublished data). This suggests an activation of the salt overly sensitive transduction signal leading to the activation of the Na^+/H^+ antiporter responsible for extrusion of Na^+ from the cytosol. We further compared the extent of cell death induced 6 h after the addition of various NaCl concentrations on *C. maritima* and *A. thaliana* suspension cells. The cell death began to increase from 100 mM NaCl to close to 100% at 600 mM for *A. thaliana*. For *C. maritima*, the increase in cell death was only significant at 400 mM NaCl and reached only 60% at 800 mM, remaining largely inferior when compared to *A. thaliana*. It is noteworthy that the 40% of surviving cells probably could go on dividing, explaining, although reduced, the growth of the culture at 800 mM NaCl.

12.5 *Cakile maritima*: Model Halophyte for Future Research in Salt-Stress Physiology

Cakile maritima can be used as a model plant to learn about plant reactions that are not observed or experimentally reproduced in traditional glycophytic models, for the following reasons: (i) The seeds of *C. maritima*, where a set of constitutively present proteins involved in several metabolic pathways like glycolysis, amino acid metabolism, and photosynthesis was identified, have the capacity to recover from salinity and start germination once salinity is reduced, which may happen following rain. Seed dimorphism is another strategy whereby two different seeds are produced that germinated consecutively at suitable intervals. (ii) By contrasting stress responses of *C. maritima* to those of stress-sensitive models such as *Arabidopsis*, we are convinced that the differences between both kinds of plants lie mainly in the mechanisms that control how stress signals are perceived, transduced and how adaptive processes are controlled within the plant.

The attributes of using *C. maritima* as a good model plant over other halophytes were addressed by Debez et al. (2013). First, *C. maritima* shows traits suitable for a model system (small genome, rapid life cycle, diploid genetics with nine chromosomes and gene duplication, large number of seeds produced, well positioned in phylogeny and easily transformed). It is a Na^+ includer plant like the majority of halophytes.

In contrast to the *Thellungiella* model, *C. maritima* can be especially useful for beach dune fortification. Despite these attributes, little is known about the molecular biology of *Cakile*—only cystatin genes have been studied in relation to drought and salt stress (Megdiche et al., 2009). Certainly this might be a constraint for its use as a model system.

References

Aleman, F., Nieves-Cordones, M., Martinez, V., Rubio, F., 2009. Potassium/sodium steady-state homeostasis in *Thellungiella halophila* and *Arabidopsis thaliana* under long-term salinity conditions. Plant Sci. 176, 768–774.

Amtmann, A., 2009. Learning from evolution: *Thellungiella* generates new knowledge on essential and critical components of abiotic stress tolerance in plants. Mol. Plant. 2, 3–12.

Barbour, M.G., Rodman, J.E., 1970. Saga of the westcoast sea rockets: *Cakile edentula* ssp. *californica* and *C. maritima*. Rhodora. 70, 370–386.

Ben Amor, N., Jiménez, A., Megdiche, W., Lundqvist, M., Sevilla, F., Abdelly, C., 2006. Responses of antioxidant systems to NaCl stress in the halophyte Cakile maritima. Physiol. Plant. 126, 446–457.

Ben Hamed, I., Biligui, B., Arbelet-Bonnin, D., Abdelly, C., Ben Hamed, K., Bouteau, F., 2014. Establishment of a cell suspension culture of the halophyte *Cakile maritima*. Adv. Hortic. Sci. 28, 43–48.

Bohnert, H.J., Cushman, J., 2000. The ice plant Cometh: lessons in abiotic stress tolerance. J. Plant Growth Regul. 19, 334–346.

Bose, J., Rodrigo-Moreno, A., Shabala, S., 2014. ROS homeostasis in halophytes in the context of salinity stress tolerance. J. Exp. Bot. 65, 1241–1257.

Boyd, R.S., Barbour, M.G., 1993. Replacement of *Cakile edentula* by *C. maritima* in the strand habitat of California. Am. Midl. Nat. 130, 209–228.

Breckon, G.J., Barbour, M.G., 1974. Review of North American Pacific coast beach vegetation. Madrono. 22, 333–360.

Clausing, G., Vickers, K., Kadereit, J.W., 2000. Historical biogeography in a linear system: genetic variation of Sea Rocket (*Cakile maritima*) and Sea Holly (*Eryngium maritimum*) along European coasts. Mol. Ecol. 9, 1823–1833.

Cousens, R.D., Cousens, J.M., 2011. Invasion of the New Zealand coastline by european sea-rocket (*Cakile maritima*) and American sea-rocket (*Cakile edentula*). Invasive Plant Sci. Manage. 4, 260–263.

Dassanayake, M., Oh, D.-H., Haas, J.S., Hernandez, A., Hong, H., Ali, S., et al., 2011. The genome of the extremophile crucifer *Thellungiella parvula*. Nat. Genet. 43, 913–918.

Debez, A., Ben, H.K., Grignon, C., Abdelly, C., 2004. Salinity effects on germination, growth, and seed production of the halophyte *Cakile maritima*. Plant Soil. 262, 179–189.

Debez, A., Saadaoui, D., Ramani, B., Ouerghi, Z., Koyro, H.-W., Huchzermeyer, B., et al., 2006. Leaf H^+-ATPase activity and photosynthetic capacity of *Cakile maritima* under increasing salinity. Environ. Exp. Bot. 57, 285–295.

Debez, A., Koyro, H.W., Grignon, C., Abdelly, C., Huchzermeyer, B., 2008. Relationship between the photosynthetic activity and the performance of *Cakile maritima* after long-term salt treatment. Physiol. Plant. 133, 373–385.

Debez, A., Braun, H.P., Pich, A., Taamalli, W., Koyro, H.-W., Abdelly, C., et al., 2012. Proteomic and physiological responses of the halophyte *Cakile maritima* to moderate salinity at the germinative and vegetative stages. J. Proteomics. 75, 5667−5694.

Debez, A., Ben Rejeb, K., Ghars, M.A., Gandour, M., Megdiche, W., Ben Hamed, K., et al., 2013. Ecophysiological and genomic analysis of salt tolerance of *Cakile maritima*. Environ. Exp. Bot. 92, 64−72.

Ellouzi, H., Ben Hamed, K., Cela, J., Munné-Bosch, S., Abdelly, C., 2011. Early effects of salt stress on the physiological and oxidative status of *Cakile maritima* (halophyte) and *Arabidopsis thaliana* (glycophyte). Physiol. Plant. 142, 128−143.

Flowers, T.J., Colmer, T.D., 2008. Salinity tolerance in halophytes. New Phytol. 179, 945−963.

Gandour, M., Hessini, K., Abdelly, C., 2008. Understanding the population genetic structure of coastal species (*Cakile maritima*): seed dispersal and the role of sea currents in determining population structure. Genet. Res. 90, 167−178.

Heyligers, P.C., 1985. The impact of introduced plants on foredune formation in southeastern Australia. Proc. Ecol. Soc. Aust. 14, 23−41.

Lokhande, V.H., Nikam, T.D., Ghane, S.G., Suprasanna, P., 2010. *In vitro* culture, plant regeneration and clonal behavior of *Sesuvium portulacastrum* (L.) L.: a prospective halophyte. Physiol. Mol. Biol. Plants. 16, 187−193.

Lokhande, V.H., Gor, B.K., Desai, N.S., Nikam, T.D., Suprasanna, P., 2013. *Sesuvium portulacastrum*, a plant for drought, salt stress, sand fixation, food and phytoremediation. A review. Agron. Sustainable Dev. 33, 329−348.

Maun, M.A., Payne, A.M., 1989. Fruit and seed polymorphism and its relation to seedling growth in the genus *Cakile*. Can. J. Bot. 67, 2743−2750.

Megdiche, W., Passaquet, C., Zourrig, W., Zuily-Fodil, Y., Abdelly, C., 2009. Molecular cloning and characterization of novel cystatin gene in leaves *Cakile maritima* halophyte. J. Plant Physiol. 166, 739−749.

Pakeman, R.J., Lee, J.A., 1991a. The ecology of the strandline annuals *Cakile maritima* and *Salsola kali*, 1. Environmental factors affecting plant performance. J. Ecol. 79, 143−153.

Pakeman, R.J., Lee, J.A., 1991b. The ecology of the strandline annuals *Cakile maritima* and *Salsola kali*, II. The role of nitrogen in controlling plant performances. J. Ecol. 79, 155−165.

Panta, S., Flowers, T., Lane, P., Doyle, R., Haros, G., Shbala, S., 2014. Halophyte agriculture: success stories. Environ. Exp. Bot. 107, 71−83.

Rodman, J.E., 1986. Introduction, establishment, and replacement of sea rockets (*Cakile*: Cruciferae) in Australia. J. Biogeogr. 13, 159−171.

Subudhi, P.K., Baisakh, N., 2011. *Spartina alterniflora* Loisel., a halophyte grass model to dissect salt stress tolerance. In Vitro Cell Dev. Biol. Plant. 47, 441−457.

Westberg, E., Kadereit, J.W., 2009. The influence of sea currents, past disruption of gene flow and species biology on the phylogeographical structure of coastal flowering plants. J. Biogeogr. 36, 398−1410.

Willis, C.G., Hall, J.C., Rubio de Casas, R., Wang, T.Y., Donohue, K., 2014. Diversification and the evolution of dispersal ability in the tribe Brassiceae (Brassicaceae). Ann. Bot. 114, 1675−1686.

EXOGENOUS CHEMICAL TREATMENTS HAVE DIFFERENTIAL EFFECTS IN IMPROVING SALINITY TOLERANCE OF HALOPHYTES

Abdul Hameed[1], Bilquees Gul[1] and M. Ajmal Khan[1,2]
[1]Institute of Sustainable Halophyte Utilization, University of Karachi, Karachi, Pakistan [2]Centre for Sustainable Development, College of Arts and Sciences, Qatar University, Doha, Qatar

13.1 Introduction

Soil salinity is one of the major threats to food production for the fast-growing human population, which is expected to surpass the 9 billion mark by 2050 (FAO, 2011; Panta et al., 2014). Salinity has already destroyed about 950 Mha of land (Flowers and Yeo, 1995; Ruan et al., 2010) and resulted in economic penalties of over US$12 billion at the global scale (Qadir et al., 2008; Flowers et al., 2010). Salinity has destroyed about 6.30 Mha of land in Pakistan that results in annual economic losses of about US$230 million (Qureshi et al., 2008). For an agricultural country like Pakistan, increasing soil salinization is a serious problem, which needs to be dealt with on an urgent basis. According to some estimates, nearly half of the country's population is already food insecure ($>$2000 cal $-$ person^{-1} day^{-1}) and about 28% is severely food insecure ($>$1700 cal person^{-1} day^{-1}) (IFPRI Food Security Portal, 2014). Thus, the situation demands instant and innovative measures to cope with such food shortages in sustainable ways.

Most of the conventional crops used today are salinity-sensitive and cannot withstand even low ($\geq$40 mM NaCl) salinity levels (Munns and Tester, 2008; Cabot et al., 2014). In contrast, halophytes can grow in salinity that is as high as seawater

M.A. Khan, M. Ozturk, B. Gul, & M.Z. Ahmed (Eds): Halophytes for Food Security in Dry Lands.
DOI: http://dx.doi.org/10.1016/B978-0-12-801854-5.00013-3

(~600 mM NaCl) and some can even grow better under saline conditions due to their specialized adaptations (Flowers and Colmer, 2008; Rozema and Schat, 2013; Shabala, 2013). Therefore, cultivation of halophytes on barren saline lands for food/fodder, fuel, and other economic uses is now widely recognized as an environmental friendly way to support the burgeoning human population in the future (Glenn et al., 1999; Khan et al., 2009; Abideen et al., 2011; Cassaniti et al., 2013; Rozema et al., 2013; Panta et al., 2014). However, a thorough understanding of salinity tolerance limits and mechanisms of potential cash crop halophytes is essential to better exploit their economic utilities (Koyro et al., 2013; Abideen et al., 2014).

Despite being naturally tolerant, nearly all halophytes show reduction in both seed germination and growth under highly saline conditions (Gul et al., 2013; Moinuddin et al., 2014). Therefore, low-cost techniques to improve both seed germination and growth under saline conditions need to be explored. Exogenous application of different chemicals, such as glycine betaine (GB; Demiral and Türkan, 2004; Raza et al., 2007; Abbas et al., 2010), ascorbic acid (Shalata and Neumann, 2001; Athar et al., 2008; Salama, 2009), and hydrogen peroxide (Gong et al., 2001; Li et al., 2011; Neto et al., 2005; Wahid et al., 2007; Gondim et al., 2010), is reported to improve the salt tolerance of many crop plants. However, such studies on halophytes are few and inconclusive, thus warranting their importance (Khan et al., 2006; Hameed et al., 2012). Keeping this in mind, we investigated the potential of the exogenous application of different chemicals for improving the salinity tolerance of potential cash crop halophytes *Suaeda fruticosa*, *Limonium stocksii*, and *Atriplex stocksii*.

13.2 Materials and Methods

13.2.1 Test Species

Suaeda fruticosa Forssk is a leaf succulent perennial halophyte distributed in coastal and inland saline areas of Pakistan and most Saharo-Sindian and southern Irano-Turanian regions (http://www.efloras.org). This plant belongs to the family Amaranthaceae and can be utilized for obtaining high-quality edible oil from seeds (Weber et al., 2007), domestic soap from burnt leaves (Freitag et al., 2001), and as forage for camels (Towhidi et al., 2011). Different plant parts also possess anti-ophthalmic, hypolipidemic, and hypoglycemic effects (Chopra et al., 1986; Bennani-Kabachi et al., 1999; Benwahhoud et al.,

2001). In addition, cultivation of *S. fruticosa* may help in the bioremediation and reclamation of soils contaminated with toxic metals (Bareen and Tahira, 2011) and salinity (Khan et al., 2009). *Limonium stocksii* is a low-branched perennial halophyte from the family Plumbaginaceae. This plant is restricted to the coastal areas of Gujarat (India), Sindh, and Balochistan (Pakistan). It is a shrubby salt-secreting halophyte that produces purple-pink flowers twice a year (Zia and Khan, 2004). This plant has enormous potential to become an ornamental or cut-flower crop for saline coastal areas. *A. stocksii* Boiss is a perennial halophyte from the family Amaranthaceae, which is endemic to the coastal belt of the Arabian sea. There are many medicinal uses for this shrub. For instance, decoctions/infusions of the leaves of this plant are used to cure fever, jaundice, edema, and liver diseases by locals (Qureshi et al., 2009).

13.2.2 Seed Collection Sites

Seeds of *S. fruticosa* and *L. stocksii* were collected from Hawks Bay, Karachi, Pakistan (24°52′21.87″N, 66°51′24.58″E, about 1.5 km away from the seafront), while those of *A. stocksii* were from saline land located in the University of Karachi campus (24°93′81.72″N, 67°12′42.92″E; about 25 km away from the seafront). Seeds were transported to the laboratory, separated from the inflorescence, surface sterilized with 1% sodium hypochlorite for 1 min, followed by thorough rinsing with distilled water and air-drying. Seeds were stored at room temperature in dry clear petri plates until use.

13.2.3 Experiment 1: Effect of Salinity and Exogenous Chemical Treatments on Seed Germination

Germination experiments were conducted in programmed incubators (Percival, USA) with a 30/20°C day/night temperature regimen and 12-h photoperiod (Philips fluorescent lamps, $25\ \mu mol\ m^{-2}\ s^{-1}$, 400−700 nm). Tight-fitting clear lid petri plates (50 mm diameter) with 5 mL of test solution were used. There were three NaCl concentrations (0, 200, and 400 mM) with and without GB (10 mM), ascorbic acid (40 mM), and hydrogen peroxide (100 mM). These concentrations of exogenous treatments were selected on the basis of a preliminary trial (data not given). There were four replicates of 25 seeds for each treatment. Percent germination (embryo protrusion from seeds; Bewley and

Black, 1994) was recorded at 2-day intervals for a period of 20 days. The rate of germination was calculated using the Timson index of germination velocity (Khan and Ungar, 1984).

13.2.4 Experiment 2: Effect of Salinity and Exogenous Chemical Treatments on Growth

Seedlings were raised from seeds in plastic pots (12 cm) containing sandy soil and sub-irrigated with half-strength Hoagland's nutrient solution in a green net house under semi-ambient conditions. After 2 months, salinity (0, 300 and 600 mM NaCl) was introduced gradually at the rate of 150 mM NaCl after every 2 days to avoid osmotic shock. All salinity concentrations were achieved on the same day. Fresh water was added daily to compensate for evaporative loss and treatment solutions were renewed after every 5 days. Ascorbic acid (20 mM), GB (10 mM), and hydrogen peroxide (100 μM) each with 0.1% Tween-20 were sprayed, 7 days after achieving final salinity concentrations, on the leaves of plants till dripping twice a week until harvest. Concentrations of chemicals sprayed were determined in preliminary trials. Unsprayed plants served as controls for studying the effects of foliar sprays. There were five plants per pot and four pots per treatment. Plants were harvested after 30 days of treatments.

Fresh weight (FW) of plants was noted immediately after harvest. Dry weight (DW) was determined after drying vegetative parts at 60°C for 48 h in a forced-draft oven. Tissue water was calculated by subtracting DW from FW. Osmolality (-c) of the leaf sap was determined using the press sap method with the help of a vapor pressure osmometer (VAPRO 5520, Wescor Inc., USA). Leaf sap osmolality was converted into osmotic potential (OP) using Van't Hoff equation. Electrolyte leakage (EL) from membranes was assessed according to the method of Dionisio-Sese and Tobita (1998).

13.2.5 Experiment 3: Effect of Water-Spray on Salinity Tolerance

Two-month-old seedlings of *S. fruticosa* and *L. stocksii* were sprayed on leaves till dripping with distilled water containing 0.1% Tween-20, twice a week under nonsaline and saline (gradually introduced 600 mM NaCl) conditions for a period of 30 days. Unsprayed plants served as controls. After 30 days FW, DW, tissue water, leaf sap osmolality, and EL were examined, according to the methods described above. These data were compared with unsprayed (control) and ascorbic acid (AsA)-sprayed plants, to check the efficacy of water spray in improving the salinity tolerance of plants.

13.2.6 Statistical Analyses

Data were analyzed with the help of SPSS version 11.0 for windows (SPSS Inc., 2011). Analysis of variance (ANOVA) was used to determine whether treatments had significant effect on germination, growth, or other parameters. Student t-test was conducted to determine whether significant ($P < 0.05$) differences existed among means.

13.3 Results

13.3.1 Experiment 1: Effect of Salinity and Exogenous Chemical Treatments on Seed Germination

Seeds of all test species lacked innate dormancy and germinated maximally under nonsaline conditions in the following order of magnitude: *L. stocksii* > *S. fruticosa* > *A. stocksii* (Figure 13.1). Increases in NaCl concentration decreased mean final germination (MFG) of all test species; however a few (~10%) seeds of all species could germinate in as high as 400 mM NaCl. The rates of germination of all species followed a pattern similar to that of their seed germination percentages (Figure 13.1).

Exogenous application of AsA could improve seed germination in *S. fruticosa* (under both nonsaline and saline conditions) and *A. stocksii* (only under saline conditions) but not in *L. stocksii* (Figure 13.1). On the other hand, exogenously applied hydrogen peroxide (H_2O_2) ameliorated seed germination of *S. fruticosa* (under both nonsaline and saline conditions) and *L. stocksii* (only under saline conditions) but not of *A. stocksii* (Figure 13.1). GB did not improve seed germination of any test species in this study (data not given).

13.3.2 Experiment 2: Effect of Salinity and Exogenous Chemical Treatments on Growth

Test species responded differently to salinity treatments at growth stage. FW and tissue water (TW) of *S. fruticosa* and *L. stocksii* remained unaffected, while TW of *A. stocksii* decreased

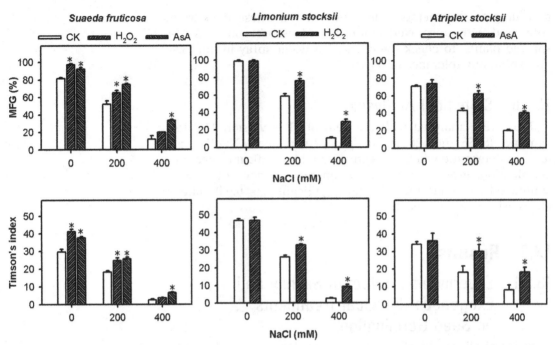

Figure 13.1 Effects of salinity and chemical treatments on mean final germination (MFG) and rate (Timson's index) of germination of *Suaeda fruticosa*, *Limonium stocksii*, and *Atriplex stocksii* seeds. Bars represent mean ± standard error. Asterisks (*) indicate significant difference between a particular exogenous treatment bar and nil-exogenous control bar within a salinity level (*t*-test, $P < 0.05$).

at moderate (300 mM NaCl) salinity treatment in comparison to nonsaline control (Figure 13.2). DW of *S. fruticosa* improved and that of both *L. stocksii* and *A. stocksii* remained similar to nonsaline control in response to moderate salinity treatment (Figure 13.2). High (600 mM) salinity generally inhibited growth parameters (except DW of *A. stocksii*) of all test species (Figure 13.2). Leaf sap OP decreased linearly with increases in salinity, with more negative values in *A. stocksii* (Figure 13.3). EL values in *S. fruticosa* were comparable under nonsaline and moderately saline conditions but increased slightly at high salinity, increased linearly with increases in salinity in *L. stocksii*, and increased substantially at high salinity in *A. stocksii* (Figure 13.3).

Exogenous AsA and GB improved growth of *S. fruticosa* and *L. stocksii* under saline conditions in the following order of magnitude: AsA > GB (Figure 13.2). Exogenous GB could marginally improve FW and tissue water of *A. stocksii* under saline conditions, while AsA ameliorated its growth under nonsaline conditions only (Figure 13.2). Exogenous AsA and GB decreased

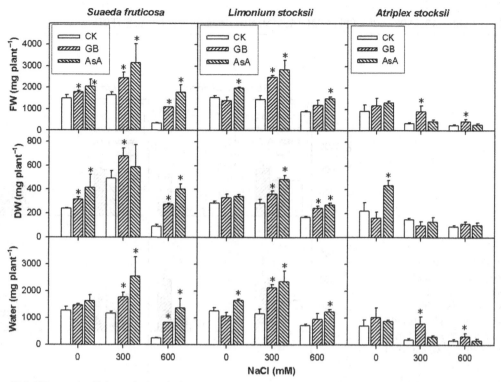

Figure 13.2 Effects of salinity and chemical treatments on fresh weight (FW), dry weight (DW), and tissue water of *Suaeda fruticosa*, *Limonium stocksii*, and *Atriplex stocksii* plants. Bars represent mean ± standard error. Asterisks (*) indicate significant difference between a particular exogenous treatment bar and nil-exogenous control bar within a salinity level (*t*-test, $P < 0.05$).

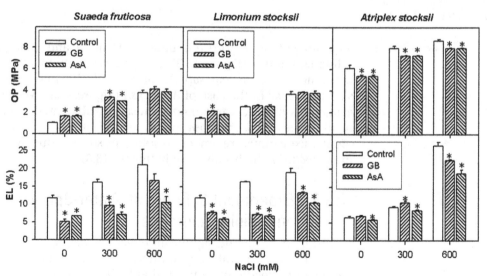

Figure 13.3 Effects of salinity and chemical treatments on osmotic potential (OP) and electrolyte leakage (EL) of *Suaeda fruticosa*, *Limonium stocksii*, and *Atriplex stocksii* leaves. Bars represent mean ± standard error. Asterisks (*) indicate significant difference between a particular exogenous treatment bar and nil-exogenous control bar within a salinity level (*t*-test, $P < 0.05$).

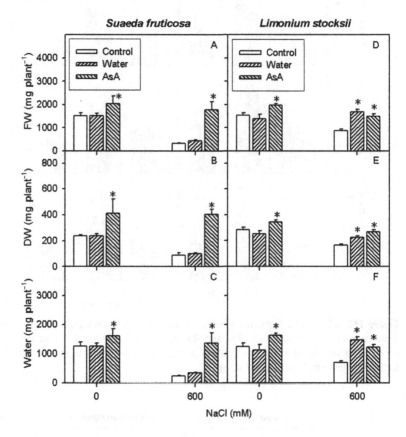

Figure 13.4 Comparison of water-sprayed plants of *Suaeda fruticosa* and *Limonium stocksii* with unsprayed and AsA-sprayed plants, respectively, for fresh weight (FW), dry weight (DW), and tissue water in response to increasing salinity. Bars represent mean ± standard error. Asterisks (*) indicate significant difference between a particular exogenous treatment bar and nil-exogenous control bar within a salinity level (t-test, $P < 0.05$).

leaf sap OP of *S. fruticosa* under nonsaline and moderately saline conditions (Figure 13.3). Exogenous treatments had generally no effect on leaf sap OP of *L. stocksii*, while causing a slight increase in the case of *A. stocksii* (Figure 13.3). Meanwhile exogenous chemical treatments caused a substantial reduction in EL values of *S. fruticosa* and *L. stocksii* in all salinity treatments. Exogenous treatments resulted in some reduction in EL in *A. stocksii* at high salinity only (Figure 13.3).

13.3.3 Experiment 3: Effect of Water-Spray on Salinity Tolerance

Water-spray had no significant effect on the growth of *S. fruticosa*, however significant improvement in all growth parameters of *L. stocksii* was observed at high salinity in response to water-spray (Figure 13.4). Water-spray resulted in a decrease in leaf sap OP of *S. fruticosa* in all salinity treatments, which was comparable to exogenous AsA treatment (Figure 13.5). Likewise,

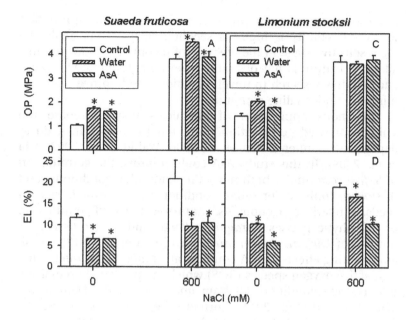

Figure 13.5 Comparison of water-sprayed plants of *Suaeda fruticosa* and *Limonium stocksii* with unsprayed and AsA-sprayed plants, respectively, for osmotic potential (OP) and electrolyte leakage (EL) responses to increasing salinity. Bars represent mean ± standard error. Asterisks (*) indicate significant difference between a particular exogenous treatment bar and nil-exogenous control bar within a salinity level (*t*-test, $P < 0.05$).

effects of water-spray on leaf sap OP of *L. stocksii* were also similar to the effects of exogenous AsA. Water-spray led to a reduction in EL in both *S. fruticosa* and *L. stocksii* with a greater reduction in the former (Figure 13.5).

13.4 Discussion

13.4.1 Seed Germination Responses to Salinity and Chemical Treatments

Seeds of *S. fruticosa, L. stocksii,* and *A. stocksii* were nondormant and germinated maximally in distilled water, similar to most subtropical perennial halophytes (Gul et al., 2013). Seed germination of test species decreased with increases in NaCl concentration and only ~10% of seeds could germinate in the highest (400 mM NaCl) salinity treatment used. Likewise, seed germination of many other co-occurring halophytes, such as *Arthrocnemum indicum* (Saeed et al., 2011), *Arthrocnemum macrostachyum* (Khan and Gul, 1998), *Salsola imbricata* (Mehrunnisa et al., 2007), *Aeluropus lagopoides* (Gulzar and Khan, 2001), *Urochondra setulosa* (Gulzar et al., 2001), and *Halopyrum mucronatum* (Khan and Ungar, 2001) also decreased with increases in salinity. According to Khan and Gul (2006) this decline in seed germination under saline conditions is an adaptation of halophytes to prevent seedling formation when soil salinity is high, as seedling stage is reportedly the most sensitive stage in the lifecycle of halophytes. However, this adaptive feature is not desirable from an agronomic viewpoint and requires research to unveil techniques to improve seed germination under saline conditions.

Exogenous application of chemicals may overcome the salinity-imposed germination reduction in species in a dose-dependent manner (Khan et al., 2006; Wahid et al., 2007; Atia et al., 2009). In this study, AsA could improve the germination of *S. fruticosa* (under both nonsaline and saline conditions) and *A. stocksii* (only under saline conditions) but not of *L. stocksii* seeds. Similarly, exogenous AsA improved the seed germination of halophytic grasses *Phragmites karka* and *Eragrostis ciliaris* but not of *Dichanthium annulatum* (Zehra et al., 2013). These ameliorating effects of AsA could be attributed to the quenching of reactive oxygen species (ROS) which are produced excessively under stress conditions (Sairam and Tyagi, 2004; Neto et al., 2005; Hameed et al., 2014). Furthermore, AsA is also a cofactor for enzymes which synthesize the germination-regulating hormones gibberellins and ethylene (De Tullio and Arrigoni, 2003) and also promotes cell division in germinated seeds for seedling formation (Noctor and Foyer, 1998; De Tullio et al., 1999; De Tullio and Arrigoni, 2003).

Exogenously applied hydrogen peroxide (H_2O_2) ameliorated seed germination of *S. fruticosa* (under both nonsaline and saline conditions) and *L. stocksii* (only under saline conditions) but not of *A. stocksii*. H_2O_2 treatments also improved seed germination in *Zinnia elegans* (Ogawa and Iwabuchi, 2001), *Vitis rotundifolia* (Conner, 2008), *Zea mays* (Wahid et al., 2008), *Citrullus lanatus* (Jaskani et al., 2006), *Avena fatua* (Hsiao and Quick, 1984), and *Tripsacum dactyloides* (Klein et al., 2006). Such improvement due to exogenous H_2O_2 could be a consequence of oxidation of germination inhibitors (Ogawa and Iwabuchi, 2001), testa scarification (Ching, 1959), activation of H_2O_2-mediated oxidative pentose phosphate pathway (Fontaine et al., 1994), or abscisic acid antagonism (Çavusoglu and Kabar, 2010).

13.4.2 Growth Responses to Salinity and Chemical Treatments

The effects of salinity on growth varied among species. Moderate (300 mM NaCl) salinity treatment was not inhibitory for the growth of *S. fruticosa* or *L. stocksii*. *Suaeda fruticosa* produced even greater dry biomass under moderately saline condition than the nonsaline control, like many other *Suaeda* species such as *S. salsa* (100–200 mM NaCl, Song et al., 2009) and *Suaeda maritima* (170–340 mM NaCl; Flowers, 1972). High (600 mM NaCl) salinity was, however, inhibitory for the growth of all test species. Generally, *A. stocksii* was relatively less tolerant to salinity than *S. fruticosa* and *L. stocksii*.

Osmotic adjustment by sequestering inorganic ions (mainly Na^+ and Cl^-) in cell vacuoles and organic osmolytes (such as GB, proline, sugars, etc.) in cytoplasm is an important adaptation of halophytes to withstand physiological drought imposed by salinity (Flowers and Colmer, 2008). Leaf sap OP is a commonly used indicator of osmotic adjustment in plants under saline conditions (Aziz and Khan, 2001; Katschnig et al., 2013; Moinuddin et al., 2014). In this study, leaf sap OP decreased linearly with increases in salinity, with more negative values in *A. stocksii*. A similar reduction in leaf sap OP value with increase in salinity was also observed in *Salicornia dolichostachya* (Katschnig et al., 2013), *Suaeda physophora*, *Haloxylon ammodendron*, *Haloxylon persicum* (Song et al., 2006) and in four subtropical halophytic grasses (Moinuddin et al., 2014).

EL is an inverse measure of membrane integrity and a commonly used indicator of plant salinity tolerance (Mansour, 2013). A growing body of evidence indicates that EL values markedly increase in salt-sensitive plants in comparison to tolerant plant species under saline conditions (Mansour et al., 1993; Ashraf and Ali, 2008; Tuna et al., 2009; Munns, 2010). In this study, EL values of *S. fruticosa* leaves under moderately saline conditions were comparable to values in unstressed plant leaves but increased significantly in response to high-salinity treatments. EL values in *L. stocksii* and *A. stocksii* increased with increasing salinity, with the highest value in *A. stocksii* at 600 mM NaCl. The highest EL value in *A. stocksii* at 600 mM NaCl coincides with higher leaf OP which might be indicative of high solute accumulation approaching toxic levels. The appearance of necrotic patches on *A. stocksii* in response to 600 mM NaCl also indicates the occurrence of ionic toxicity.

AsA is a water-soluble antioxidant, which can quench ROS directly or indirectly and is widely implicated in stress tolerance of plants (Jithesh et al., 2006; Shabala and Mackay, 2011). It has direct or indirect involvement in almost every plant process (Debolt et al., 2007). Exogenous application of AsA improved growth of *S. fruticosa* and *L. stocksii* but not of *A. stocksii* under saline conditions. Exogenous AsA reportedly improved growth by way of increasing soluble sugars in wheat (Al-Hakimi and

Hamada, 2001), lowering lipid peroxidation in tomato (Shalata and Neumann, 2001), enhancing antioxidant enzyme activities and proline in *Saccharum* spp. Hybrid cv. HSF-240 (Ejaz et al., 2012), and increasing Ca^{2+} content in maize and two wheat cultivars (Athar et al., 2008; Bassuony et al., 2008). In this study, growth improvement by exogenous AsA coincided with a decrease in EL, indicating a role for AsA in plant defense that prevented membrane damage. AsA is known to facilitate recycling of important lipid-soluble antioxidant α-tocopherol that protects membranes from ROS-induced peroxidation (Lushchak and Semchuk, 2012).

GB is a water-soluble quaternary ammonium compound that accumulates in osmotically significant levels in many salt-tolerant plants (Rhodes and Hanson, 1993) and is also involved in many protective functions such as protection of photosynthetic machinery and antioxidant activities (Chen and Murata, 2011). Exogenous application of GB is also known to improve the growth of crop plants in a number of ways. For instance, exogenous GB improved growth by enhancing SOD activity in two wheat cultivars and *Prosopis ruscifolia* (Raza et al., 2007; Meloni and Martínez, 2009) and decreasing Na^+ and increased Ca^{2+} in wheat and eggplant (Raza et al., 2007; Abbas et al., 2010). In this study, exogenous GB improved the growth of our two test species, *S. fruticosa* and *L. stocksii*, which accompanied lower EL, thus indicating a protective role for GB in response to salinity.

13.4.3   Involvement of Water-Spray in Salinity Injury Mitigation

Studies involving foliar application of various chemicals to alleviate salinity effects on plants use aqueous solutions of those chemicals. Any improvement in growth and/or tolerance thus might be a consequence of water rather than the chemical itself. We, therefore, investigated the possible involvement of water-spray in plant salinity tolerance. In this study water-spray improved all growth parameters of nonsucculent *L. stocksii* but not of leaf succulent *S. fruticosa*. However, eco-physiological parameters (EL and leaf sap OP) of both species improved under saline conditions, which indicates some role of water-spray in sub-cellular defense for long-term survival under saline conditions. This may also hint at the possible role of dew in survival of plants in arid and semiarid environments, where rains are insufficient and unpredictable.

13.5 Conclusions

Seed germination of all test species decreased with increases in NaCl concentration and exogenous chemical treatments differentially improved seed germination under saline conditions in a species-specific manner. High (600 mM NaCl) salinity inhibited the growth of all species. Exogenous AsA and GB ameliorated growth of two test species, *S. fruticosa* and *L. stocksii*, probably by protective activities. Exogenous chemicals tested in this study have differential and species-specific effects in improving germination and growth of halophytes under saline conditions.

Acknowledgments

The authors would like to thank Higher Education of Commission of Pakistan for provision of funds and Dr. Irfan Aziz for his help during determination of leaf sap OPs of the plants.

References

Abbas, W., Ashraf, M., Akram, N.A., 2010. Alleviation of salt-induced adverse effects in eggplant (*Solanum melongena* L.) by glycinebetaine and sugarbeet extracts. Sci. Hortic. 125, 188–195.

Abideen, Z., Ansari, R., Khan, M.A., 2011. Halophytes: potential source of lignocellulosic biomass for ethanol production. Biomass Bioenergy. 35, 1818–1822.

Abideen, Z., Koyro, H.W., Huchzermeyer, B., Ahmed, M.Z., Gul, B., Khan, M.A., 2014. Moderate salinity stimulates growth and photosynthesis of *Phragmites karka* by water relations and tissue specific ion regulation. Environ. Exp. Bot. 105, 70–76.

Al-Hakimi, A.M.A., Hamada, A.M., 2001. Counteraction of salinity stress on wheat plants by grain soaking in ascorbic acid, thiamin or sodium salicylate. Biol. Plant. 44, 253–261.

Ashraf, M., Ali, Q., 2008. Relative membrane permeability and activities of some antioxidant enzymes as the key determinants of salt tolerance in canola (*Brassica napus* L.). Environ. Exp. Bot. 63, 266–273.

Athar, H.R., Khan, A., Ashraf, M., 2008. Exogenously applied ascorbic acid alleviates salt-induced oxidative stress in wheat. Environ. Exp. Bot. 63, 224–231.

Atia, A., Debez, A., Barhoumi, Z., Smaoui, A., Abdelly, C., 2009. ABA, GA3, and nitrate may control seed germination of *Crithmum maritimum* (Apiaceae) under saline conditions. C. R. Biol. 332, 704–710.

Aziz, I., Khan, M.A., 2001. Experimental assessment of salinity tolerance of *Ceriops tagal* seedlings and saplings from the Indus delta, Pakistan. Aquat. Bot. 70, 259–268.

Bareen, F., Tahira, S.A., 2011. Metal accumulation potential of wild plants in tannery effluent contaminated soil of Kasur, Pakistan: field trials for toxic metal cleanup using *Suaeda fruticosa*. J. Hazard. Mater. 186, 443–450.

Bassuony, F.M., Hassanein, R.A., Baraka, D.M., Khalil, R.R., 2008. Physiological effects of Nictinamide and Ascorbic acid on *Zea mays* plant grown under salinity stress. II changes in nitrogen constituents, protein profiles, protease enzymes and certain inorganic cations. Aust. J. Basic Appl. Sci. 2, 350–359.

Bennani-Kabachi, N., El-Bouayadi, F., Kehel, L., Fdhil, H., Marquié, G., 1999. Effect of *Suaeda fruticosa* aqueous extract in the hypercholesterolaemic and insulin-resistant sand rat. Therapie. 54, 725–730.

Benwahhoud, M., Jouad, H., Eddouks, M., Lyoussi, B., 2001. Hypoglycemic effect of *Suaeda fruticosa* in streptozotocin-induced diabetic rats. J. Ethnopharmacol. 76, 35–38.

Bewley, J.D., Black, M., 1994. Seeds: Physiology of Development and Germination. Plenum Press, New York, NY.

Cabot, C., Sibole, J.V., Barceló, J., Poschenrieder, C., 2014. Lessons from crop plants struggling with salinity. Plant Sci. 226, 2–13.

Cassaniti, C., Romano, D., Hop, M.E.C.M., Flowers, T.J., 2013. Growing floricultural crops with brackish water. Environ. Exp. Bot. 92, 165–175.

Çavusoglu, K., Kabar, K., 2010. Effects of hydrogen peroxide on the germination and early seedling growth of barley under NaCl and high temperature stresses. Eur. Asian J. Biosci. 4, 70–79.

Chen, T.H., Murata, N., 2011. Glycinebetaine protects plants against abiotic stress: mechanisms and biotechnological applications. Plant Cell Environ. 34, 1–20.

Ching, T.M., 1959. Activation of germination in Douglas fir seed by hydrogen peroxide. Plant Physiol. 34, 557–563.

Chopra, R.N., Nayar, S.L., Chopra, I.C., 1986. Glossary of Indian medicinal plants (including the supplement). Council of Scientific and Industrial Research. CSIR Publications, New Delhi.

Conner, P.J., 2008. Effects of stratification, germination temperature and pretreatment with gibberellic acid and hydrogen peroxide on germination of "Fry" muscadine (*Vitis rotundifolia*) seed. HortScience. 43, 853–856.

Debolt, S., Melino, V., Ford, C.M., 2007. Ascorbate as a biosynthetic precursor in plants. Ann. Bot. 99, 3–8.

Demiral, T., Türkan, I., 2004. Does exogenous glycinebetaine affect antioxidative system of rice seedlings under NaCl treatment? J. Plant Physiol. 161, 1089–1100.

De Tullio, M.C., Arrigoni, O., 2003. The ascorbic acid system in seeds: to protect and to serve. Seed Sci. Res. 13, 249–260.

De Tullio, M.C., Paciolla, C., Vecchia, F.D., Rascio, N., D'Emerico, S., De Gara, L., et al., 1999. Changes in onion root development induced by the inhibition of peptidyl-proyl hydroxylase and influence of the ascorbate system on cell division and elongation. Planta. 209, 424–434.

Dionisio-Sese, M.L., Tobita, S., 1998. Antioxidant responses of rice seedlings to salinity stress. Plant Sci. 135, 1–9.

Ejaz, B., Sajid, Z.A., Aftab, F., 2012. Effect of exogenous application of ascorbic acid on antioxidant enzyme activities, proline contents, and growth parameters of *Saccharum* spp. hybrid cv. HSF-240 under salt stress. Turk. J. Biol. 36, 630–640.

FAO, 2011. The State of the World's 690 Land and Water Resources for Food and Agriculture (SOLAW), Managing Systems at Risk. Food and Agriculture Organization of the United Nations, Rome and Earthscan, London.

Flower, T.J., Yeo, A.R., 1995. Breeding for salinity resistance in crop plants; where next. Aust. J. Plant Physiol. 22, 875–884.

Flowers, T.J., Colmer, T.D., 2008. Salinity tolerance in halophytes. New Phytol. 179, 945–963.

Flowers, T.J., Galal, H.K., Bromham, L., 2010. Evolution of halophytes: multiple origins of salt tolerance in land plants. Funct. Plant Biol. 37, 604–612.

Fontaine, O., Huault, C., Pavis, N., Billard, J.-P., 1994. Dormancy breakage of *Hordeum vulgare* seeds: effects of hydrogen peroxide and scarification on glutathione level and glutathione reductase activity. Plant Physiol. Biochem. 32, 677–683.

Freitag, H., Hedge, I.C., Jaffri, S.M.H., Kothe, H.G., Omer, S., Uotila, P., 2001. Flora of Pakistan No. 204 Chenopodiaceae. Department of Botany, University of Karachi, Karachi, Pakistan.

Glenn, E.P., Brown, J.J., Blumwald, E., 1999. Salt tolerance and crop potential of halophytes. Crit. Rev. Plant Sci. 18, 227–255.

Gondim, F.A., Gomes-Filho, E., Lacerda, C.F., Prisco, J.T., Neto, A.D.A., Marques, E.C., 2010. Pretreatment with H_2O_2 in maize seeds: effects on germination and seedling acclimation to salt stress. Braz. J. Plant Physiol. 22, 103–112.

Gong, M., Chen, B., Li, Z.-G., Guo, L.-H., 2001. Heat-shock-induced cross adaptation to heat, chilling, drought and salt in maize seedlings and involvement of H_2O_2. J. Plant Physiol. 158, 1125–1130.

Gul, B., Ansari, R., Flowers, T.J., Khan, M.A., 2013. Germination strategies of halophyte seeds under salinity. Environ. Exp. Bot. 92, 4–18.

Gulzar, S., Khan, M.A., 2001. Seed germination of a halophytic grass *Aeluropus lagopoides*. Ann. Bot. 87, 319–324.

Gulzar, S., Khan, M.A., Ungar, I.A., 2001. Effect of salinity and temperature on the germination of *Urochondra setulosa* (Trin.) CE Hubbard. Seed Sci. Technol. 29, 21–30.

Hameed, A., Hussain, T., Gulzar, S., Aziz, I., Gul, B., Khan, M.A., 2012. Salt tolerance of a cash crop halophyte *Suaeda fruticosa*: biochemical responses to salt and exogenous chemical treatments. Acta Physiol. Plant. 34, 2331–2340.

Hameed, A., Rasheed, A., Gul, B., Khan, M.A., 2014. Salinity inhibits seed germination of perennial halophytes *Limonium stocksii* and *Suaeda fruticosa* by reducing water uptake and ascorbate dependent antioxidant system. Environ. Exp. Bot. 107, 32–38.

Hsiao, A., Quick, W.A., 1984. Actions of sodium hypochlorite and hydrogen peroxide on seed dormancy and germination of wild oats, *Avena fatua* L. Weed Res. 24, 411–419.

IFPRI Food Security Portal, 2014. <http://www.foodsecurityportal.org/pakistan?print>.

Jaskani, M.J., Kwon, S.W., Kim, D.H., Abbas, H., 2006. Seed treatments and orientation affects germination and seedling emergence in tetraploid watermelon. Pak. J. Bot. 38, 89–98.

Jithesh, M.N., Prashanth, S.R., Sivaprakash, K.R., Parida, A.K., 2006. Antioxidative response mechanisms in halophytes: their role in stress defence. J. Genet. 85, 237–254.

Katschnig, D., Broekman, R., Rozema, J., 2013. Salt tolerance in the halophyte *Salicornia dolichostachya* Moss: growth, morphology and physiology. Environ. Exp. Bot. 92, 32–42.

Khan, M.A., Gul, B., 2006. Halophyte seed germination. In: Khan, M.A., Weber, D. J. (Eds.), Ecophysiology of High Salinity Tolerant Plants. Springer, Netherlands, pp. 11–30.

Khan, M.A., Gul, B., 1998. High salt tolerance in germinating dimorphic seeds of *Arthrocnemum indicum*. Int. J. Plant Sci. 5, 826–832.

Khan, M.A., Ungar, I.A., 1984. Effects of salinity and temperature on the germination and growth of *Atriplex triangularis* Willd. Am. J. Bot. 71, 481–489.

Khan, M.A., Ungar, I.A., 2001. Alleviation of salinity stress and the response to temperature in two seed morphs of *Halopyrum mucronatum* (Poaceae). Aust. J. Bot. 49, 777–783.

Khan, M.A., Ahmed, M.Z., Hameed, A., 2006. Effect of sea salt and L-ascorbic acid on the seed germination of halophytes. J. Arid Environ. 67, 535–540.

Khan, M.A., Ansari, R., Ali, H., Gul, B., Nielsen, B.L., 2009. *Panicum turgidum*, a potentially sustainable cattle feed alternative to maize for saline areas. Agric. Ecosyst. Environ. 129, 542–546.

Klein, J.D., Wood, L.A., Geneve, R.L., 2006. Hydrogen peroxide and color sorting improves germination and vigor of eastern gamagrass (*Tripsacum dactyloides*) seeds. In: IV International Symposium on Seed, Transplant and Stand Establishment of Horticultural Crops; Translating Seed and Seedling, vol. 782, pp. 93–98.

Koyro, H.W., Hussain, T., Huchzermeyer, B., Khan, M.A., 2013. Photosynthetic and growth responses of a perennial halophytic grass *Panicum turgidum* to increasing NaCl concentrations. Environ. Exp. Bot. 91, 22–29.

Li, J.-T., Qiu, Z.-B., Zhang, X.-W., Wang, L.-S., 2011. Exogenous hydrogen peroxide can enhance tolerance of wheat seedlings to salt stress. Acta Physiol. Plant. 33, 835–842.

Lushchak, V.I., Semchuk, N.M., 2012. Tocopherol biosynthesis: chemistry, regulation and effects of environmental factors. Acta Physiol. Plant. 34, 1607–1628.

Mansour, M.M.F., 2013. Plasma membrane permeability as an indicator of salt tolerance in plants. Biol. Plant. 57, 1–10.

Mansour, M.M.F., Lee-Stadelmann, O.Y., Stadelmann, E.J., 1993. Salinity stress and cytoplasmic factors—a comparison of cell permeability and lipid partiality in salt sensitive and salt resistant cultivars and lines of *Triticum aestivum* and *Hordeum vulgare*. Physiol. Plant. 88, 141–148.

Mehrunnisa, Khan, M.A., Weber, D.J., 2007. Dormancy, germination and viability of *Salsola imbricata* seeds in relation to light, temperature and salinity. Seed Sci. Technol. 35, 595–606.

Meloni, D.A., Martínez, C.A., 2009. Glycinebetaine improves salt tolerance in vinal (*Prosopis ruscifolia* Griesbach) seedlings. Braz. J. Plant Physiol. 21, 233–241.

Moinuddin, M., Gulzar, S., Ahmed, M.Z., Gul, B., Koyro, H.W., Khan, M.A., 2014. Excreting and non-excreting grasses exhibit different salt resistance strategies. AoB Plants 6-plu038.

Munns, R., 2010. Plant salt tolerance. In: Sunkar, R. (Ed.), Methods in Molecular Biology. Springer, Berlin, pp. 25–38.

Munns, R., Tester, M., 2008. Mechanisms of salinity tolerance. Annu. Rev. Plant Biol. 59, 651–681.

Neto, A.D.D.A., Prisco, J.T., Enéas-Filho, J., Medeiros, J.V.R., Gomes-Filho, E., 2005. Hydrogen peroxide pre-treatment induces salt-stress acclimation in maize plants. J. Plant Physiol. 162, 1114–1122.

Noctor, G., Foyer, C.H., 1998. Ascorbate and glutathione: keeping active oxygen under control. Annu. Rev. Plant Physiol. 49, 249–279.

Ogawa, K., Iwabuchi, M., 2001. A mechanism for promoting the germination of *Zinnia elegans* seeds by hydrogen peroxide. Plant Cell Physiol. 42, 286–291.

Panta, S., Flowers, T., Lane, P., Doyle, R., Haros, G., Shabala, S., 2014. Halophyte agriculture: success stories. Environ. Exp. Bot. 107, 71–83.

Qadir, M., Tubeileh, A., Akhtar, J., Larbi, A., Minhas, P.S., Khan, M.A., 2008. Productivity enhancement of salt-affected environments through crop diversification. Land Degrad. Dev. 19, 429–453.

Qureshi, A.S., McCornick, P.G., Qadir, M., Aslam, Z., 2008. Managing salinity and waterlogging in the Indus Basin of Pakistan. Agric. Water Manage. 95, 1–10.

Qureshi, R., Waheed, A., Arshad, M., Umbreen, T., 2009. Medico-ethnobotanical inventory of Tehsil Chakwal, Pakistan. Pak. J. Bot. 41, 529–538.

Raza, S.H., Athar, H.R., Ashraf, M., Hameed, A., 2007. Glycinebetaine-induced modulation of antioxidant enzymes activities and ion accumulation in two wheat cultivars differing in salt tolerance. Environ. Exp. Bot. 60, 368–376.

Rhodes, D., Hanson, A.D., 1993. Quaternary ammonium and tertiary sulfonium compounds in higher plants. Annu. Rev. Plant Biol. 44, 357–384.

Rozema, J., Schat, H., 2013. Salt tolerance of halophytes, research questions reviewed in the perspective of saline agriculture. Environ. Exp. Bot. 92, 83–95.

Rozema, J., Muscolo, A., Flowers, T., 2013. Sustainable cultivation and exploitation of halophyte crops in a salinising world. Environ. Exp. Bot. 92, 1–3.

Ruan, C.J., da Silva, J.A.T., Mopper, S., Qin, P., Lutts, S., 2010. Halophyte improvement for a salinized world. Crit. Rev. Plant Sci. 29, 329–359.

Saeed, S., Gul, B., Khan, M.A., 2011. Comparative effects of NaCl and sea salt on seed germination of *Arthrocnemum indicum*. Pak. J. Bot. 43, 1091–1103.

Sairam, R.K., Tyagi, A., 2004. Physiology and molecular biology of salinity stress tolerance in plants. Curr. Sci. Bangalore. 86, 407–421.

Salama, K.H.A., 2009. Amelioration of nacl-induced alterations on the plasma membrane of *Allium cepa* L. by ascorbic acid. Aust. J. Basic Appl. Sci. 3, 990–994.

Shabala, S., 2013. Learning from halophytes: physiological basis and strategies to improve abiotic stress tolerance in crops. Ann. Bot. 112 (7), 1209–1221.

Shabala, S., Mackay, A., 2011. Ion transport in halophytes. In: Kader, J.C., Delseny, M. (Eds.), Advances in Botanical Research, vol. 57. Elsevier, San Diego, pp. 151–199.

Shalata, A., Neumann, P.M., 2001. Exogenous ascorbic acid (vitamin C) increases resistance to salt stress and reduces lipid peroxidation. J. Exp. Bot. 52, 2207–2211.

Song, J., Feng, G., Tian, C.Y., Zhang, F.S., 2006. Osmotic adjustment traits of *Suaeda physophora*, *Haloxylon ammodendron* and *Haloxylon persicum* in field or controlled conditions. Plant Sci. 170, 113–119.

Song, J., Chen, M., Feng, G., Jia, Y., Wang, B.-S., Zhang, F., 2009. Effect of salinity on growth, ion accumulation and the roles of ions in osmotic adjustment of two populations of *Suaeda salsa*. Plant Soil. 314, 133–141.

SPSS, 2011. SPSS version 11.0 for Windows. SPSS Inc.

Towhidi, A., Saberifar, T., Dirandeh, E., 2011. Nutritive value of some herbage for dromedary camels in the central arid zone of Iran. Trop. Anim. Health Prod. 43, 617–622.

Tuna, A.L., Kaya, C., Ashraf, M., Altunlu, H., Yokas, I., Yagmur, B., 2009. The effects of calcium sulphate on growth, membrane stability and nutrient uptake of tomato grown under salt stress. Environ. Exp. Bot. 59, 173–178.

Wahid, A., Perveen, M., Gelani, S., Basra, S.M., 2007. Pretreatment of seed with H_2O_2 improves salt tolerance of wheat seedlings by alleviation of oxidative damage and expression of stress proteins. J. Plant Physiol. 164, 283–294.

Wahid, A., Sehar, S., Perveen, M., Gelani, S., Basra, S.M.A., Farooq, M., 2008. Seed pre-treatment with hydrogen peroxide improves heat tolerance in maize at germination and seedling growth stages. Seed Sci. Technol. 36, 633–645.

Weber, D.J., Ansari, R., Gul, B., Khan, M.A., 2007. Potential of halophytes as source of edible oil. J. Arid Environ. 68, 315–321.

Zehra, A., Shaikh, F., Ansari, R., Gul, B., Khan, M.A., 2013. Effect of ascorbic acid on seed germination of three halophytic grass species under saline conditions. Grass Forage Sci. 68, 339–344.

Zia, S., Khan, M.A., 2004. Effect of light, salinity, and temperature on seed germination of *Limonium stocksii*. Can. J. Bot. 82, 151–157.

FOOD AND WATER SECURITY FOR DRY REGIONS: A NEW PARADIGM

M. Ajmal Khan

Institute of Sustainable Halophyte Utilization, University of Karachi, Karachi, Pakistan; Centre for Sustainable Development, College of Arts and Sciences, Qatar University, Doha, Qatar

14.1 Introduction

The availability of safe drinking water, access to sufficient, safe and nutritious food to meet the dietary needs and food preferences for an active and healthy life and sanitation, in other words "food and water security," is a human right (FAO, 1996). Closely linked with these two sectors is energy security, which has been defined as "access to clean and reliable energy sources for productive uses at a price which is affordable, while respecting environment" (Anonymous, 2010a). Three sectors—water, food, and energy security—are closely linked, hence actions in one area have impacts in one or both of the others (Holger, 2011).

14.2 Water and Food Production

Water is the fountain of life and at the center of economic and social development; it is vital to maintain health, grow food, manage the environment, and create jobs. Irrigated agriculture largely caters to the food needs of the inhabitants of our planet, but as the world population hurtles towards 8 billion with increasing demands for basic services and growing desires for higher living standards, food production will have to be increased accordingly. The world population is predicted to double by the year 2050 (Kendall and Pimentel, 1994) and food production will require twice as much water as it does today, whereas the quantity of fresh water on the planet remains

M.A. Khan, M. Ozturk, B. Gul, & M.Z. Ahmed (Eds): Halophytes for Food Security in Dry Lands.
DOI: http://dx.doi.org/10.1016/B978-0-12-801854-5.00014-5

almost constant (<2% of total) with no check on soil degradation. An important deterrent is climate change which, in the long run, threatens to alter the rate of aquifer recharge, making the availability of water even less predictable and render the ambient conditions unfavorable for crop production.

Mismanaged use of water for crop production on the other hand has led to widespread soil salinization and caused a decline in crop yields. The problem is of worldwide occurrence and is on the increase, however it is of special significance for arid regions. These areas characterized by scarcity of water have a further disadvantage that the ability of soils to recover from negative changes is lower than in humid areas, hence the menace here is more severe. A similar situation exists in the region that spans from India up to Morocco covering Pakistan, the Middle East and countries of North Africa, which is suffering from serious problems of food and water security. Here, natural water resources do not exist in most places and in some countries water storage is available only for a short period, endangering the water and food security of the entire region. Despite the importance of water, a substantial number of people globally are, however, still without access to safe drinking water. Not surprisingly, world leaders now rank water as one of their top critical issues and it has also become a cause of conflict between many countries, such as India–Pakistan, Turkey–Iraq, Sudan–Egypt, USA–Mexico (Qadir et al., 2008).

14.3 Conventional Solutions

Irrigated agriculture has so far remained the major source of food production and the largest consumer of global fresh water supplies. This has also adversely impacted the soil and water resources through land degradation, changes in runoff, disruption of ground water discharge, water quality, and the availability of land for agriculture and other purposes such as natural habitats. Furthermore, the increased yields of conventional crops have come at a high energy cost, as land preparation, fertilizer production, irrigation and the sowing, harvesting and transportation of crops are energy-intensive. There is hence a need to search for better and sustainable methods to achieve the desirable objectives.

14.4 Nonconventional Solutions

Keeping in view the fact that degraded land area is increasing, particularly in many arid and semi-arid countries, due to natural causes and anthropogenic activities (UNEP, 2010) with a concomitant reduction in the yield of crops used for human and animal consumption, it has become imperative to search for suitable alternatives which leads us to halophytic plants capable of completing their lifecycle in a saline environment. These plants have a history of multifarious usages and there is a need to identify suitable halophytic flora that will be of benefit for mankind.

Of some 400,000 flowering plant species around the world, ~2600 are halophytes belonging to a class of plants which can thrive in soils with varying levels of salinity, some even tolerate seawater salinity (Aronson, 1989; Joppa et al., 2011; Menzel and Lieth, 1999) and saline agriculture is hardly a new concept because salinity and plants are not incompatible. Salt-tolerant crops have been cultivated on marginal lands for hundreds and even thousands of years. For instance, nomads in the Euphrates region used to cultivate *Alhaji maurorum* to improve arable land that had become saline due to intense agriculture and reuse the land after a few years of halophyte cultivation (Khan et al., 2006).

14.5 Potential Uses of Halophytes

14.5.1 Food

Conventional agriculture has so far managed to meet demand, despite the water scarcity and loss of good-quality land to salinization. Almost 50% of our food comes from just four commodities—rice, maize, wheat, and potato—which illustrates the narrow base and fragility of our food resources (Khan et al., 2006). It is further noteworthy that all agricultural crops have a low tolerance to salt and are subject to becoming redundant in the future due to high soil salinity.

Halophytes, because they are nonconventional crops, may not easily find a place on our dining table due to our established culinary preferences, however there is an array of salt-tolerant halophytic species with potential for use as human food and many of them are used for this purpose, albeit not on a very large scale. For instance, species of *Salicornia*, *Sessuvium*, *Sporobolus*, *Chenopodium*, *Portulaca*, *Suaeda*, and many others have been used as vegetables, salads, and in pickles. Fruits of *Salvadora oleoides* and *Avicennia marina* are used as food among the coastal populations in Pakistan. *Suaeda fruticosa* is used to prepare a kind of baking soda, which is used in the preparation of food. Radicles of *Rhizophora, Bruguiera,* and *Ceriops,* tender leaves of *Thespesia populnea,* and fruits or kernels of littoral species such as *Terminalia catappa* are consumed by local populations (Dagar, 1995; Khan and Qaiser, 2006; Leith et al., 1999).

14.5.2 Fodder, Feed, Forage

A great potential of halophytes rests with their utilization as forage and fodder (Pasternak, 1990). The vast coastal areas around the world lie barren because of the shortage of good-quality water and there is no dearth of similar tracts inland, particularly in arid regions where halophytes could be grown. Some promising species include *Cenchrus ciliaris, Dactyloctenium scindicum, Dichanthium annulatum, Kochia scoparia, Phragmites karka,* and *Sporobolus ioclados,* while *Urochondra setulosa* and *Desmostachya bipinnata* can be used after cutting and drying. Similarly, there are several other species (e.g., *A. maurorum, Athrocnemum macrostachyum, Atriplex stocksii, Haloxylon salicornicum, Halocnemum strobilaceum, Kochia indica, Limonium stocksii, Salsola drummondii, Salsola imbricata, S. oleoides, S. fruticosa, Zygophyllum simplex, Zygophyllum propinqum,* etc.) which have high ash and/or toxic chemicals but have potential to be used as fodder after processing. The methods include chopping, soaking, and washing (even with brackish water if the foliage salt contents are higher) and air drying. The halophytic feed can also be supplemented with high-energy fodder and protein rations (Gul et al., 2014). Young leaves of these plants are usually better in comparison to old, and during the winter compared to summer. The suitable species should carefully be selected and the animals conditioned to adapt to the halophytic diet gradually (Khan, unpublished).

14.5.3 Medicine

Modern medical help is generally inadequate in rural areas where herbal therapy plays a significant role in human health care and many wild and cultivated plants are used for this purpose. To face the abiotic stress of their saline habitats, halophytes synthesize secondary metabolites like flavonoids, alkaloids, tannins, terpenoids, and so on, which can be rich sources of medicinally important compounds. Hence, in addition to their medicinal uses halophytes represent a rich source of natural antioxidants and could be considered as potential alternatives to synthetic antioxidants for food, pharmaceutical, cosmetics, and other industrial products. These plants would not only provide sufficient biomass at a commercial scale using brackish water and saline land but would spare arable land for regular agricultural practices. There is a need for proper cataloging of relevant halophytic plants of a particular area with knowledge of the specific useful compounds for deriving maximum benefit from these resources.

14.5.4 Edible Oil

The composition of cooking oils has an impact on health. Studies support the concept that diets high in saturated fats pose greater risk to heart diseases (Hu, 1997; Lang, 1997). In contrast, oilseeds high in polyunsaturated fatty acids and tocochromanols, which are important lipid-soluble antioxidants and are an essential part of the mammalian diet, are considered much healthier (Matthaus and Angelini, 2005). In addition to the degree of unsaturation, individual lipid fractions of the oil have also to be taken into consideration before making any recommendations. A high level of erucic acid (C 22:1), exceeding 25% for instance, is not considered fit for animal or human consumption (Karleskind, 1996).

There is potential to extract high-quality edible oil from seed-bearing halophytes, which should not have any problems of acceptability from users. Very little information about the composition of halophytic oilseed plants is, however, available. Seeds of various halophytes, for example, species of *Suaeda, Arthrocnemum, Salicornia, Halogeton, Kochia, Haloxylon* and so on possess sufficient quantities of high-quality edible oil with unsaturation ranging from 70% to 80% (Weber et al., 2001). He et al. (2003) reported that the oil and crude protein contents in the seed of a halophytic species, *Kosteletzkya virginica*, were 11.28% and 8.17%, respectively. The oil was composed largely of unsaturated fatty acids, with high potassium and low sodium levels.

14.5.5 Biofuel

The generally inflationary trends of world oil prices have an adverse influence on the economies of many countries. The energy demand met by oil, gas, and coal for industry, transportation and domestic purposes has increased manifold recently, with most of it happening in heavily populated Asia (Abideen et al., 2011). It has been predicted that energy requirements are likely to increase another 1.5-fold globally and that they will almost double in Asian countries by the year 2035 (Matsuo et al., 2013). The nonrenewable nature and consumption at an increasing rate of these fuel sources indicates that they are not going to last for very long. A further problem with fossil fuels is their adverse environmental impact because burning them releases a great deal of carbon into the atmosphere that had been locked away for millions of years. Hence, a bio-alternative may be a better choice because biofuels are made from plants and burning them releases almost as much carbon as was recently taken from the air, thus they do not contribute to the atmospheric carbon count. The only catch is that, if they come from plants that have to be grown and cultivated on arable lands with fresh-water irrigation, they would compete with food crops. This could be avoided by using non-food alternatives for biofuel production (Abideen et al., 2014). Halophytes grow on saline lands and thrive in salty water, producing large quantities of ligno-cellulosic biomass that can be converted into biofuel while the oil from seed of many halophytes may serve as a suitable source of biodiesel (Gul et al., 2013) and could decrease the dependence of many countries on foreign fuel sources (Anonymous, 2010b). However, challenges, such as the efficient conversion of biomass into biofuel, still exist (Zhu and Ketola, 2012).

14.6 What We Have Done

14.6.1 Fodder

The Institute of Sustainable Halophyte Utilization, University of Karachi, has identified halophytic grasses which can partly or fully replace regular animal fodder grasses. Initial studies in this regard included two halophytic grasses which were selected on the basis of high crude protein, low ash content compared to other halophytic shrubs of this area as well as their high yield and availability throughout the year. It was observed that the perennial grass, *Panicum antidotale*, with a high potential yield of $>50,000$ kg ha^{-1} year^{-1} when raised with brackish water on saline land, can be used both green and dry as a replacement for maize and wheat straw, respectively. It is comparable to conventional fodder in terms of body weight gain of calves and quality of meat and its secondary metabolites are not in high enough quantity to be harmful to animals (Khan et al., 2009). Dried and chopped *D. bipinnata*, another perennial halophytic grass, could also replace 75% wheat straw as a dry fodder when used in combination with green maize or *P. antidotale.* The absence of energy ration in feed sustained weight only for about 6−7 weeks, after which it started to decrease and was restored on supplementing the diet with the missing ingredients. *Prosopis juliflora* pods and *Manilkara zapota* fruit could replace cottonseed cake and molasses, respectively, as cheaper ingredients of the concentrate (Khan et al., unpublished).

14.6.2 Medicine

Extensive ethno-medicinal surveys were conducted amongst the rural communities of the coastal and inland areas of the Sindh−Balochistan provinces of Pakistan for the presence of halophytes of medicinal significance, as well as using the experience of local herbalists in this regard, which established the medicinal importance of 100 plant species from 31 families in this area. The majority (54%) of these plants were xerophytes, followed by halophytes (46%). The aboveground plant parts were most commonly used as decoctions to cure various ailments such as gastrointestinal diseases, arthritis, asthma, cough, body pain, skin diseases, respiratory and sexual problems, and so on (Qasim et al., 2011, 2014). Screening of these plants showed high antioxidant and phenolic content in aboveground parts as well as in roots. The highest antioxidant capacity was recorded in 80% methanolic extract of *Thespesia populneoides*, *Salvadora persica, Ipomoea pes-caprae, S. fruticosa*, and *Pluchea lanceolata* (Qasim et al., unpublished). The herbal practitioners of the area are well aware of the medicinal potential of these plants but the knowledge is generally not being transferred to the next generation, thus endangering the future use and safety of this valuable resource.

14.6.3 Edible Oil

Preliminary studies on oilseeds of local halophytic species indicate that their oil quality is comparable with those of conventionally used oils, such as corn and canola oils (Khan et al., unpublished). Screening trials to find alternate sources of edible oil from halophytic plants distributed along the coastal areas of Pakistan have shown large variations in oil and other constituents (protein, carbohydrate, crude fiber, phenols, moisture, organic matter, ash and ion contents) of these species. Of the 45 species analyzed so far, 26 show >20% oil recovery. Gas chromatographic analysis of oil has shown the diversity of fatty acid fractions with predominantly unsaturated fatty acids (generally C16, C18, and C20) in many of these species (Weber et al., 2007). There is a need to find more such species and conduct detailed studies regarding their suitability as a source of oil. Making use of this untapped resource will open vast opportunities to utilizing barren saline coastal lands and poor-quality irrigation water to meet the edible oil shortage and save valuable foreign exchange.

14.6.4 Biofuel

Chemical tests were conducted on a number of halophytic species, largely from family Poaceae distributed in a 100-km radius around Karachi (Abideen et al., 2011). A wide variation in their height and the quantity of cellulose, hemi-cellulose, and lignin was recorded. *Suaeda fruticosa*, *Suaeda monoica*, *Arthrocnemum indicum*, and *S. imbricata* contained low (around 8−11%) cellulose followed by *Calotropis procera*, *Tamarix indica*, *Ipomea pes-caprea*, and *Aerva javanica* (12−15%), while others contained >20% cellulose and the highest amount (37%) was found in *Halopyrum mucronatum*. Hemi-cellulose was low (11−13%) in *A. javanica*, *A. indicum*, *C. procera*, *S. persica*, and *S. monoica*, while others had >20% hemi-cellulose with *Typha domingensis* containing an exceptionally high (38.67%) level of hemi-cellulose. The lignin content in all the species was generally less than 10%. From this study, *H. mucronatum*, *D. bipinnata*, *P. karka*, *T. domingensis*, and *P. antidotale* appear promising potential sources of lignocellulosic biomass which could be converted into ethanol.

This research illustrates the potential of halophytes for biofuel production and provides an option of selecting high-biomass salt-tolerant plants that contain suitable lignocellulosic material for conversion into ethanol. These plants, which are outside the human food chain, are tolerant to saline conditions and relatively inexpensive to grow. Investigations are, however, needed to explore the possibility of breeding plants having desirable characters like low lignin content and high biomass yield, which will help in enhancing and improving the feedstock availability and efficiency of biofuel production.

14.7 Future Directions, Pitfalls, and Possibilities

About one-third of the earth's land surface is affected with desertification due to the spread of salinity, especially in arid and semi-arid regions, resulting in reduced crop yields. To prevent this situation from further deteriorating, crop diversification through halophyte utilization to suit particular soil conditions seems a viable option. Although halophytes offer multifarious usages, research on their utilization has gained momentum only recently. For instance, halophytic grasses (*Paspalum vaginatum* and *Sporobolus virginicus*) have been used for landscaping and turf development (DePew and Tillman, 2006), while *Atriplex* spp. and some other shrubs are used as animal fodder (Leith et al., 2000), although the high oxalate content of *Atriplex* spp. poses serious health problems. Aggressive marketing of *Salicornia bigelovii* by Arizonians (O'Leary, 1979; O'Leary et al., 1986) has opened an avenue to introducing other halophytes as oilseed crops. Halophytes may be used as a land cover which is not only aesthetically pleasing but that also checks land erosion and desertification. Their potential in reforestation or replanting and ecological recovery of saline areas that have fallen into disuse, coastal development and protection, production of cheap biomass for renewable energy, environment conservation through carbon sequestration, cannot be denied. They may serve as a source of edible oil and various chemicals (including those of medicinal importance), provide fodder for animals, wood for building purposes and furniture, boat- and canoe-making, etc.

The saline culture of halophytes however means that they need careful handling as we are dealing with soils which are already saline and irrigation water that is brackish. It is imperative that the knowledge of the mechanisms adopted by plants to cope with salinity and the methods to deal with salt buildup in the root zone are studied in detail and more information is accumulated on the related aspects. Only then will we be able to efficiently manage and manipulate plant growth according to our requirements and in a manner suited to the prevailing conditions. It is emphasized that creating a demand in the marketplace for halophyte food, fodder, and other products is a daunting task. Halophytes certainly deserve our attention, but their acceptance as a commercial entity requires convincing all stakeholders—grass root farmer to highest policy planner—through extensive field trials demonstrating the potential uses and benefits of adopting this novel approach. It should also be kept in mind that agriculture is as old as human civilization and a very slow-moving industry. The current conventional crops took thousands of years to reach the present state whereas commercial utilization of halophytes is still in its infancy.

References

Abideen, Z., Ansari, R., Khan, M.A., 2011. Halophytes: potential source of ligno-cellulosic biomass for ethanol production. Biomass Bioenergy. 35, 1818–1822.

Abideen, Z., Hameed, A., Koyro, H.W., Gul, B., Ansari, R., Khan, M.A., 2014. Sustainable biofuel production from non-food sources—an overview. Emir. J. Food Agric. 26 (12).

Anonymous, 2010a. UN Secretary General's Advisory Group on Energy and Climate Change (AGECC), Summary Report and Recommendations, April 28, 2010, p. 13.

Anonymous, 2010b. Biofuels 2020. A policy driven logistics and business challenge. Research and Innovation, Position Paper 02. DNV Climate Change and Environmental Services.

Aronson, J., 1989. HALOPH: Salt Tolerant Plants for the World—A Computerized Global Data Base of Halophytes with Emphasis on Their Economic Uses. University of Arizona Press, Tucson, AZ.

Dagar, J.C., 1995. Characteristics of halophytic vegetation in India. In: Khan, M.A., Ungar, I.A. (Eds.), Biology of Salt Tolerant Plants. University of Karachi, Pakistan, pp. 255–276.

DePew, M.W., Tillman, P.H., 2006. Commercial application of halophytic turf for golf and landscape developments utilizing hyper-saline irrigation. In: Khan, M.A., Weber, D.J. (Eds.), Ecophysiology of High Salinity Tolerant Plants. Springer, The Netherlands, pp. 255–278.

FAO, 1996. Rome Declaration on World Food Security and World Food Summit Plan of Action. World Food Summit November 13–17, 1996. Rome.

Gul, B., Abideen, Z., Ansari, R., Khan, M.A., 2013. Halophytic biofuels revisited. Biofuels. 4, 575–577.

Gul, B., Ansari, R., Ali, II., Adnan, M.Y., Weber, D.J., Nielsen, B.L., et al., 2014. The sustainable utilization of saline resources for livestock feed production in arid and semi-arid regions: a model from Pakistan. Emir. J. Food Agric. 26 (12).

He, Z., Ruana, C., Qin, P., Seliskar, D.M., Gallagher, J.L., 2003. *Kosteletzkya virginica*, a halophytic species with potential for agro-ecotechnology in Jiangsu Province, China. Ecol. Eng. 21, 271–276.

Holger, H., 2011. Understanding the Nexus. Background Paper for the Bonn 2011 Conference: The Water, Energy and Food Security Nexus. Stockholm Environment Institute, Stockholm.

Hu, F., 1997. Dietary fat intake and the risk of heart diseases in women. N. Engl. J. Med. 337, 1491–1499.

Joppa, L.N., Roberts, D.L., Pimm, L.S., 2011. How many species of flowering plants are there? Proc. Biol. Sci. 278 (1705), 554–559, London, UK.

Karleskind, A., 1996. Oils and Fats Manual, A Comprehensive Treatise, vol. 1. Lavoisier Publishing, Paris, France, p. 806.

Kendall, H., Pimentel, D., 1994. Constraints on the expansion of the global food supply. AMBIO J. Hum. Environ. 23, 198–205.

Khan, M.A., Qaiser, M., 2006. Halophytes of Pakistan: distribution, ecology, and economic importance. In: Khan, M.A., Barth, H.-J., Kust, G.C., Boer, B. (Eds.), Sabkha Ecosystems Volume II: The South and Central Asian Countries. Springer, The Netherlands, pp. 135–160.

Khan, M.A., Ansari, R., Gul, B., Qadir, M., 2006. Crop diversification through halophyte production on salt prone land resources. CAB Reviews: Perspectives in Agriculture, Veterinary Science, Nutrition and Natural Resources, No. 48.

Khan, M.A., Ansari, R., Ali, H., Gul, B., Nielsen, B.L., 2009. *Panicum turgidum*, a potentially sustainable cattle feed alternative to maize for saline areas. Agric. Ecosyst. Environ. 129, 542–546.

Lang, L., 1997. Types of Dietary Fats: Key to Heart Risk. Reuters Health Information Services Inc.

Leith, H., Moschencko, M., Lohmann, M., Koyro, H.-W., Hamdy, A., 1999. Halophyte uses in different climates I: ecological and eco-physiological Studies, Progress in Biometeriology, vol. 13. Backhuys Publishers, The Netherlands, pp. 77–88.

Leith, H., Lohmann, M., Guth, M., Menzel, U., 2000. Cash Crop Halophytes for Future Halophytes Growers. Institute of Environmental System Research, University of Osnabruck, Germany.

Matsuo, Y., Yanagisawa, A., Yamashita, Y., 2013. A global energy outlook to 2035 with strategic considerations for Asia and Middle East energy supply and demand inter dependencies. Energy Strategy Rev. 2, 79–91.

Matthaus, B., Angelini, L.G., 2005. Anti-nutritive constituents in oilseed crops from Italy. Ind. Crops Prod. 21, 89–99.

Menzel, U., Leith, H., 1999. Halophyte database vers. 2.0. In: Leith, H., Moschenko, M., Lohman, M., Koyro, H.–W., Hamdy, A. (Eds.), Halophyte Uses in Different Climates I: Ecological and Eco-Physiological Studies. Progress in Biometeriology, vol. 13. Backhuys Publishers, The Netherlands, pp. 77–88.

O'Leary, J.W., 1979. Yield potential of halophytes and xerophytes. In: Goodin, J.R., Northington, D.K. (Eds.), Arid Land Plant Resources. Texas Tech University, Lubbock, TX, pp. 574–581.

O'Leary, J.W., Glenn, E.P., Watson, M.C., 1986. Agricultural production of halophytes irrigated with seawater. Plant Soil. 89, 311–322.

Pasternak, D., 1990. Fodder production with saline water. The Institute for Applied Research, Ben-Gurion University of the Negev. Project Report.

Qadir, M., Tubeileh, A., Akhtar, J., Larbi, A., Minhas, P.S., Khan, M.A., 2008. Productivity enhancement of salt-prone land and water resources through crop diversification. Land Degrad. Dev. 19, 429–453.

Qasim, M., Gulzar, S., Khan, M.A., 2011. Halophytes as medicinal plants. In: Ozturk, M., Mermut, A.R., Celik, A. (Eds.), Urbanization, Land Use, Land Degradation and Environment. Daya Publishing House, Delhi, pp. 330–343.

Qasim, M., Abidin, Z., Adnan, M.Y., Ansari, R., Gul, B., Khan, M.A., 2014. Traditional ethobotanical uses of medicinal plants from coastal areas of Pakistan. J. Coast. Life Med. 2, 22–30.

UNEP, 2010. UNEP Year Book. pp. 66. ISBN: 9789280730449.

Weber, D.J., Gul, B., Khan, M.A., Williams, T., Wayman, P., Warner, S., 2001. Comparison of vegetable oil from seeds of native halophytic shrubs. In: Mc Arthur, E.D., Fairbanks, D.J. (Eds.), Proceeding Shrub Land Ecosystem Genetics and Biodiversity. RMTS-P-21. USDA Forest Service, Ogden, UT; Rocky Mountain Research Station, USA, pp. 287–290.

Weber, D.J., Ansari, R., Gul, B., Khan, M.A., 2007. Potential of halophytes as source of edible oil. J. Arid Environ. 68, 315–321.

Zhu, L., Ketola, T., 2012. Microalgae production as a biofuel feedstock: risks and challenges. Int. J. Sustainable Dev. World Ecol. 19, 268–274.

15

GENETIC AND ENVIRONMENTAL MANAGEMENT OF HALOPHYTES FOR IMPROVED LIVESTOCK PRODUCTION

David G. Masters[1,2] **and Hayley C. Norman**[2]
[1]*School of Animal Biology, The University of Western Australia, Crawley, WA, Australia* [2]*CSIRO Agriculture, Wembley, WA, Australia*

15.1 Introduction

The interest in halophytes for food security in dry lands is a consequence of predicted climate change and increasing soil and water salinity. Some halophytes have the ability to grow productively in environments that are dry, saline, or a combination of both. These plants have been used opportunistically for agriculture with little or no plant improvement. This will need to change.

Climate change is expected to significantly decrease rainfall in many dry farming areas in Australia, Africa, and the Middle East (NOAA Geophysical Fluid Dynamics Laboratory (GFDL), 2007). This will be combined with increasing soil and water salinity as a consequence of unsustainable agricultural practices. The increase in salinity in agricultural land is resulting from:

- The elevation of the water table resulting from unsustainable irrigation practices. Flood or sprinkler irrigation even with low-saline water can lead to salinization of surface soil and crop failure. Where irrigation water is sourced from river systems, salinity impact tends to be closer to the end of the river. Salinization of irrigated land is found in all continents (Ghassemi et al., 1995) and recent reviews suggest up to 50% of irrigated land in South and Central Asia, North Africa, and

M.A. Khan, M. Ozturk, B. Gul, & M.Z. Ahmed (Eds): Halophytes for Food Security in Dry Lands.
DOI: http://dx.doi.org/10.1016/B978-0-12-801854-5.00015-7

the Arabian Peninsula has become saline (Sakadevan and Nguyen, 2010) and at least 20% of all irrigated land is affected by an increase in salt content (Rozema and Flowers, 2008).

- The elevation of the water table caused by widespread clearing of native vegetation combined with planting of annual crop plants in low-rainfall dryland farming environments. This is particularly a problem where poor internal drainage and the lack of an externally draining river system cause the saline water table to rise to the surface. Dryland salinity is similar in appearance to irrigation salinity but has a different cause. Once on the surface, evaporation again results in high surface soil salinity accompanied by moderate to high salinity in the shallow water table (Clarke et al., 2002). In some instances, groundwater may be more saline than seawater. Dryland salinity also occurs in many parts of the world and is a result of an imbalance between the ratio of evapotranspiration to drainage (Zalidis et al., 2002).

- Intrusion of seawater into groundwater aquifers near the coast. Often a consequence of inappropriate management, the intrusion may lead to salinization and abandonment of irrigation and drinking water wells (Bear et al., 1999). Under these circumstances, surface soil may not be saline, particularly if the soil is highly transmissive, but groundwater is likely to approach seawater in composition.

This review will focus on the use of halophytes in saline environments but will consider both irrigated and dryland production systems.

15.2 Potential Forage and Crop Solutions

With the large and expanding areas of salinity in most continents of the world, it has become a priority to maintain food production through the development of saline agricultural options. Much of the relevant research has focused on new salt-tolerant varieties of traditional crops. There have been some optimistic reviews of yields from a range of selected edible crops grown on saline water (Sakadevan and Nguyen, 2010). Examples are provided of high production of rice, cotton, sugar beet, date palms, sorghum, barley, wheat, and maize, along with a range of vegetable and forage crops. However, in most of the examples provided, salinity was moderate (EC of 10 dS m^{-1} or less) and they would not appear to be long-term options for environments where salinity is expected to increase towards the concentration found in seawater (EC approximately 46 dS m^{-1}). Others have similarly reported that while there is some optimism in the assessment of cropping options for saline land, most of the success to date has been associated with relatively low salinity and/or irrigation (Bauder et al., 2008; Glenn et al., 2013). There is clearly a need to address the production potential of miohalophytes (plants that have their highest growth when irrigated with fresh water), but more urgency around developing improved euhalophytes (plants that show some increase in growth with salinity) (Glenn and O'Leary, 1984) for agriculture.

A significant research effort has been directed towards the identification and incorporation of genes that will provide salt tolerance in conventional crops. Candidate genes for salt tolerance are being studied with the intention of using new molecular techniques to improve salt tolerance in high-value crops and pastures (Munns, 2005; Ahmed et al., 2013). At the same time traditional breeding strategies are being used to introduce salt-tolerant characteristics into cropping plants (e.g., Colmer et al., 2005). The application of these technologies has not yet delivered salt-tolerant cultivars of these plants, possibly because salt tolerance is a complex trait determined by many different genes (Rozema and Flowers, 2008). Overall, recent publications on cropping options for saline land indicate progress has been modest and the domestication of halophytes for food and bioenergy may provide better opportunities for the future than the modification of traditional crops (Colmer et al., 2005; Ruan et al., 2010). Domestication of halophytes and the changes in sowing, harvesting, and processing that will be required are exciting prospects but not a short-term solution. Progress is likely to be best achieved through the large and well-supported and coordinated research and development groups that have made spectacular progress in the production of traditional crop varieties.

Halophytes for forage are already available; they do not offer a solution by themselves but are likely to be an important part of the transition to saline agriculture—filling a gap while new crop options are developed and providing ongoing options for livestock in mixed farming systems. They will continue to play a role within a saline farming system, particularly where the environment is both arid and saline. The forage halophytes available are productive in both semi-arid (dryland) and saline environments. They can be managed using conventional agricultural practices and do not come with a requirement for changes in machinery, processing, or marketing.

15.3 Halophytes for Livestock

This review focuses on the use of halophytes for livestock in dry lands. To some extent all three forms of salinity described above can be associated with dry lands, management however may be quite different depending on whether the solution involves only rainfed production or if it involves irrigation with saline water.

There are many halophytes available for grazing or conserved forage (Masters et al., 2007). Many have, historically, been used for grazing or incorporated into mixed rations to replace roughage (Le Houérou, 1992).

These plants are variable in both biomass production and nutritive value, they are characterized by slow growth, low digestibility (therefore low metabolizable energy) and high content of anti-nutritional factors. Poor feeding value has resulted from a natural selection process within native plant populations to survive herbivory and environmental stress. Hundreds or even thousands of years of opportunistic and uncontrolled grazing has resulted in higher survival of unpalatable, low-nutritive-value plants with poor livestock production as an inevitable consequence.

Limited attempts have been made to make the plants more suitable for animals either through agronomic, genetic, or environmental manipulation of the plant or through identification of the species or class of animal best suited to the plants. Recent progress in the adaption of methods for the assessment of feeding and nutritive value means that research into plant improvement, beyond simply increasing biomass, is feasible and justified.

15.4 Current Limitations in the Use of Halophytes for Livestock Production

To understand the options and requirements for genetic and environmental management of halophytes for improved livestock production, it is first essential to understand the limitations that currently exist and the best available plant options. This information has been reviewed extensively over the past 20 years and is now well described (Ben Salem et al., 2010; El Shaer, 2010; Gihad and El Shaer, 1994; Masters and Norman, 2010; Masters et al., 2001, 2005a, 2007; Norman et al., 2013; Rogers et al., 2005). It is not the purpose of this review to re-assess these limitations but they will be summarized to provide

context (more information is available in the cited reviews).
Primary problems include:

1. Low dry matter (DM) production. Forage halophytes can have wide variation in the production of edible DM for grazing or harvest, production may range from <1 tonne ha^{-1} from shrubs grown in semi-arid saline environments (Norman et al., 2008; Warren et al., 1994) up to 50 tonnes ha^{-1} for grasses irrigated with brackish water (Pasternak et al., 1993).

2. Low digestibility and metabolizable energy (ME). ME intake is the primary driver of growth and production in grazing ruminants. For maintenance of live weight, ruminants require around 7.5 MJ ME kg^{-1} DM when fed a roughage diet. For many of the halophytic grasses the fiber content is high and digestibility of that fiber low. As a consequence digestibility of the plant is low and so is the ME value. ME is usually less than 10 MJ ME kg^{-1} and may be less than 5 MJ ME kg^{-1} (Masters and Norman, 2010). Halophytic shrubs present a different concern, the plant may be highly digestible but much of the digested content is of little nutritional value (e.g., excess minerals and plant osmolytes). The component of the plants that is organic, like many halophytic grasses, is also not highly digestible. Again the ME value of the plants is low with measured or predicted ME often <7 MJ ME kg^{-1} (Norman et al., 2010b).

3. Low-protein and/or high-nonprotein nitrogen. Adult sheep and cattle require 7−9% protein in the diet to maintain weight and up to 18% for lactation and growth. Most of the halophytic grasses contain less than 8% crude protein (CP) and therefore will only support low levels of production if fed without a supplement. The halophytic shrubs tend to have high CP (10−15%) but the CP is calculated as 6.25 × nitrogen content and does therefore not represent true protein. In fact, much of the CP in halophytic shrubs is nonprotein nitrogen. This can be converted to protein by the ruminant, but only if the supply of ME is also high.

4. High mineral composition. The halophytic grasses and legumes that grow in low/moderate-saline conditions exclude minerals from absorption or in some cases secrete minerals from the leaves. This presents no major mineral nutrition problems for livestock but these plants do not grow well in either highly saline or semi-arid environments. The halophytic shrubs that do grow in semi-arid and saline environments accumulate high concentrations of many minerals, particularly sodium (often >7% of DM), but also sulfur, magnesium, calcium, and potassium (Norman et al., 2010b).

The high concentration suppresses appetite (Masters et al., 2005b) and creates a mineral imbalance within the animal (Mayberry et al., 2010).

5. High concentration of anti- or nonnutritional compounds. Again this is primarily a characteristic of shrubs and can be explained from a natural selection process within native plant populations to survive herbivory and environmental stress. Opportunistic and uncontrolled grazing has resulted in higher survival of unpalatable, low-nutritive-value plants with poor livestock production. Other compounds assist the plant to grow and survive in environments with high salinity, low moisture availability, and high evapotranspiration. The number and role of many of these secondary compounds is not well understood—there appear to be many that are still not characterized (Gihad and El Shaer, 1994).

Even though much is now understood on the factors that influence feeding value of halophytes, the ability to predict livestock production and preferences through measurement of plant composition is not reliable (Norman et al., 2004), there is still much to know.

Such a list of limitations may (and has) discouraged many livestock scientists away from grazing options for saline and semi-arid land. In many cases research has focused on how to gain some value from halophytes by feeding higher-quality supplements at the same time (Benjamin et al., 1992; Mayberry et al., 2008; Norman et al., 2008). In a world where high-quality feed is scarce and best used for feeding people or at least for livestock with a high feed conversion ratio, the sustainable use of feed supplements for grazing ruminants is questionable.

We suggest the use of high-quality supplements is short-sighted. A more sustainable long-term approach is to improve the feeding value of halophytes. With such an approach, the current limitations represent opportunities.

From an agricultural systems' perspective, improved livestock production is essential; however, this should also be seen as part of a package that can provide broader environmental benefits. Recent studies in Western Australia, for example, have shown that establishment and grazing of chenopod shrubs and improved annual pastures on a highly saline site dramatically reduced salt, nutrient, and sediment discharge in surface run-off over 7 years of observation. Both the concentration and mass of these components of run-off were consistently much lower than from an adjacent unimproved area of salt land. By comparison, one tenth of the salt was discharged from the improved catchment once the saltbush system was established

(Bennett et al., 2012). Others have highlighted the potential land rehabilitation benefits of *Atriplex* spp. in semi-arid and saline environments (Barrett-Lennard, 2002; Le Houérou, 1992), however, the study of Bennett and colleagues appears to be the first to quantify in detail the long-term benefits that may result from the cultivation and grazing of halophytes.

In a situation where plants grow and provide obvious environmental benefits the potential to use genetic and environmental manipulation to improve feeding value is significant.

15.5 Genetic Improvement of Halophytes for Livestock

To improve production from ruminants grazing halophytes, the aims are simple, these are:

1. Increased biomass production
2. Improved nutritive value and palatability
3. Reduction in anti-nutritional properties
4. Plant survival in a grazing environment.

An example of an approach to improvement is a specific project aimed at increasing the feeding value of chenopod shrubs for saline land in Australia. Seeds from 27 native populations (provenances) of old man saltbush (*Atriplex nummularia*) were collected across Australia (Norman and Masters, 2010). The collection included two subspecies (ssp. *nummularia* and ssp. *spiculata*). In the first stage of the project, 60,000 shrubs were established in three nursery sites in three different states of Australia. The plants were assessed for agronomic traits such as survival, plant form, and biomass growth. Each of the nurseries was also grazed with sheep to determine relative preference and recovery from grazing. A full suite of nutritive value measurements were conducted on a representative subset of the plants. There was significant variation in digestibility (and therefore ME content) (Figure 15.1), salt, CP, and biomass growth. There was also significant variation in grazing preference for shrubs with consistent preference for specific provenances measured across the nursery sites (Figure 15.2). Of the initial 60,000 wild shrubs, 30 from each nursery site were selected and each planted at three sites in each of three states. When compared to the average of the unselected provenances, the selected plants had 20% higher ME and produced eight times more biomass. They were also consistently preferred by grazing sheep. From these 90 clones, 12 were subsequently selected for testing under

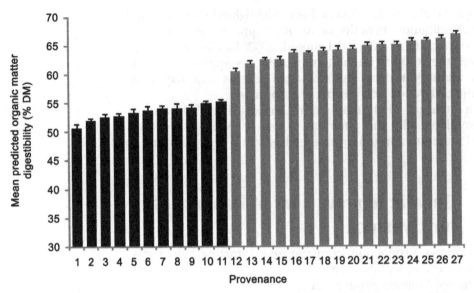

Figure 15.1 Predicted organic matter digestibility of 27 provenances of old man saltbush (mean ± SE) from subspecies *spiculata* (black columns) and subspecies *nummularia* (gray columns). Digestibility estimated using a pepsin cellulase assay, adjusted for soluble ash and a shrub correction factor as described by Norman et al. (2010a,b).

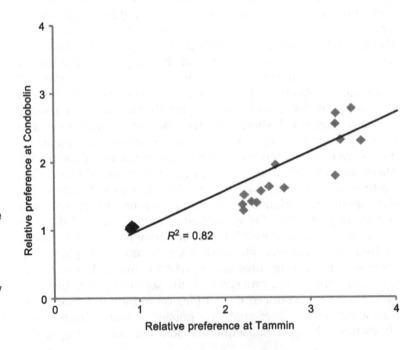

Figure 15.2 Grazing preference of 27 provenances of old man saltbush from subspecies *spiculata* (black symbols) and subspecies *nummularia* (gray symbols) at two geographically distant sites. A higher preference score indicates higher grazing preference.

commercial conditions. The potential difference in livestock production across this range of nutritive values should be put into perspective. A forage with an ME content of 7.4 MJ kg^{-1} DM (organic matter digestibility approximately 57%—the digestibility of an average old man saltbush provenance) will support maintenance but not growth in a 50-kg sheep. An increase in biomass alone will increase the number of sheep that can be carried but will still not provide any growth, no matter what stocking rate is applied. Production of meat or lambs could only be achieved through the use of feed supplements or alternative feed sources. An increase in ME to 9.3 MJ kg^{-1} DM (organic matter digestibility approximately 68%—the digestibility of the best old man saltbush provenance) would support a growth rate of approximately 36 g day^{-1} in a 50-kg sheep. A simultaneous eight-fold increase in biomass would allow more stock to be carried. Where one sheep could be grazed without growth on average unimproved old man saltbush, up to eight sheep could be grazed, each growing at 36 g day^{-1} (growth estimated using Graz Feed (Freer et al., 1997)), on the selected high feeding value old man saltbush.

This selection program has been livestock-focused from the beginning and progress has been partially attributable to the use of shrub clones to reduce the intervals between selection and testing under grazing. It is only now that a short list of plants suitable for livestock production has been identified that further research on seed yield, collection, and germinability have also commenced.

There are two characteristics of this selection program that have been critical. The first is the ability to use in vitro measurements and near infrared spectroscopy to indicate digestibility and therefore ME value of the plants. This is not a trivial achievement as traditional in vitro methods are unsuitable for plants containing salt concentrations and provide unsound in vivo estimates (Norman et al., 2010a).

The second critical characteristic of the program is the use of livestock in the selection process to screen for palatability. With the high number of secondary compounds contained in halophytes, it is an impossible task to identify and measure all of these and then select for low concentrations of each. In fact, this may be a flawed selection pathway and just lead to plants that cannot cope with the environmental conditions in which they are planted. Some of the secondary compounds that are not attractive to livestock are certainly necessary for osmoregulation and their removal will compromise plant survival in saline/semi-arid conditions. Importantly though, only some (e.g., oxalates, nitrates), but not all secondary compounds used by plants to cope with high salinity are detrimental to livestock production. Livestock may be able to metabolize some compounds (e.g., free amino acids, protein, and soluble sugars) or even benefit from consumption of others (e.g., vitamin E). In using palatability as part of the selection process, composition of selected plants may change without a reduction in salinity tolerance. This should be seen as an evolutionary step in the selection of plants for livestock—traditional selection of pasture plants is based on a prescriptive approach where the scientist makes the decision on the appropriate plant characteristics. This alone is unlikely to be successful when the plants involved have a complex composition partly developed to discourage grazing and are growing in hostile and uncontrolled environments.

The outcome of this project is that a new saltbush cultivar will shortly be released that has the potential to support modest weight gains in livestock compared with live weight loss or at best maintenance expected from most unselected provenances. This will allow for the development of a more productive farming system with less reliance on expensive supplements to support growth and reproduction.

15.6 Environmental Manipulation

Given the high variability in feeding value of the halophytic chenopods, there is also a strong case to improve our understanding of environmental impacts on plant composition and growth. There is some evidence that the feeding value of halophytes is influenced by the growing environment. While this may be the case, from a farming perspective, this is really only of interest if the environmental change is capable of manipulation through management.

15.6.1 Water

Some control over the environment is possible where unlimited water supplies are available. Using an intensive system of agronomic management, involving irrigation, fertilizer, and multiple harvesting, DM production of 35–40 tonnes DM ha^{-1} year^{-1} has been reported for salt-grass (*Distichlis spicata*) and marine couch (*Sporobolus virginicus*) (Al-Dakheel et al., 2007, 2008). The plants were irrigated with saline water of up to

30 dS m^{-1}. High levels of production of the forage *Panicum turgidum* irrigated with brackish water have also been reported (Khan et al., 2009). At even higher levels of salinity the euhalophytes are more successful. Saltbush (*Atriplex* spp.) irrigated with saline drainage water has produced forage yields of 2.2–12.3 tonnes DM ha^{-1} year^{-1} (Watson and O'Leary, 1993; Watson et al., 1987). In other experiments, a broad range of halophytes from the genera *Atriplex* and *Salicornia* produced up to 1.8 tonnes DM ha^{-1} year^{-1} even under hypersaline irrigation water (Glenn and O'Leary, 1985). These euhalophytes are highly salt-tolerant and have increased growth rates at 4–5 dS m^{-1} and still only have a 50% decrease in growth at 40 dS m^{-1} (70% seawater).

While growth of halophytes can be improved with irrigation, the potential long-term consequences of this irrigation on soil and water need consideration.

15.6.2 Salinity

There is evidence that salinity may alter nutritive value (Masters et al., 2010; Rogers et al., 2008), but under practical conditions, salinity cannot easily be managed. Similarly, the type of salts in ground and irrigation water influence both the growth response to salinity, nutritive value of the plants, and the type of plants that can be cultivated. For example, plants are significantly more tolerant to high salinity and have nutritive value sufficient to support reasonable levels of ruminant production when the ionic composition of water available to plants is dominated by sulfates rather than chlorides (Robinson et al., 2004; Rogers et al., 1998; Suyama et al., 2007).

15.6.3 Fertilizer

With higher biomass production and more intensive livestock production as a possibility, further environmental manipulation may be an option. To date there is little information on the influence of fertilizers on growth and composition of chenopods, however, there is evidence that the type of fertilizer may influence the composition of toxic compounds within the plant and therefore both palatability and production (Al Daini et al., 2013). This indicates a multidisciplinary approach that includes consideration of both the plant and animal responses to fertilizer use is logical where forage production is the primary goal.

15.7 Conclusions

The domestication of halophytes offers opportunities for the improvement of agriculture in saline and semi-arid landscapes. Domestication, particularly for the development of new food crops, will require time. Short-term options are available for the use of halophytes for livestock.

Currently many of the halophytes used for grazing have low feeding value due to low growth rates, poor digestibility, and high content of minerals and antinutritional compounds. These constraints also represent opportunities with recent research indicating significant short-term progress can be made through the identification and selection, from within natural populations, of plants that support at least modest levels of livestock production and a reduced requirement for expensive feed supplements.

References

Ahmed, M.Z., Shimazaki, T., Gulzar, S., Kikuchi, A., Gul, B., Khan, M.A., et al., 2013. The influence of genes regulating transmembrane transport of Na^+ on the salt resistance of *Aeluropus lagopoides*. Funct. Plant Biol. 40 (9), 860–871.

Al-Dakheel, A., Al-Hadrami, G., Al-Shorabi, S., AbuRumman, G., 2007. Optimizing management practices for maximum production for two salt-tolerant grasses: *Sporobolus virginicus and Distichlis spicata*. In: 7th Annual U.A.E University Research Conference, Dubai, pp. CFA-72–CFA-78.

Al-Dakheel, A., Al-Hadrami, G., Al-Shoraby, S., Shabbir, G., 2008. The potential of salt-tolerant plants and marginal resources in developing an integrated forage-livestock production system, 2nd International Salinity Forum, Adelaide.

Al Daini, H., Norman, H.C., Young, P., Barrett Lennard, E.G., 2013. The source of nitrogen (NH_4^+ or NO_3^-) affects the cocentration of oxalate in the shoots and the growth of *Atriplex nummularia* (oldman saltbush). Funct. Plant Biol. 40, 1057–1064.

Barrett-Lennard, E.G., 2002. Restoration of saline land through revegetation. Agric. Water Manage. 53, 213–226.

Bauder, J.W., Browning, L.S., Phelps, S.D., Kirkpatrick, A.D., 2008. Biomass production, forage quality, and cation uptake of quail bush, four-wing saltbush, and seaside barley irrigated with moderately saline–sodic water. Commun. Soil Sci. Plant Anal. 39, 2009–2031.

Bear, J., Cheng, A.H.-D., Sorek, S., Ouazar, D., Herrara, I., 1999. Seawater Intrusion in Coastal Aquifers—Concepts, Methods and Practices. Kluwer Academic Publishers, Dordrecht.

Benjamin, R.W., Oren, E., Katz, E., Becker, K., 1992. The apparent digestibility of *Atriplex barclayana* and its effect on nitrogen balance in sheep. Anim. Prod. 54, 259–264.

Bennett, D., George, R., Silberstein, R., 2012. Changes in Run-Off and Groundwater Under Saltbush Grazing Systems: Preliminary Results of a Paired Catchment Study. Perth, Australia, p. 38.

Ben Salem, H., Norman, H.C., Nefzaoui, A., Mayberry, D.E., Pearce, K.L., Revell, D.K., 2010. Potential use of oldman saltbush (*Atriplex nummularia* Lindl.) in sheep and goat feeding. Small Rumin. Res. 91, 13–28.

Clarke, C.J., George, R.J., Bell, R.W., Hatton, T.J., 2002. Dryland salinity in south-western Australia; its origins, remedies, and future research directions. Aust. J. Soil Res. 40, 93–113.

Colmer, T.D., Munns, R., Flowers, T.J., 2005. Improving salt tolerance of wheat and barley: future prospects. Aust. J. Exp. Agric. 45, 1425–1443.

El Shaer, H.M., 2010. Halophytes and salt-tolerant plants as potential forage for ruminants in the Near East region. Small Rumin. Res. 91, 3–12.

Freer, M., Moore, A.D., Donnelly, J.R., 1997. GRAZPLAN: Decision support systems for Australian enterprises—II. The animal biology model for feed intake, production and reproduction and the GrazFeed DSS. Agric. Syst. 54, 77–126.

Ghassemi, F., Jakeman, A.J., Nix, H.A., 1995. Salinisation of Land and Water Resources. CAB International, Wallingford, UK.

Gihad, E.A., El Shaer, H.M., 1994. Utilization of halophytes by livestock on range-lands. Problems and prospects. In: Squires, V., Ayoub, A.T. (Eds.), Halophytes as a Resource for Livestock and for Rehabilitation of Degraded Land. Kluwer Academic Publishers, Dordrecht, Netherlands, pp. 77–96.

Glenn, E.P., O'Leary, J.W., 1984. Relationship between salt accumulation and water content of dicotyledenous halophytes. Plant Cell Environ. 7, 253–261.

Glenn, E.P., O'Leary, J.W., 1985. Productivity and irrigation requirements of halo-phytes grown with seawater in the Sonoran Desert. J. Arid Environ. 9, 81–91.

Glenn, E.P., Anday, T., Chaturvedi, R., Martinez-Garcia, R., Pearlstein, S., Soliz, D., et al., 2013. Three halophytes for saline-water agriculture: an oilseed, a forage and a grain crop. Environ. Exp. Bot. 92, 110–121.

Khan, M.A., Ansari, R., Ali, H., Gul, B., Nielsen, B.L., 2009. *Panicum turgidum*, a potentially sustainable cattle feed alternative to maize for saline land. Agric. Ecosyst. Environ. 129, 542–546.

Le Houérou, H.N., 1992. The role of saltbushes (*Atriplex* spp.) in arid land reha-bilitation in the Mediterranean Basin: a review. Agroforestry Syst. 18, 107–148.

Masters, D., Norman, H., 2010. Salt tolerant plants for livestock production. In: Ahmed, M., Al-Rawahy, S.A. (Eds.), International Conference on Soils and Groundwater Salinization in Arid Environments. Sultan Qaboos University, Muscat, Oman, pp. 23–31.

Masters, D., Tiong, M., Vercoe, P., Norman, H., 2010. The nutritive value of river saltbush (*Atriplex amnicola*) when grown in different concentrations of sodium chloride irrigation solution. Small Rumin. Res. 91, 56–62.

Masters, D.G., Norman, H.C., Dynes, R.A., 2001. Opportunities and limitations for animal production from saline land. Asian-Australas. J. Anim. Sci. 14 (Special issue), 199–211.

Masters, D.G., Norman, H.C., Barrett-Lennard, E.G., 2005a. Agricultural systems for saline soil: the potential role of livestock. Asian-Australas. J. Anim. Sci. 18, 296–300.

Masters, D.G., Rintoul, A.J., Dynes, R.A., Pearce, K.L., Norman, H.C., 2005b. Feed intake and production in sheep fed diets high in sodium and potassium. Aust. J. Agric. Res. 56, 427–434.

Masters, D.G., Benes, S.E., Norman, H.C., 2007. Biosaline agriculture for forage and livestock production. Agric. Ecosyst. Environ. 119, 234–248.

Mayberry, D., Masters, D., Vercoe, P., 2010. Mineral metabolism of sheep fed salt-bush or a formulated high-salt diet. Small Rumin. Res. 91, 81–86.

Mayberry, D.E., Masters, D.G., Vercoe, J.E., 2008. What is the optimal level of bar-ley to feed sheep grazing saltbush? In: 2nd International Salinity Forum.

Salinity, water and society—global issues, local action, Adelaide, South Australia, p. CD Rom.

Munns, R., 2005. Genes and salt tolerance: bringing them together. New Phytol. 167, 645–663.

NOAA Geophysical Fluid Dynamics Laboratory (GFDL), 2007. NOAA GFDL Climate Research Highlights Image Gallery. Will the Wet Get Wetter and the Dry Drier?

Norman, H.C., Masters, D.G., 2010. Predicting the nutritive value of saltbushes (*Atriplex* spp) with near infrared reflectance spectroscopy, Management of soil and groundwater salinization in arid regions, Oman.

Norman, H.C., Freind, C., Masters, D.G., Rintoul, A.J., Dynes, R.A., Williams, I.H., 2004. Variation within and between two saltbush species in plant composition and subsequent selection by sheep. Aust. J. Agric. Res. 55, 999–1007.

Norman, H.C., Masters, D.G., Wilmot, M.G., Rintoul, A.J., 2008. Effect of supplementation with grain, hay or straw on the performance of weaner Merino sheep grazing oldman (*Atriplex nummularia*) or river (*Atriplex amnicola*) saltbush. Grass Forage Sci. 63, 179–192.

Norman, H.C., Revell, D.K., Mayberry, D.E., Rintoul, A.J., Wilmot, M.G., Masters, D.G., 2010a. Comparison of *in vivo* organic matter digestibility of native Australian shrubs to *in vitro* and *in sacco* predictions. Small Rumin. Res. 91, 69–80.

Norman, H.C., Wilmot, M.G., Rintoul, A.J., Hulm, E., Revell, D.K., 2010b. Nutritive value of 5 genera of West Australian chenopods for livestock. In: Proceedings of the 15th Australian Society of Agronomy Conference, Lincoln, New Zealand.

Norman, H.C., Masters, D.G., Barrett Lennard, E.G., 2013. Halophytes as forages in saline landscapes: interaction between gentotype and environment change their feeding value to ruminants. Environ. Exp. Bot. 92, 96–109.

Pasternak, D., Nerd, A., De Malach, Y., 1993. Irrigation with brackish water under desert conditions IX. The salt tolerance of six forage crops. Agric. Water Manage. 24, 321–334.

Robinson, P.H., Grattan, S.R., Getachew, G., Grieve, C.M., Poss, J.A., Suarez, D.L., et al., 2004. Biomass accumulation and potential nutritive value of some forages irrigated with saline-sodic drainage water. Anim. Sci. Technol. 111, 175–189.

Rogers, M., Colmer, T., Frost, K., Henry, D., Cornwall, D., Hulm, E., et al., 2008. Diversity in the genus *Melilotus* for tolerance to salinity and waterlogging. Plant Soil. 304, 89–101.

Rogers, M.E., Grieve, C.M., Shannon, M.C., 1998. The response of lucerne (*Medicago sativa* L.) to sodium sulphate and chloride salinity. Plant Soil. 202, 271–280.

Rogers, M.E., Craig, A.D., Munns, R., Colmer, T.D., Nichols, P.G.H., Malcolm, C.V., et al., 2005. The potential for developing fodder plants for the salt-affected areas of southern and eastern Australia: an overview. Aust. J. Exp. Agric. 45, 301–329.

Rozema, J., Flowers, T., 2008. Crops for a salinized world. Science. 322, 1478–1480.

Ruan, C.J., da Silva, J.A.T., Mopper, S., Qin, P., Lutts, S., 2010. Halophyte improvement for a salinized world. Crit. Rev. Plant Sci. 29, 329–359.

Sakadevan, K., Nguyen, M.L., 2010. Chapter two—extent, impact, and response to soil and water salinity in arid and semiarid regions. In: Donald, L.S. (Ed.), Advances in Agronomy. Academic Press, pp. 55–74.

Suyama, H., Benes, S.E., Robinson, P.H., Getachew, G., Grattan, S.R., Grieve, C.M., 2007. Biomass yield and nutritional quality of forage species under long-term irrigation with saline-sodic drainage water: field evaluation. Anim. Feed Sci. Technol. 135, 329–345.

Warren, B.E., Casson, T., Ryall, D.H., 1994. Production from grazing sheep on revegetated saltland in Western Australia. In: Squires, V.R., Ayoub, A.T. (Eds.), Halophytes as a Resource for Livestock and for Rehabilitation of Degraded Lands. Kluwer Academic Publishers, Dordrecht, Netherlands, pp. 263–265.

Watson, M.C., O'Leary, J.W., 1993. Perfomance of *Atriplex* species in the San Joaquin Valley, California, under irrigation and with mechanical harvests. Agric. Ecosyst. Environ. 43, 255–266.

Watson, M.C., O'Leary, J.W., Glenn, E.P., 1987. Evaluation of *Atriplex lentiformis* (Torr.) S. Wats. and *Atriplex nummularia* Lindl. as irrigated forage crops. J. Arid Environ. 13, 293–303.

Zalidis, G., Stamatiadis, S., Takavakoglou, V., Eskridge, K., Misopolinos, N., 2002. Impacts of agricultural practices on soil and water quality in the Mediterranean region and proposed assessment methodology. Agric. Ecosyst. Environ. 88, 137–146.

16

DROUGHT AND SALINITY DIFFERENTLY AFFECT GROWTH AND SECONDARY METABOLITES OF "*CHENOPODIUM QUINOA WILLD*" SEEDLINGS

Adele Muscolo[1], Maria Rosaria Panuccio[1], Angelo Maria Gioffrè[2] and Sven-Erik Jacobsen[2]

[1]*Department of Agriculture, Mediterranea University, Reggio Calabria, Italy*
[2]*Department of Plant and Environmental Sciences, Faculty of Science, University of Copenhagen, Tåstrup, Denmark*

16.1 Introduction

Drought and salinity are the most limiting factors for agricultural productivity worldwide, causing crop failure and decreasing average yields by more than 50% (Buchanan et al., 2000; Bartels and Sunkar, 2005; Mittler, 2006). The majority of food crops do not generally adapt quickly to changes resulting from drought and/or salinity stress (Lane and Jarvis, 2007). Hence, there is the need to find alternative food crops with a major constitutive tolerance to these environmental constraints.

Chenopodium quinoa is a seed crop and an important food source (Wilson et al., 2002; Jacobsen et al., 2003; Trognitz, 2003; Koyro and Eisa, 2008) that has been cultivated in the Andean regions for thousands of years (Jacobsen et al., 2009; Hariadi et al., 2011). Considered a pseudo-cereal, it is a broadleaf plant that, recently, has had serious worldwide attention because of its extraordinary tolerance to different environmental stresses.

M.A. Khan, M. Ozturk, B. Gul, & M.Z. Ahmed (Eds): Halophytes for Food Security in Dry Lands.
DOI: http://dx.doi.org/10.1016/B978-0-12-801854-5.00016-9

It is a facultative halophyte and its grains are a rich source of a wide range of minerals (Ca, P, Mg, Fe, and Zn), vitamins (B1, B9, C, and E), linolenate, natural antioxidants, and high-quality proteins containing ample amount of essential amino acids such as lysine and methionine (Koyro and Eisa, 2008; Abugoch et al., 2009). Its grains are gluten-free, representing a marketable cultivar for commercial health food producers, and its leaves are widely used as a food for humans and livestock. Due to its high nutritional quality quinoa is considered one of the major alternative crops to meet food shortage in this century (FAO, 2012).

Abiotic stresses affect plant growth, leading to the overproduction of reactive oxygen species (ROS) which are highly reactive and toxic. The ROS comprise both free radicals (superoxide radicals, hydroxyl radicals, perhydroxy radicals, and alkoxy radicals) and nonradical molecular forms (hydrogen peroxide and singlet oxygen). The excess of ROS causes damage to proteins, lipids, carbohydrates, and DNA, which ultimately results in the disruption of cellular homeostasis and subsequently in cellular death. Thus, to maintain growth and productivity under stress conditions, plants have to activate several strategies to scavenge the enhanced generation of ROS. Recent studies have demonstrated that halophytes, in respect to glycophytes, usually have a robust constitutive anti-oxidative defense system able to alleviate oxidative damage during salt or drought stress, showing halophytes as a source of valuable secondary metabolites with potential economic value (Ksouri et al., 2007; Buhmann and Papenbrock, 2013). Most of the beneficial foods for human health contain a higher amount of compounds, with good antioxidant properties like phenols, ascorbic acid, flavonoids, and others (Arts and Hollman, 2005; Scalbert et al., 2005; Gallie, 2013). However, direct evidence of increasing the concentration of valuable secondary metabolites and the antioxidant capacity as a consequence of increasing abiotic stress still remains ambiguous. In consideration of this background, the general objective of this research was to test salt- and drought-induced responses in *C. quinoa* seedlings to get precise insights into the individual physiological mechanisms that may be involved in salt and/or drought tolerance.

The specific aim was to identify the optimal saline concentration and the limit of water deficiency to increase the biosynthesis of valuable antioxidant compounds whilst maintaining acceptable biomass production. A growth chamber experiment was performed in the presence of different concentrations of

NaCl and PEG, to assess seedling growth in terms of biomass, root and shoot length, root–shoot ratio, and the antioxidant responses of quinoa as total antioxidant capacity, total phenols, total carotenoids, proline, and antioxidant enzyme activities. In this study we used Danish-bred quinoa (cv. Titicaca) (Adolf et al., 2012; Jacobsen et al., 2010), a variety well adapted to European climatic conditions mainly because quinoa production may be a viable option for farmers interested in a high-value crop with regional production and local markets in Mediterranean countries where drought and salinity are major risks for the future of agricultural development.

16.2 Materials and Methods

16.2.1 Plant Material

Mature seeds of the Danish-bred quinoa (cv. Titicaca) (provided by the Department of Plant Environmental Science of the University of Copenhagen) were stored at 5°C until the start of the experiments. Two different experiments were carried out in a growth chamber (temperature of $25 \pm 1°C$ in the dark with a relative humidity of 70%) to characterize the responses of quinoa to salt stress.

16.2.2 Plantlet Growth in Pots

Seeds were germinated in petri dishes. After 3 days from the beginning of germination, germinated seeds were grown for 21 days in plastic pots (10 cm diameter $\times$ 7 cm height), in a growth chamber Green line WRS 96–85, KW apparecchi scientifici, Italy, under white light (80 W m^{-2}, Osram HQI halogen vapor W lamp, PAR 1055 μmol m^{-2} s^{-1}), in a 16/8 h photoperiod, 70% relative humidity and at 21°C. All pots were filled with Perlite that had been equilibrated, before transplanting the germinated seeds, with PEG or NaCl (at 0, -0.62, -0.80, -1.54, and -2.43 MPa). All reagents used were of the highest analytical grade and were purchased from Sigma Chemical Co. (St. Louis, MO). All pots were watered with a quarter-strength Murashige and Skoog medium (MS/4) containing macro- and micronutrients at pH 5.8. The pots were weighed daily, and watered when their weight decreased by 30% (corresponding to water that was lost by evapotranspiration). The control pots were watered with

MS/4 alone. Leaf and root length were evaluated 21 days after the beginning of the stress, using six plants for each treatment.

16.2.3 Determination of Enzyme Activities

After 21 days of growth, fresh leaves (0.5 g) were ground using a chilled mortar and pestle and homogenized in 0.1 M phosphate buffer solution (pH 7.0) containing 0.1% 100 mg soluble polyvinylpolypyrrolidone (PVPP) and 0.1 mM ethylene-diamine tetra acetic acid (EDTA). The homogenate was centrifuged at 15,000 g for 15 min at 4 °C. The resulting supernatant was used to evaluate the activity of catalase (CAT, EC 1.11.1.6), peroxidase (POX, EC 1.11.1.7), ascorbate peroxidase (APX, EC 1.11.1.11), and superoxide dismutase (SOD EC 1.15.1.1). All enzyme activities were measured at 25°C by a UV:visible light spectrophotometer (UV-1800 CE, Shimadzu, Japan).

Catalase activity was determined by monitoring the disappearance of H_2O_2 at 240 nm, calculated by using its extinction coefficient $(\varepsilon) = 0.036 \, \text{mM}^{-1} \, \text{cm}^{-1}$. The reaction mixture contained 1 mL potassium phosphate buffer (50 mM, pH 7.0), 40 μL enzyme extract, and 5 μL H_2O_2 (Beaumont et al., 1990).

Ascorbate peroxidase activity was assayed according to Nakano and Asada (1981). The reaction mixture (1.5 mL) contained 50 mM phosphate buffer (pH 6.0), 0.1 μM EDTA, 0.5 mM ascorbate, 1.0 mM H_2O_2, and 50 μL enzyme extract. The reaction was started by the addition of H_2O_2 and ascorbate oxidation measured at 290 nm and the activity was quantified using the molar extinction coefficient for ascorbate $(2.8 \, \text{mM}^{-1} \, \text{cm}^{-1})$.

O-guaiacol-peroxidase activity was measured on the basis of determination of guaiacol oxidation at 436 nm for 90 s (Panda et al., 2003). The reaction mixture contained 1 mL potassium phosphate buffer (0.1 M, pH 7.0), 20 μL guaiacol, 40 μL enzyme extract, and 15 μL H_2O_2. POX activity was quantified by the amount of tetraguaiacol formed using its extinction coefficient $(\varepsilon) = 25.5 \, \text{mM}^{-1} \, \text{cm}^{-1}$.

Superoxide dismutase activity was estimated by recording the decrease in absorbance of superoxide nitro-blue tetrazolium complex by the enzyme (Gupta et al., 1993). The reaction mixture (2 mL) contained 0.1 mL of 200 mM methionine, 0.1 mL of 2.25 mM nitro-blue tetrazolium (NBT), 0.1 mL of 3 mM EDTA, 1.5 mL of 100 mM potassium phosphate buffer, 1 mL distilled water, and 0.05 mL of enzyme extract. The assay was performed in duplicate for each sample. Two tubes without enzyme extract were used as controls. The reaction was started by adding 0.1 mL riboflavin (60 μM) and placing the tubes below a light source of two 15-W florescent lamps for 15 min. The reaction was stopped by switching off the light and covering the tubes with black cloth. Tubes without enzyme developed maximal color. A nonirradiated complete reaction mixture, which did not develop color, served as the blank. Absorbance was recorded at 560 nm and one unit of enzyme activity was taken as the quantity of enzyme which reduced the absorbance reading of samples to 50% in comparison with tubes lacking enzymes.

For CAT, APX, SOD, and POX activities, the results were expressed as enzyme units (U) per mg fresh weight. One unit of enzyme was defined as the amount of enzyme necessary to decompose 1 μmol of substrate per min at 25°C.

16.2.4 Determination of Total Antioxidant Capacity

Leaves of seedlings (3 weeks treated) were homogenized in a chilled mortar with distilled water 1:4 (w/v) and centrifuged at 14,000 g for 30 min. All steps were performed at 4°C. The supernatants were filtered through two layers of muslin cloth and were used to determine the total antioxidant capacity by the spectrophotometric method of Prieto et al. (1999). Aqueous extracts were combined in Eppendorf tubes with 1 mL of reagent solution (0.6 M H_2SO_4, 28 mM sodium phosphate, 4 mM ammonium molybdate mixture). The tubes were incubated for 90 min at 95°C. The mixture was cooled to room temperature and the absorbance read at 695 nm against a blank (mixture without leaf extract). The assay was conducted in triplicate and the total antioxidant activity was expressed as μg ascorbic acid g^{-1} FW. The higher absorbance value indicated higher antioxidant activity (Prasad et al., 2009).

16.2.5 Determination of Total Phenolic Content

Total phenolic content was determined with the Folin–Ciocalteu reagent according to a modified procedure described by Singleton and Rossi (1965). Briefly, 0.50 mL of the aqueous extract of the leaves was reacted with 2.5 mL of 0.2 mol L^{-1} Folin–Ciocalteu reagent for 4 min, and then 2 mL saturated sodium carbonate solution (about 75 g L^{-1}) was added into the reaction mixture. The absorbance readings were taken at 760 nm after incubation at room temperature for 2 h. Tannic acid was used as a reference standard, and the results were expressed as milligram tannic acid equivalent (mg TAET g^{-1} fresh weight).

16.2.6 Detection of Proline

To detect free proline content, leaf samples (0.3 g) were homogenized with 5 mL of 3% sulfosalicylic acid and centrifuged at 3000 rpm for 20 min. The supernatant was added to 2 mL of glacial acetic acid with 2 mL acidic ninhydrin. The mixture was heated at 100°C for 25 min. After the liquid was cooled, the mixture was added to 4 mL toluene. The absorbance of the extracts was read at 520 nm (Bates et al., 1973).

16.2.7 Cation and Anion Detection

Cations (Na^+, K^+, Ca^{2+} Mg^{2+}, and NH_4^+) and anions (Cl^- and SO_4^{2-}) were determined in the water extracts of treated seedlings by ion chromatography (DIONEX ICS-1100).

16.2.8 Statistical Analysis

All data were analyzed by one-way analysis of variance (ANOVA). We performed separate ANOVAs using PEG or NaCl at different osmotic potential (0, −0.62, −0.80, −1.54, and −2.43 MPa) as the grouping factor. The response variables for these ANOVAs were: seedling growth, enzyme activities, ion content, and antioxidants. Since salt and PEG concentrations had five levels, on all significant ANOVAs we performed a Tukey multiple comparison test to compare all pairs of means. All data collected were statistically analyzed using SYSTAT 8.0 software (SPSS Inc.).

16.3 Results

16.3.1 Seedling Growth

Fresh and dry weights of the seedlings decreased with increasing concentrations of NaCl and PEG (Table 16.1), although PEG was the most detrimental at all concentrations. Leaf and root mass ratio (LMR and RMR) were differently affected by NaCl and PEG; NaCl decreased significantly LMR at the lowest concentration and increased it significantly at the highest one (Table 16.1), conversely, RMR increased at the lowest NaCl concentrations, and decreased when NaCl increased. Under PEG, LMR and RMR decreased in a concentration-dependent manner suggesting that the osmotic effect was more detrimental than ionic toxicity at this plant stage (Table 16.1).

Table 16.1 Growth Parameters: Fresh Weight (FW), Dry Weight (DW), Dry Matter (DM), Leaf Mass Ratio (LMR), and Root Mass Ratio (RMR) of 21-Day-Old Seedlings Treated with NaCl and PEG at Increasing Iso-Osmotic Potentials (0, −0.62, −0.80, −1.54, and −2.43 MPa)

Treatment	FW (g plant^{-1})	DW (g plant^{-1})	DM (%)	LMR (g plant^{-1})	RMR (g plant^{-1})
NaCl					
0	10.5 ± 1.0^a	0.92 ± 0.04^a	8.76 ± 0.2^b	0.71 ± 0.01^b	0.29 ± 0.03^b
−0.62	11.0 ± 0.5^a	0.98 ± 0.05^a	8.91 ± 0.2^b	0.61 ± 0.02^c	0.39 ± 0.02^a
−0.80	10.5 ± 0.7^a	0.90 ± 0.04^a	8.57 ± 0.3^b	0.67 ± 0.02^b	0.33 ± 0.02^b
−1.54	7.2 ± 0.6^b	0.82 ± 0.03^b	11.39 ± 0.5^a	0.70 ± 0.01^b	0.30 ± 0.02^b
−2.43	5.7 ± 0.4^c	0.41 ± 0.06^c	7.01 ± 0.6^c	0.75 ± 0.01^a	0.25 ± 0.01^c
PEG					
0	10.5 ± 1.0^a	0.92 ± 0.04^a	8.76 ± 0.2^a	0.71 ± 0.01^b	0.29 ± 0.03^a
−0.62	10.0 ± 0.3^a	0.90 ± 0.05^a	8.50 ± 0.2^a	0.51 ± 0.02^c	0.21 ± 0.02^b
−0.80	10.0 ± 0.4^a	0.88 ± 0.04^a	8.33 ± 0.3^a	0.47 ± 0.02^b	0.18 ± 0.02^b
−1.54	5.2 ± 0.6^b	0.61 ± 0.03^b	7.39 ± 0.5^b	0.42 ± 0.01^b	0.16 ± 0.02^b
−2.43	3.7 ± 0.4^c	0.21 ± 0.06^c	5.01 ± 0.6^c	0.35 ± 0.01^a	0.13 ± 0.01^c

Values are means ± SD of four different experiments. Within each column and for each measurement a different superscript letter indicates that any difference between control and saline or PEG treatment is significant. Both analyses according to the Tukey's Studentized Range test ($P \leq 0.05$).

16.3.2 Enzyme Activities

As expected, all the antioxidant enzymatic activities increased in NaCl- and PEG-treated seedlings compared to controls (Figure 16.1). SOD activities increased in a concentration-dependent manner in NaCl- and PEG-treated seedlings, but the highest activities were observed under NaCl, at all concentrations. CAT activity was increased by both NaCl and PEG, and the greatest increase was observed with PEG already at the lower concentrations (Figure 16.1). No significant variation in POX activity was observed in seedlings under NaCl compared to controls (Figure 16.1). Conversely, increasing PEG

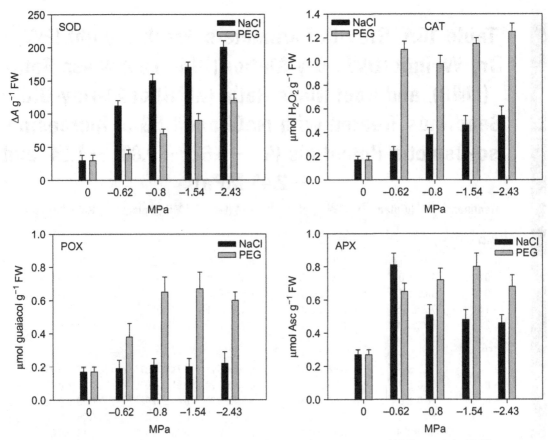

Figure 16.1 Effect of different NaCl and PEG treatments on superoxide dismutase (SOD), catalase (CAT), peroxidase (POX), and ascorbate peroxidase (APX) of 21-day-old quinoa seedlings. Values are means ± SD of four different experiments.

concentrations, the POX activity increased compared to control and NaCl-treated seedlings (Figure 16.1). APX activity, significantly higher in PEG-treated seedlings than in NaCl-treated seedlings and controls, was not dependent on PEG concentrations (Figure 16.1).

16.3.3 Total Antioxidant Activity, Carotenoids, Photosynthetic Pigments, and Proline Contents

Phenols increased significantly with increasing osmotic potential of NaCl and PEG with respect to controls (Table 16.2). The greatest increase was under NaCl compared to PEG at the

Table 16.2 Phenols and Total Antioxidant Capacity in Untreated and NaCl- and PEG-Treated Quinoa Seedlings

Treatments (MPa)	Phenols (mg g^{-1} DW)		Total Antioxidant Activity (μg ascorbic acid g^{-1} FW)	
	NaCl	PEG	NaCl	PEG
0	100 ± 5^e	100 ± 5^e	2.07 ± 0.2^e	2.07 ± 0.1^d
−0.62	150 ± 8^d	132 ± 7^d	2.99 ± 0.1^d	2.35 ± 0.1^c
−0.80	210 ± 10^c	170 ± 10^c	4.35 ± 0.3^c	3.41 ± 0.2^b
−1.54	318 ± 8^b	215 ± 8^b	5.66 ± 0.2^b	3.95 ± 0.3^{ab}
−2.43	358 ± 9^a	252 ± 11^a	6.05 ± 0.2^a	4.44 ± 0.4^a

Values are means $\pm$ SD of four different experiments. Within each column and for each measurement a different letter indicates that any difference between control and saline or PEG treatment is significant. Both analyses according to the Tukey's Studentized Range test ($P \leq 0.05$).

same iso-osmotic potential. NaCl also increased the total anti-oxidant capacity of seedlings in a concentration-dependent manner and much more than PEG (Table 16.2). Chlorophyll a and b slightly decreased under NaCl and the Chl a/b ratio slightly increased (Table 16.3). On the contrary, with increasing PEG concentration, the amount of chlorophyll a and b significantly lowered and the Chl a/b ratio was also lowered. No significant differences compared to controls in carotenoid content were observed in the presence of NaCl, instead proline content increased in a concentration-dependent manner. Carotenoids did not show significant differences with respect to controls, while proline increased with increasing PEG level, but to a lesser extent with NaCl at the same iso-osmotic potentials (Table 16.3).

16.3.4 Ion Content

In 21-day-old seedlings, we observed a different ion accumulation between shoot and root. More cations were accumulated in shoots, conversely more anions were in roots (Tables 16.4 and 16.5). In shoots, the total percentage of ions increased considerably in the presence of NaCl, but not in the presence of PEG. The cation more present in leaves of NaCl-treated seedlings was Na$^+$ (Table 16.4). K$^+$ was instead much higher in

Table 16.3 Chlorophyll a (Chl a), Chlorophyll b (Chl b), Chl a/b Ratio, Carotenoids, and Proline in Control, NaCl-, and PEG-Treated Seedlings

Treatments	Chl a (μg g^{-1} FW)	Chl b (μg g^{-1} FW)	Chl a/b	Carotenoids (μg g^{-1} FW)	Proline (μg g^{-1} FW)
0	33.4[b]	11.2[a]	2.98[a]	4.52[b]	0.37[d]
NaCl −0.62 MPa	34.8[a]	11.1[a]	3.13[a]	4.32[c]	0.51[c]
NaCl −0.80 MPa	35.2[a]	11.5[a]	3.06[a]	4.46[c]	0.67[b]
NaCl −1.54 MPa	31.0[b]	9.6[b]	3.20[a]	4.59[b]	0.77[a]
NaCl −2.43 MPa	29.3[b]	9.2[b]	3.18[a]	4.54[b]	0.79[a]
PEG −0.62 MPa	31.2[b]	9.8[b]	3.18[a]	4.59[b]	0.41[d]
PEG −0.80 MPa	24.4[c]	9.5[b]	2.61[b]	4.69[b]	0.45[d]
PEG −1.54 MPa	19.3[c]	7.4[c]	2.56[b]	4.79[ab]	0.49[c]
PEG −2.43 MPa	15.1[d]	5.9[d]	2.55[b]	4.89[a]	0.53[c]

Values are means of four different experiments. Within each column and for each measurement a different superscript letter indicates that any difference between control and saline or PEG treatment is significant. Both analyses according to the Tukey's Studentized Range test ($P \leq 0.05$).

shoots of untreated and PEG-treated seedlings. Mg^{2+} content did not change significantly in shoots of seedlings treated with PEG and NaCl compared to controls (Table 16.4), instead NH_4^+ increased in leaves of stressed seedlings and the greatest amount was under PEG at all concentrations (Table 16.4). Ca^{2+} content increased only in shoots in the presence of the highest concentrations of NaCl. Chloride and phosphate ions were undetectable in shoots of treated and untreated seedlings. Sulfate was accumulated in shoots of treated seedlings and mainly under PEG treatment (Table 16.4). In roots the greatest accumulation of Na^{2+} was detected in NaCl compared to untreated and PEG-treated seedlings. NH_4^+ and PO_4^{3-} were below the detection limit in roots of both treated and untreated seedlings (Table 16.5), K^+ content increased in the presence of NaCl, conversely no significant differences were observed between the control and the PEG-treated seedlings. Mg^{2+} content did not vary in roots of treated and untreated seedlings, Ca^{2+} levels with NaCl were lower than controls, but were similar to controls under PEG. Cl^- ions were accumulated only in roots of NaCl-treated seedlings and SO_4^{2-} was higher in roots of seedlings treated with PEG in comparison with NaCl and control (Table 16.5).

Table 16.4 Cation and Anion Content in Shoots of 21-Day-Old Seedlings of Quinoa Treated with NaCl and PEG at Increasing Iso-Osmotic Potential

Treatments	Cl^-	PO_4^{3-}	SO_4^{2-}	Na^+	NH_4^+	K^+	Mg^{2+}	Ca^{2+}	Total Ions	Total Cations	Total Anions
0	nd	nd	nd	0.324[e]	0.014[f]	3.223[c]	0.123[b]	0.0085[c]	3.692[d]	3.692[c]	0.00[d]
NaCl −0.62 MPa	nd	nd	0.133[e]	0.976[d]	0.022[e]	1.992[d]	0.125[b]	0.0099[c]	3.258[d]	3.125[d]	0.133[d]
NaCl −0.80 MPa	nd	nd	0.215[d]	2.451[c]	0.045[c]	1.331[d]	0.125[b]	0.0138[a]	4.180[c]	3.965[c]	0.215[c]
NaCl −1.54 MPa	nd	nd	0.216[d]	2.987[b]	0.049[c]	1.222[d]	0.123[b]	0.0132[a]	4.675[b]	4.459[b]	0.216[c]
NaCl −2.43 MPa	nd	nd	0.220[d]	4.150[a]	0.056[b]	1.012[e]	0.110[c]	0.0121[b]	5.570[a]	5.348[a]	0.220[c]
PEG −0.62 MPa	nd	nd	0.235[d]	0.344[e]	0.037[d]	3.332[b]	0.128[a]	0.0093[c]	4.086[c]	3.850[c]	0.235[c]
PEG −0.80 MPa	nd	nd	0.316[c]	0.332[e]	0.053[b]	3.435[a]	0.123[b]	0.0088[c]	4.268[c]	3.952[c]	0.316[b]
PEG −1.54 MPa	nd	nd	0.398[b]	0.321[e]	0.065[a]	3.243[c]	0.131[a]	0.0091[c]	4.167[c]	3.769[c]	0.398[a]
PEG −2.43 MPa	nd	nd	0.455[a]	0.311[e]	0.072[a]	3.329[b]	0.133[a]	0.0095[c]	3.855[d]	3.400[d]	0.455[a]

The ions are expressed as g g^{-1} DW. Values are means of four different experiments. Within each column and for each measurement a different superscript letter indicates that any difference between control and saline or PEG treatment is significant. Both analyses according to the Tukey's Studentized Range test ($P \leq 0.05$).

16.4 Discussion

At present, many regions are not available to agriculture because of soil salinity and drought that result in losses of cultivation areas. Utilization of halophytic plants is a well-ascertained option for the reclamation of saline lands but it is not a well-proved opportunity for recovering soil in drought conditions. In saline environments, the vulnerability of plants is determined not only by salt concentrations but also by salt type, so adaptation to salinity may be affected by ion toxicity and/or osmotic gradient while both are crucial for the establishment of species (Koyro and Eisa, 2008). During seedling establishment, abiotic stress modifies many biological processes

Table 16.5 Cation and Anion Content in Roots of 21-Day-Old Seedlings of Quinoa Treated with NaCl and PEG at Increasing Iso-Osmotic Potential

Treatments	Cl^-	PO_4^{3-}	SO_4^{2-}	Na^+	NH_4^+	K^+	Mg^{2+}	Ca^{2+}	Total Ions	Total Cations	Total Anions
0	0.00^d	nd	0.363^a	0.472^e	nd	1.033^c	0.093^a	0.056^a	2.017^e	1.654^c	0.363^d
NaCl −0.62 MPa	1.2^c	nd	0.061^c	0.765^d	nd	1.321^b	0.099^a	0.014^b	3.460^d	2.199^b	1.261^c
NaCl −0.80 MPa	1.5^b	nd	0.063^c	1.249^c	nd	1.45^a	0.107^a	0.014^b	4.383^c	2.820^a	1.563^b
NaCl −1.54 MPa	1.6^b	nd	0.065^c	1.367^b	nd	1.47^a	0.105^a	0.014^b	4.621^b	2.956^a	1.665^b
NaCl −2.43 MPa	1.9^a	nd	0.063^c	1.760^a	nd	1.020^c	0.089^b	0.010^b	4.842^a	2.879^a	1.963^a
PEG −0.62 MPa	0.00^d	nd	0.344^b	0.466^e	nd	1.041^c	0.097^a	0.059^a	2.007^e	1.663^c	0.344^e
PEG −0.80 MPa	0.00^d	nd	0.335^b	0.469^e	nd	1.037^c	0.099^a	0.061^a	2.001^e	1.666^c	0.335^e
PEG −1.54 MPa	0.00^d	nd	0.366^a	0.478^e	nd	1.040^c	0.094^a	0.059^a	2.037^e	1.671^c	0.366^d
PEG −2.43 MPa	0.00^d	nd	0.368^a	0.476^e	nd	1.036^c	0.095^a	0.058^a	2.033^e	1.665^c	0.368^d

The ions are expressed as g g^{-1} DW. Values are means of four different experiments. Within each column and for each measurement a different superscript letter indicates that any difference between control and saline or PEG treatment is significant. Both analyses according to the Tukey's Studentized Range test ($P \leq 0.05$).

such as growth, osmotic homeostasis, photosynthesis, carbon partitioning, carbohydrate and lipid metabolism, protein synthesis, and gene expression (Prado et al., 2000; Munns, 2002). Quinoa, because of its high-quality nutrient content as well as resistance to diverse abiotic stresses, has been suggested by the FAO as a future candidate in food security programs. In addition, the comparative results on seedling growth persistence to drought and salinity need to distinguish quinoa for cultivation in arid and/or saline regions. Data presented here evidence a good resistance to both stresses, even if it shows a more *advantageous* tolerance to salinity, both from a physiological and a metabolic point of view. The growth parameters during the experimental period showed a different trend in respect to treatment. Fresh and dry mass were mainly reduced by PEG compared to salinity at the same iso-osmotic potential, and both LMR and RMR, indicators of dry matter partitioning, were more affected by PEG, suggesting that this cultivar of quinoa is more sensitive to osmotic than ionic effects. In most plants, especially halophytes, the solute content of cells at high salinity is higher than in non-saline conditions, due largely to the accumulation of ions (e.g., Na^+ and Cl^-) and organic solutes. The higher salt than drought tolerance in quinoa, during seedling growth, may be explained by the existence of a significant gradient in the accumulation of potentially toxic (Na^+ and Cl^-) and nontoxic (K, Mg, Ca, P, and S) essential elements, useful for osmotic adjustment, and particularly by the different distributions of ions between shoots and roots of treated seedlings, as already demonstrated by Koyro and Eisa (2008). The data from the partitioning of ions between roots and shoots showed differences between treatments. Specifically, with NaCl, we observed a significant accumulation of Na^+, and no Cl^- in shoots. In accordance with previous investigations (Eisa et al., 2012), Na^+ was shown to be preferentially accumulated in shoots, thereby the plants avoid its excessive accumulation in the root tissues (Koyro, 2000; Ashraf et al., 2006). Considering the high energy cost of the de novo synthesis of organic osmolytes (Raven, 1985), we can suppose that the use of Na^+ also for osmotic adjustment enables seedlings to make efficient use of water under our experimental conditions. A further role in the maintenance of turgor was also attributed to Cl^- ions accumulated in roots (James et al., 2006). Our results showed that the Cl^- concentration was more than enough to contribute to osmotic adjustment maintaining root turgor as was previously demonstrated in seedlings of *Stylosanthes guianensis* by Veraplakorn et al. (2013). Under PEG the major amount of SO_4^{2-} and NH_4^+ could be the result of an accelerated protein catabolism generally activated in high-stress conditions. The deleterious effects of the accumulation of sulfate in PEG-treated seedlings were sufficient to cause metabolic disorders such as chlorophyll decrease and consequent seedling growth depression. Our results are consistent with previous observations (Reginato et al., 2014) showing that the deleterious effects of sulfate accumulation on leaf development, and on root and shoot elongation, are presumably a consequence of several metabolic reactions, for example, sulfide formation in the process of sulfate assimilation in the chloroplast, wherein sulfite reductase catalyzes the reduction of sulfite to sulfide using reduced ferredoxin as an electron donor (De Kok et al., 2005). Thus we can assume that the active osmotic adjustment by using ions and proline as osmolytes in NaCl-treated seedlings may be the key to the different resistances of this quinoa cultivar to the two abiotic stresses. The proline content is generally considered an adaptive strategy to salt stress; a previous study has demonstrated that *Prosopis strombulifera* cells synthesized high concentrations of compatible solutes, such as proline and pinitol, for charge balance and osmotic adjustment (Llanes et al., 2012). In addition, the tolerance of quinoa to salinity may also be associated with the different protective mechanisms activated by seedlings under NaCl and PEG. Under PEG we observed an increase in the enzymatic anti-oxidative system; however, in NaCl to prevent ROS damage seedlings have mainly developed a nonenzymatic antioxidant mechanism with a major production of phenols and a higher total antioxidant capacity (in terms of ascorbic acid). These results suggest that even high NaCl concentrations are not a negative stress factor for this species but may represent an inducer of valuable compounds with nutraceutical properties (Arts and Hollman, 2005). Phenolic compounds have been found to exert many biological functions (Pandey and Rizvi, 2009) and in particular the nutraceutical properties of dietary plants are commonly considered in terms of phenolic content. In addition, previous works have demonstrated an anticancer and antioxidant activity of *Chenopodium quinoa* leaf extract, due to the content of phenols.

The explanation for the better adaptation of quinoa titicaca to salinity than drought conditions seems to be related to: (i) a delicate balance between Na^+ accumulation (and its use for osmotic adjustment) compared to other cations, (ii) the ability of the plant to store Cl^- in roots as inorganic osmolyte, (iii) osmotic balance and protection by accumulation of compatible solutes such as proline, and (iv) an increase in enzymatic and nonenzymatic systems in particular in phenolic compounds and ascorbic acid content.

16.5 Conclusion

In conclusion our results evidence that the high adaptability to salinity of quinoa seedlings is a consequence of better metabolic control under NaCl compared to PEG, based mainly on a balanced accumulation of inorganic osmolytes (Na^+ and Cl^-), compatible organic solutes (proline), and an increase in nonenzymatic anti-oxidative system (phenolic compounds and ascorbic acid content). The present study helps to understand the response of *Chenopodium quinoa* to simulated drought and salinity stresses, evidencing that even if high salinity levels decrease the biomass of seedlings the concomitant increase in phenols, proline, and ascorbic acid in this leafy vegetable species used as food can compensate for the reduced yield.

References

Abugoch, L., Castro, E., Tapia, C., Anōn, M.C., Gajardo, P., Villarroel, A., 2009. Stability of quinoa four proteins (*Chenopodium quinoa* Willd.) during storage. Int. J. Food Sci. Technol. 44, 2013–2020.

Adolf, V.I., Shabala, S., Andersen, M.N., Razzaghi, F., Jacobsen, S.-E., 2012. Varietal differences of quinoa's tolerance to saline conditions. Plant Soil. 357, 117–129.

Arts, I.C.W., Hollman, P.C.H., 2005. Polyphenols and disease risk in epidemiologic studies. Am. J. Clin. Nutr. 81, 317–325.

Ashraf, M., Hameed, M., Arshad, M., Ashraf, Y., Akhtar, K., 2006. Salt tolerance of some potential forage grasses from Cholistan desert of Pakistan. In: Khan, M. A., Weber, D.J. (Eds.), Ecophysiology of High Salinity Tolerant Plants. Task Vegetation Science, vol. 40. Springer Verlag, Dordrecht, pp. 31–54.

Bartels, D., Sunkar, R., 2005. Drought and salt tolerance in plants. Crit. Rev. Plant Sci. 24, 23–58.

Bates, L.S., Waldren, R.P., Teare, I.D., 1973. Rapid determination of the free proline in water stress studies. Plant Soil. 38, 205–208.

Beaumont, F., Jouve, H.M., Gagnon, J., Gaillard, J., Pelmont, J., 1990. Purification and properties of a catalase from potato tubers (*Solanum tuberosum*). Plant Sci. 72, 19–26.

Buchanan, B.B., Gruissem, W., Jones, R.L., 2000. Subsequent effect on embryonic protein synthesis in barley. J. Plant Physiol. 136, 621–625.

Buhmann, A., Papenbrock, J., 2013. An economic point of view, secondary compounds in halophytes. Funct. Plant Biol. 40, 952–967.

De Kok, L.J., Castro, A., Durenkamp, M., Koralewska, A., Posthumus, F.S., Stuiver, C.E., et al., 2005. Pathways of plant sulfur uptake and metabolism—an overview. Landbauforschung VEolkenrode. 283, 5–13 (Special issue).

Eisa, S., Hussin, S., Geissler, N., Koyro, H.W., 2012. Effect of NaCl salinity on water relations, photosynthesis and chemical composition of Quinoa (*Chenopodium quinoa* Willd.) as a potential cash crop halophyte. Aust. J. Crop Sci. 6, 357–368.

FAO. 2012. Quinoa. Available at: <http://www.fao.org/docrep/t0646e/T0646E0f.htm>.

Gallie, D.R., 2013. L-ascorbic acid: a multifunctional molecule supporting plant growth and development. Scientifica 1, 1–24.

Gupta, S.A., Webb, R.P., Holaday, A.S., Allen, R.D., 1993. Overexpression of superoxide dismutase protects plants from oxidative stress. Plant Physiol. 103, 1067–1073.

Hariadi, Y., Marandon, K., Tian, Y., Jacobsen, S.E., Shabala, S., 2011. Ionic and osmotic relations in quinoa (*Chenopodium quinoa* Willd.) plants grown at various salinity levels. J. Exp. Bot. 62, 185–193.

Jacobsen, S.-E., Mujica, A., Jensen, C.R., 2003. The resistance of quinoa (*Chenopodium quinoa* Willd.) to adverse abiotic factors. Food Rev. Int. 19, 99–109.

Jacobsen, S.-E., Liu, F., Jensen, C.R., 2009. Does root-sourced ABA play a role for regulation of stomata under drought in quinoa (*Chenopodium quinoa* Willd.). Sci. Hortic. 122, 281–287.

Jacobsen, S.-E., Christiansen, J.L., Rasmussen, J., 2010. Weed harrowing and inter-row hoeing in organic grown quinoa (*Chenopodium quinoa* Willd.). Outlook Agric. 39, 223–227.

James, J.J., Alder, N.N., Muhling, K.H., Läuchli, A.E., Shackel, K.A., Donovan, L.A., et al., 2006. High apoplastic solute concentrations in leaves alter water relations of the halophytic shrub, *Sarcobatus vermiculatus*. J. Exp. Bot. 57, 139–147.

Koyro, H.W., 2000. Effect of high NaCl-salinity on plant growth, leaf morphology, and ion composition in leaf tissues of *Beta vulgaris* ssp. *maritima*. J. Appl. Bot. 74, 67–73.

Koyro, H.-W., Eisa, S.S., 2008. Effect of salinity on composition, viability and germination of seeds of *Chenopodium quinoa* Willd. Plant Soil. 302, 79–90.

Ksouri, R., Megdiche, W., Debez, A., Falleh, H., Grignon, C., Abdelly, C., 2007. Salinity effects on polyphenol content and antioxidant activities in leaves of the halophyte *Cakile maritima*. Plant Physiol. Biochem. 45, 244–249.

Lane, A., Jarvis, A., 2007. Changes in climate will modify the geography of crop suitability: agricultural biodiversity can help with adaptation. e-Journal. 4, 1–12.

Llanes, A., Bertazza, G., Palacio, G., Luna, V., 2012. Different sodium salts cause different solute accumulation in the halophyte *Prosopis stombulifera*. Plant Biol. 15, 118–125.

Mittler, R., 2006. Abiotic stress, the field environment and stress combination. Trends Plant Sci. 11, 15–19.

Munns, R., 2002. Comparative physiology of salt and water stress. Plant Cell Environ. 25, 239–250.

Nakano, Y., Asada, K., 1981. Hydrogen peroxide is scavenged by ascorbate-specific peroxidase in spinach chloroplasts. Plant Cell Physiol. 22, 867–880.

Panda, S.K., Singha, L.B., Khan, M.H., 2003. Does aluminium phytotoxicity induce oxidative stress in greengram (*Vigna radiate*)? Bulg. J. Plant Physiol. 29, 77–86.

Pandey, B.K., Rizvi, S.I., 2009. Plant polyphenols as dietary antioxidants in human health and disease. Oxid. Med. Cell. Longev. 2, 270–278.

Prado, F.E., Boero, C., Gallarodo, M., Gonzalez, J.A., 2000. Effect of NaCl on germination, growth and soluble sugar content in *Chenopodium quinoa* Willd seeds. Bot. Bull. Acad. Sin. 41, 27–34.

Prasad, K.N., Yang, B., Yang, S.Y., Chen, Y.L., Zhao, M.M., Ashraf, M., 2009. Identification of phenolic compounds and appraisal of antioxidant and anti-tyrosinase activities from litchi (*Litchi sinensis* Sonn.) seeds. Food Chem. 116, 1–7.

Prieto, P., Pineda, M., Aguilar, M., 1999. Spectrophotometric quantization of antioxidant capacity through the formation of a phosphomolybdenum complex: specific application to the determination of Vitamin E. Anal. Biochem. 269, 337–341.

Raven, J.A., 1985. Regulation of pH and generation of osmolality in vascular plants: a cost-benefit analysis in relation to efficiency of use of energy, nitrogen and water. Phytologist. 101, 25–77.

Reginato, M.A., Castagna, A., Furlán, A., Stella Castro, S., Ranieri, A., Luna, V., 2014. Physiological responses of a halophytic shrub to salt stress by Na_2SO_4 and NaCl: oxidative damage and the role of polyphenols in antioxidant protection. Ann. Bot. Plants. 6, plu042.

Scalbert, A., Manach, C., Morand, C., Rémésy, C., 2005. Dietary polyphenols and the prevention of diseases. Crit. Rev. Food Sci. Nutr. 45, 287–306.

Singleton, V.L., Rossi, J.A., 1965. Colorimetry of total phenolics with phosphomolybdic-phosphotungstic acid reagents. Am. J. Enol. Viticult. 16, 144–158.

Trognitz, B.R., 2003. Prospects of breeding quinoa for tolerance to abiotic stress. Food Rev. Int. 19, 129–137.

Veraplakorn, V., Nanakorn, M., Kaveeta, L., Suwanwong, S., Bennett, I.J., 2013. Variation in ion accumulation as a measure of salt tolerance in seedling and callus of *Stylosanthes guianensis*. Theor. Exp. Plant Physiol. 25, 106–115.

Wilson, C., Read, J.J., Abo-Kassem, E., 2002. Effect of mixed-salt salinity on growth and ion relations of a quinoa and a wheat variety. J. Plant Nutr. 25, 2689–2704.

17

GERMINATION ECO-PHYSIOLOGY AND PLANT DIVERSITY IN HALOPHYTES OF SUNDARBAN MANGROVE FOREST IN BANGLADESH

A.K.M. Nazrul Islam

Ecology Laboratory, Department of Botany, University of Dhaka, Dhaka, Bangladesh

17.1 Introduction

There are 16 million hectares of halophytic mangrove forest in the world but these are only 1% of the total area of tropical forests and form important ecosystems along the tropical coasts of the world. Halophytes of mangrove ecosystems have a wide diversity of plant species (Field, 1996). The Sundarban mangrove forest of Bangladesh is one of the single largest tracts in the world and is situated in the southwest corner of the country (Agro-ecological region 13; FAO, 1988; Figure 17.1); between 21°.30′ and 22°.30′N latitude; and 89° to 90°E longitude. The total forest area is 577,220 ha (Figure 17.2); of these, river channels and other watercourses consist of 175,600 ha. Plant communities occupy approximately 401,600 ha (Chaffey et al., 1985). The low-lying areas are occupied by a network of small creeks and the riverbanks are dominated by *Nypa fruticans*, *Acanthus ilicifolius*, *Phragmites karka*, *Porteresia coarctata*, *Typha elephantina*, and *Sonneratia apetala*. Soil physico-chemical properties and plant community types of various ecological zones were evaluated (Nazrul-Islam, 1994, 1995). The influence of salinity on germination response was studied in halophytes and nonhalophytes (glycophytes) by Rozema (1975).

M.A. Khan, M. Ozturk, B. Gul, & M.Z. Ahmed (Eds): Halophytes for Food Security in Dry Lands.
DOI: http://dx.doi.org/10.1016/B978-0-12-801854-5.00017-0

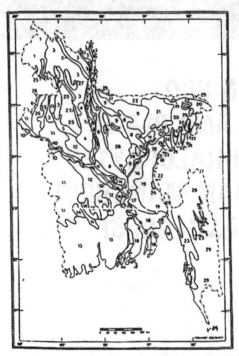

Figure 17.1 Agro-ecological regions. 1. Old Himalayan Piedmont Plain, 2. Active Tista Floodplain, 3. Tista Mander Floodplain, 4. Karatoya-Bangali Floodplain, 5. Lower Atrai Basin, 6. Lower Purnabhaba Floodplain, 7. Active Brahmaputra-Jamuna Floodplain, 8. Young Brahmaputra and Jamuna Floodplains, 9. Old Brahmaputra Floodplain, 10. Active Ganges Floodplain, 11. High Ganges River Floodplain, 12. Low Ganges River Floodplain, 13. Ganges Tidal Floodplain, 14. Gopalganj-Khulna Bils, 15. Arial Bil, 16. Middle Meghna River Floodplain, 17. Lower Meghna River Floodplain, 18. Young Meghna Estuarine Floodplain, 19. Old Meghna Estuarine Floodplain, 20. Eastern Surma-Kusiyara Floodplain, 21. Sylhet Basin, 22. Northern and Eastern Piedmont Plains, 23. Chittagong Coastal Plain, 24. St. Martin's Coral Island, 25. Level Barind Tract, 26. High Barind Tract, 27. North-Eastern Barind Tract, 28. Madhupur Tract, 29. Northern and Eastern Hills, 30. Akhaura Terrace.

However, inhibition of germination by salt could play a decisive role in limiting the geographical distribution of a species and determining their position in the vegetation.

For optimal growth, *Atriplex halimus* and other halophilic species of *Atriplex* require relatively high concentrations of sodium chloride in the soil or culture medium (Black, 1956; Greenway, 1968; Brownell, 1965) indicating that small quantities of sodium are required for the growth of some *Atriplex* species. Comparison and distribution of plant species and species diversity and changes in soil, water and plant diversity of different ecological zones were also reported (Nazrul-Islam, 1994, 1995, 2003). Germination is an important factor for seedling development and establishment in mangrove ecosystems and salt is

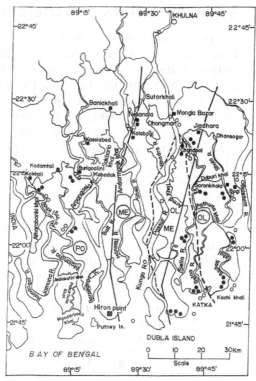

Figure 17.2 Map of Sundarbans mangrove forest showing various rivers and locations from where soil samples (solid circles) and water samples (open circles) were collected. Solid squares indicate the places where administrative offices are located within the forest. The whole forest is divided into three ecological zones (Walter, 1968; vertical lines) on the basis of salinity; OL, Oligohaline; ME, Mesohaline, and PO, Polyhaline zones; OL & ME = Oligo—Mesohaline zone demarcated by dashed lines which becomes Oligohaline during monsoon season when there is enough freshwater supply from the Ganges through Baleshwar River (extreme east) and becomes Mesohaline during winter when water in the Ganges River is diverted through the Farakka Barage.

a regulatory factor for survival, distribution, growth, reproduction, and zonation of mangroves.

Patterns of species diversity were studied by MacArthur (1965) and Laxton (1978). Edge diversity and cover type diversity for breeding avian communities in freshwater coastal marshes were evaluated by Harris et al. (1983). The diversity within habitat types and the amount of turnover in the species composition between the habitat types were described by Routledge (1977, 1979). In the present paper, the germination ecology of *Xylocarpus mekongensis* and *Heritiera fomes* and plant diversity analysis are discussed.

17.2 Materials and Methods

Seeds of *X. mekongensis* were collected from the Sundarban Forest in 2004. The germination experiment was done in plastic dishes (diameter 40 cm and depth 15 cm) filled with sand; with four salinity treatments, i.e., 0, 5, 10, and 15 ppt; ten seeds were sown in each dish and there were four replicates. The experiment continued for 10 weeks. In the case of *H. fomes*, seeds were placed naturally by tide on the sand surface in the Oligohaline zone. Seedlings of *S. apetala* were planted in the plot of the Oligohaline zone; establishment and growth were recorded during the field visit.

For diversity studies the sampling plot was selected 10 m away from the riverbank. A circular plot of 11 m radius was laid out. At each of the four cardinal points (north, south, east, and west), at a distance of 11.0 m from the plot point center, a 2.0 m circular plot was also prepared. Species were counted in both 2.0 m circular plots (including seedlings) and 11.0 m circular (trees and seedlings up to breast height only) quadrats to provide maximum habitat diversity (Figure 17.3). The modified method of Hurlbert (1971) was used to produce an unbiased estimate of species diversity from the following formula:

$$E(S) = \sum \{1 - [(N - {}_nN_i)/(N_n)]\}$$

where $E(S)$ = the expected number of species in the rarefied sample; n = the standardized sample size; N = the total number of individuals recorded in the sample to be rarefied; N_i = the number of individuals in the *i*th species recorded in the sample to be rarefied.

In order to fit the geometric series it is necessary to begin by estimating the constant K and this was done by iterating the equation of Magurran (1988) and Pielou (1983).

17.3 Results and Discussion

The germination of *X. mekongensis* showed an increasing tendency at 5 ppt NaCl salinity and declined with the further increases in salinity up to 15 ppt. This may have been due to the combined effects of osmotic pressure and toxicity of the salt (Bernstein, 1962). The response of salt to halophytic and glyco-phytic *Juncus* species was studied by Rozema (1976) and it was found that with respect to the degree of growth reduction, *Juncus geradii* (halophyte) appeared to be more salt-tolerant than *Juncus bufonius* (glycophyte). *Heritiera fomes* showed more than 80% germination under natural conditions with healthy seedlings. At the end of April all the seedlings died because of a lack of freshwater supply (Figure 17.4A and B) from the upstream Ganges River due to the Farakka barrage.

Plot layout

Distance between plots   100.0 m
Radius of large circle   11.0 m
Radius of small circle   2.0 m
The centre of plot B is the site for the
Sampling point or Gradsect.

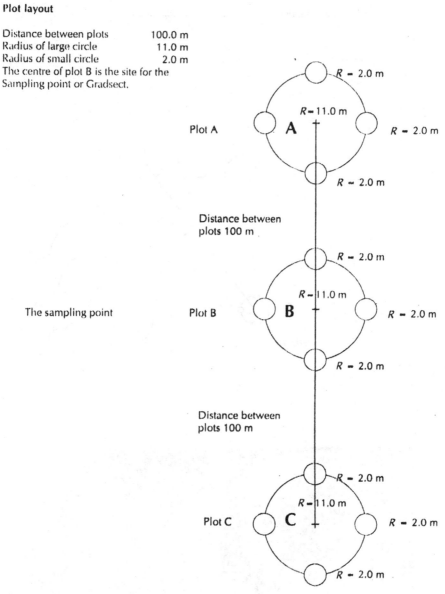

Figure 17.3 Procedure for the measurements of plant density from quadrats.

Forest Division: ....

Range: ....

Beat: ....

Sampling Point or Gradsect Number: ....

Location: Longitude/Latitude: ....

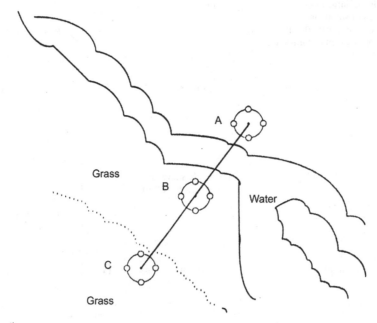

Figure 17.3 (Continued)

Figure 17.4 *Heritiera fomes* under natural habitat. (A) Healthy plants (B) Dead plants.

17.3.1 Nature of the Vegetation

The vegetation of the Sundarbans can be classified in two ways:
1. Vegetation of clearing spaces: This type of vegetation is present in the Oligohaline zone and the dominant plant species are *Pandanus foetidus*, *Flagellaria indica*, and *Achrostium aureum*, along with *Sphaeranthus africanus* and some of the other species such as *Dalbergia spinosa* and *Dalbergia candenatensis*.
2. Vegetation of the forest proper: vegetation appears undisturbed on the edge of the forest, that is the river bank, and is occupied by *S. apetala* and *N. fruticans*; and in the eastern side of the Oligohaline zone *H. fomes* occupies major areas mixed with *Excoecaria agallocha*, *Avicennia officinalis*, *Hibiscus tiliaceous*, and *Cynometra ramiflora*. Inside the forest the open places are occupied by *A. aureum*; the western side is dominated by *Bruguiera sexangula*, *E. agallocha*, *Rhizophora apiculata*, *Rhizophora mucronata* and to the extreme west mainly two species, *Ceriops decandra* and *Aegiceras corniculatum*, are found to form consociation.

The plant communities in the Sundarban mangrove forest vary in the three distinct ecological zones known as the Oligohaline, Mesohaline, and Polyhaline zones (Table 17.1). The types of plant communities of these zones are:
1. Oligohaline Zone
   a. *Heritiera fomes–C. ramiflora–H. tiliaceous*
   b. *Heritiera fomes–Avicennia alba–Barringtonia racemosa*
   c. *Heritiera fomes–E. agallocha–Amoora cucullata*
   d. *Rhizophora mucronata–Cerbera manghas–Bruguiera gymnorrhiza*
   e. *Acrostichum aureum* in open places of highland only
2. Mesohaline Zone
   a. *Excoecaria agallocha–Heritiera fomes–A. officinalis*
   b. *Sonneratia caseolaris–Tamarindus indica–B. gymnorrhiza*
   c. *Sonneratia apetala–N. fruticans–R. apiculata*
   d. *Porteresia coarctata–P. karka–A. ilicifolius* (in canal and river edges)
   e. *Xylocarpus granatum–B. sexangula–Brownlowia tersa*
3. Polyhaline Zone
   a. *Ceriops decandra* (forming consociation)
   b. *Aegiceras corniculatum* (forming consociation)
   c. *Ceriops decandra–B. gymnorrhiza–X. mekongensis*
   d. *Kandelia candel–Phoenix paludosa–A. aureum*
   e. *Phoenix paludosa–A. ilicifolius.*

Table 17.1 Dominant Vegetation of Three Ecological Zones

Oligohaline Zone	Mesohaline Zone	Polyhaline Zone
Heritiera fomes (Ab)	Excoecaria agallocha (Ab)	Aegiceras corniculatum (Ab)
Excoecaria agallocha (Ab)	Rhizophora mucronata (Ab)	Ceriops decandra (Ab)
Porteresia coarctata (Ab)	Rhizophora apiculata (Ab)	Bruguiera gymnorrhiza (Ab)
Hibiscus tiliaceous (F)	Heritiera fomes (Ab)	Xylocarpus granatum (F)
Avicennia alba (F)	(Shows top dying)	Kandelia candel (F)
Avicennia officinalis (F)	Bruguiera sexangula (F)	Phoenix paludosa (F)
Sonneratia apetala (F)	Sonneratia apetala (F)	Acanthus ilicifolius (F)
Bruguiera gymnorrhiza (F)	Porteresia coarctata (Oc)	Acrostichum aureum (F)
Sonneratia caseolaris (F)	Dalbergia spinosa (Oc)	Xylocarpus mekongensis (F)
Amoora cucullata (F)	Petunga roxburghii (R)	
Barringtonia racemosa (F)	Pandanus foetidus (R)	
Acanthus ilicifolius (F)	Sarcolobus globosus (R)	
Cynometra ramiflora (Oc)	Mucuna gigantea (VR)	
Cerbera manghans (Oc)	Bruguiera gymnorrhiza (F)	
Pandanus foetidus (Oc)		
Phragmites karka (Oc)		
Nypa fruticans (Oc)		
Flagellaria indica (R)		
Mucuna gigantea (VR)		
Ceriops decandra (Oc) dwarf		
Sphaeranthus africanus (Ab) in cleared sites		

Soil salinity (Table 17.2) in the Polyhaline zone was more than twice that in the Oligohaline zone, which is also reflected in the vegetation pattern (Table 17.1). In the Polyhaline zone freshwater supply from upstream is limited, hence the salinity value is high. Oxidation–reduction potentials (redox potentials) during different months are given in Table 17.3. In some places the values were very low, where top dying of *H. fomes* was observed. In these places, tide water reaches only during high tide and most of the time the conditions remain waterlogged and the low redox potential (+30 mv) is possibly one of the factors which are responsible for the top dying of this plant species. Armstrong and Boatman (1967) measured oxygen diffusion rate (ODR) in a small valley bog flushed by a small stream supporting the healthy growth of *Molinia* sp.; along the sides of this flush, where the surface flow was much

Table 17.2 Soil Salinity (0–10 cm depth) Micromhos cm^{-2} at Different Locations in Three Ecological Zones

Ecological Zone Oligohaline Location		Mesohaline	Polyhaline	
Bogi 3200	Supati 3450	Chandpai 4610	Kassiabad 6480	Burigoalini 6900

Table 17.3 Oxidation-Reduction Potentials (mV) in Soils Expressed as E6

Zone	Location	January	April	July
Oligohaline	Sharankhola Range	+325	+290	+325
Oligo—mesohaline	Chandpai Range	+168	+100	+90
Mesohaline	Khulna Range	+100	+80	+30

reduced, the *Molinia* sp. was considerably smaller and beyond this edge the plants were very purple and extremely stunted. ODR was very high in the flushed site and more or less zero at the edge. The authors suggested that aeration (high redox potential value) was responsible for healthy growth of *Molinia* sp. and poor growth was due to lack of aeration (low redox potential).

Values of indices of plant diversity are given in Table 17.4 (Kempton and Taylor, 1978). For comparison of diversity, values of deciduous forest are also shown. The results show that diversity is higher in halophyte mangrove forest than the deciduous forest. Soil of deciduous forest (dominated by *Shorea robusta*) is acidic and supports only acid-tolerant plant species. The species richness components are even trickier to estimate. To estimate species richness alone it is necessary to postulate some form of species abundance distribution (e.g., Pielou, 1969).

Table 17.4 Values of Indices of Plant Diversity for Mangrove and Deciduous Forests

Index	Formula	Mangrove	Deciduous
Simpson's (Inverse)	$\dfrac{1}{\sum\limits_i^s \rho_i^2}$	81.2	30.2
Shannon–Wiener (Exponent)	$\exp\left(-\sum\limits_i^s \rho_i \log_2 \rho_i\right)$	147.5	64.2
Q Statistic	$\dfrac{\frac{1}{2}S}{\log R_2/R_1}$	98.6	34.5
	Log normal gamma	98.0	38.2
	Log series ($\propto$)	101.0	27.2
Whittaker's	$\dfrac{S}{\sqrt[4]{\left[\left(\sum\limits_i^s (\log \rho_i - \log \rho)^2\right)\right]}}$	75.1	26.8
Log cycle	$\dfrac{S}{\log \rho_l - \log \rho_s}$	78.2	20.3

Values of indices of plant diversity were related to the diversity of vegetation types and also with salinity. Harris et al. (1983) have shown that the Shannon–Wiener information theory formula for various marshes the bird diversity was related to vegetation and also mosaic arrangement and distribution of cover type.

The abundance of species ranked from most to least abundant (geometric series, May, 1975) was also calculated as;

$$n_i = NC_K K(1-K)^{i-1} \quad C_K = [1-(1-K)^S]^{-1}$$

where K is a constant which ensures that $\Sigma n_i = N$; ecological diversity was calculated according to Magurran (1988). Common and rare species were also calculated and are shown in Figures 17.5 and 17.6. There were about 18 species and this was also fitted to log normal distribution and the hidden portion is the amount of rare species.

With K estimated as 0.1857 it is now possible to obtain the values of C_K

$$C_K = [1-(1-K)^S]^{-1}$$
$$= 1.020594802$$

The expected number of individuals for each of the 19 species was calculated. Thus for the most abundant species

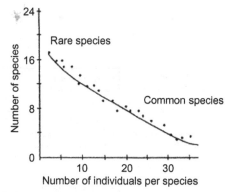

Figure 17.5 Relative abundance of plant species.

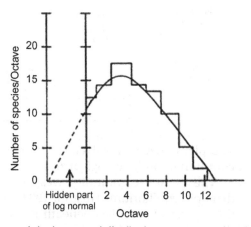

Figure 17.6 Species abundances and the log normal distribution are truncated at the point where species are represented by a single individual.

$$N_i = NC_K K(1 - K)^{i-1}$$
$$= 40.37$$

The total number of individuals is N; and when the proportion of "K" is taken from the total sample; and from the remaining sample, when the proportion "K" is taken up to Sth species, the mathematical equations takes the following form.

For the Sth species;

$$n_s = NK(1-K)^{S-1}$$
$$\Sigma n_s = N \cdot C_K$$

Species involved in various succession patterns and reaching the process of climax vegetation are given in Figure 17.7.

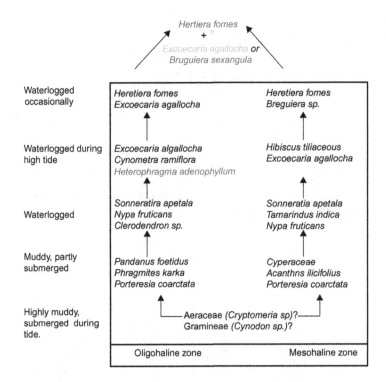

Figure 17.7 Species involved in the succession pattern of the Oligohaline and Mesohaline zones.

The plant species involved in the beginning of succession are common in both ecological zones.

The climax community was considered to be a state of dynamic equilibrium (Huston, 1994). The overstory may inhibit the growth of juveniles beneath it (juveniles of its own kind or those of any species) and in such cases local, cyclic succession will occur when the overstory plant dies. The succession in Sundarban mangrove forest follows the interpretation as discussed by Roberts and Richardson (1985) and takes five pattern changes from herbs to early dominance followed by persistence, progressive recruitment and then climax vegetation.

References

Armstrong, W., Boatman, D.J., 1967. Some field observations relating to the growth of bog plants. J. Ecol. 55, 101–110.

Bernstein, L., 1962. Salt affected soils and plants. In: Proceedings of the Paris Symposium; Arid Zone Research vol. 18. UNESCO, Paris, pp. 139–174.

Black, R., 1956. Effects of NaCl in water culture on the ion uptake and growth of *Atriplex hastata*. Aust. J. Biol. Sci. 9, 67–80.

Brownell, P.F., 1965. Sodium as an essential micronutrient element for a higher plant (*Atriplex vesilaria*). Plant Physiol. 40, 460–468.

Chaffey, D.R., Miller, F.R., Sandom, J.H., 1985. A Forest Inventory of Sundarbans. Forest Department, Bangladesh.

FAO, 1988. Agroecological Regions, vol. 2. UNDP, Rome.

Field, C., 1996. Restoration of Mangrove Ecosystems. ISME, Okinawa, Japan.

Greenway, H., 1968. Growth stimulation by high chloride concentrations in halophytes. Isr. J. Bot. 17, 169–177.

Harris, H.J., Milligan, M.S., Fewless, G.A., 1983. Diversity: quantification and ecological evaluation in freshwater marshes. Biol. Conserv. 27, 99–110.

Hurlbert, S.H., 1971. The non-concept of species diversity: a critique and alternative parameters. Ecology. 52, 577–586.

Huston, M.A., 1994. Biological Diversity: The Coexistence of Species in Changing Landscapes. Cambridge University Press, Cambridge, England.

Kempton, R.A., Taylor, l.R., 1978. The Q statistic and diversity of floras. Nature. 275, 252–253.

Laxton, R.R., 1978. The measure of diversity. J. Theor. Biol. 70, 51–67.

MacArthur, R., 1965. Patterns of species diversity. Biol. Rev. 40, 510–533.

Magurran, A.E., 1988. Ecological Diversity and Its Measurements. Chapman and Hall, London.

May, R.M., 1975. Patterns of species abundance and diversity. In: Code, M.L., Diamokd, J.M. (Eds.), Ecology and Evolution of Communities. Harvard University Press, Cambridge, MA, pp. 81–120.

Nazrul-Islam, A.K.M., 1994. Environment and vegetation of Sundarban mangrove forest. In: Leith, H. (Ed.), Towards Rational Use of High Salinity Tolerant Plants, vol. 1. Kluwer Acad. Publ., Dordrecht, pp. 81–88.

Nazrul-Islam, A.K.M., 1995. Ecological conditions and species diversity in Sundarban mangrove forest community, Bangladesh. In: Khan, M.A., Unger, I.A. (Eds.), Biology of Salt Tolerant Plants. Book Grafters, Chelsea, MI, pp. 294–305.

Nazrul-Islam, A.K.M., 2003. Mangrove forest ecology of Sundarbans: the study of change in water, soil and plant diversity. In: Ghosh, A.K., Ghosh, J.K., Mukhopadhaya, M.K. (Eds.), Sustainable Environment: A Statistical Analysis. Oxford University Press, New Delhi, pp. 126–147.

Pielou, E.C., 1969. An Introduction to Mathematical Ecology. Wiley, New York, NY.

Pielou, E.C., 1983. Population and Community Ecology: Principles and Methods. Ecological Diversity. Gordan and Breach Science, London, pp. 289–315.

Roberts, M.R., Richardson, C.J., 1985. Forty one years of population change and community succession in aspen forests on four soil types, northern lower Michigan, USA. Can. J. Bot. 63, 1641–1651.

Routledge, R.D., 1977. On Whittaker's components of diversity. Ecology. 58, 1120–1127.

Routledge, R.D., 1979. Diversity indices: which ones are admissible? J. Theor. Biol. 76, 503–515.

Rozema, J., 1975. The influence of salinity, inundation and temperature on the germination of halophytes and non-halophytes. Oecol. Plant. 10, 341–353.

Rozema, J., 1976. An ecophysiological study on the response to salt of four halophytic and glycophytic *Juncus* species. Flora. 165, 197–209.

Walter, H., 1968. Die Vegetation der Erde in Oiko-Physiologischen Betrachtung. Band II. Fischer Verlag, Stuttgart.

18

HALOPHYTIC PLANT DIVERSITY OF UNIQUE HABITATS IN TURKEY: SALT MINE CAVES OF ÇANKIRI AND IğDIR

Münir Öztürk[1], Volkan Altay[2], Ernaz Altundağ[3] and Salih Gücel[4]

[1]Botany Department, Science Faculty, Ege University, Bornova-Izmir, Turkey
[2]Biology Department, Science and Arts Faculty, Mustafa Kemal University, Antakya-Hatay, Turkey [3]Biology Department, Science and Arts Faculty, Duzce University, Duzce, Turkey [4]Institute of Environmental Sciences, Near East University, Lefkoşa, Northern Cyprus

18.1 Introduction

There are approximately 955 million hectares of salt-affected land in arid and semi-arid climatic zones. Almost 10% of the world's total land surface is salt-affected in different ways and about 20 million ha are no longer productive due to this (Malcolm, 1993). One-third of the existing irrigated areas are also affected by salt and are not good for plant production. According to Qadir et al. (2014) the irrigated land covers around 310 million ha globally, and an estimated 20% of it is salt-affected (62 million ha). Salt degradation is visible in areas of dry irrigated land with little rainfall. Recent studies have revealed that every day an average of 2000 ha of irrigated land in arid and semi-arid areas across 75 countries is degraded by salt (Qadir et al., 2014). The well-known salt-degraded land areas include: Aral Sea basin, Central Asia; Indo-Gangetic basin, India; Indus basin, Pakistan; Yellow River basin, China; Euphrates basin, Syria and Iraq; Murray-Darling basin, Australia; and San Joaquin Valley, United States (Qadir et al., 2014).

Major demographic explosions are expected in the semi-arid and arid regions of the world where salinity is very common

M.A. Khan, M. Ozturk, B. Gul, & M.Z. Ahmed (Eds): Halophytes for Food Security in Dry Lands.
DOI: http://dx.doi.org/10.1016/B978-0-12-801854-5.00018-2

and this poses a serious threat to agriculture and food security (Temel and Şimşek, 2011; Qadir et al., 2014). The saline areas of the world are increasing daily and there is currently insufficient information to overcome the problems originating from this (Hasanuzzaman et al., 2014).

Halophytes are plants that grow all over the world, in different climatic regions in soils with high salinity levels (salt marshes, terrestrial saline soils, saline deserts, arid lands, and tidal areas and sand dunes) (Akbaş et al., 1999; Adıguzel et al., 2005; Guvensen et al., 2006; Tuğ et al., 2011; Altay and Ozturk, 2012). They can complete their lifecycle at places with salt concentrations of at least 200 mM NaCl (Flowers et al., 1986) or on areas with salt concentrations greater than 0.5% (Khan and Duke, 2001). They constitute approximately 1% of all flora, although under high salinity levels these plants are capable of completing their lifecycle easily and are thus also regarded as plant indicators adapted to special ecological conditions (Stuart et al., 2012; Hasanuzzaman et al., 2014).

Approximately 2500—3000 halophytes belonging to 117 families and 550 genera are distributed throughout the world (Aronson, 1984, 1989; Khan and Duke, 2001). Nearly 700 taxa from these are distributed in the Mediterranean basin (Choukr-Allah, 1991; Chedlly et al., 2008). A total of 307 taxa were identified during a study on the halophytic taxa in Turkey (Birand, 1960, 1961; Atamov et al., 2006; Aydoğdu et al., 2002, 2004; Hamzaoğlu and Aksoy, 2006, 2009; Ozturk et al., 2014).

The halophytic taxa are also observed to grow around salt mines. In Turkey there are large salt deposits belonging to the third period of the Eocene, Oligocene, and Miocene. These originated from these sedimentations in the inland seas subjected to extensive drought and evaporation. The rock salt deposits are mainly concentrated in the northeastern and central Anatolian regions (Barutoğlu, 1961). There are more than 30 rock salt caves distributed in the country extending from Çankırı to Iran through Corum, Yozgat, Sivas, Erzincan, Erzurum, and Kars. Some large underground salt formations have also been reported from the basin in Adana and Siirt (MTA, 2007). The northeast Anatolian deposits are found at Kağızman, Tuzluca-Kulp, east of Erzurum, Oltu, Sağırkaya, with the largest of these being the deposits found in Tuzluca Kağızman; the central Anatolian deposits are scattered mostly on or near the line between Çankiri and Kırşehir. These are recorded as "Cayan," "Sekili," "Tepesidelik," and "Hacıbektaş-Tuzköy-Gülşehir." Other deposits are found in the east of central Anatolia, around Manastıraltı, in the vicinity of Sivas-Erzincan

(Barutoğlu, 1961). The areas encircling these deposits abound in saline habitats covered by halophytes.

In this study halophytic plant taxa were distributed around the Çan-Kaya Salt Mine and Potuk Salt Mine caves near Çankiri in central Anatolia and Tuzluca Salt Mine cave area located in Igdir in East Anatolia. This study attempts to evaluate some ecological characteristics of these halophytic taxa (life forms, ecotypes, and chorological features), together with the existing potential as food, fodder, and other economic uses, as well as for medicinal purposes.

18.2 Study Areas

18.2.1 Çankırı

Çankırı is located in central Anatolia near Çorum; which lies on its eastern side; between north of Kızılırmak and the west Black Sea main basin, situated between 40° 36" 410 north latitude and 32° 30" and 340 east longitude (Figure 18.1). Alluvial soils suitable for all kinds of agriculture in the lowlands are available. The limestone, clay, marl, gypsum, and rock salt beds are available in thick layers under the soil of central districts. In particular, the hilly tracts on the east of the city are rich in rock salt reserves, and pose a degraded appearance (PDEF, 2010; MTA, 1978, 2010; Kırıkoğlu, 2012). In the higher parts of this area groundwater is not present in sufficient quantities in the red-colored formations, whereas highly saline groundwater is observed in gypsum formation. We generally come across brown steppe soils, chestnut-colored steppe soils, and saline soils in the region, and alluvial soils are found alongside the streams. The plant cover is sparse on saline lands in the east.

Due to the continental climate, summers in Çankırı are hot and dry and winters are cold and harsh. The average temperature is 11.2°C, with the mean maximum being 17.9°C and the average minimum 4.8°C. The hottest months are July and August, while the coldest months are January and February. Maximum rainfall is recorded in May (53.6 mm m^{-2}), and the least precipitation in September (16.4 mm m^{-2}). Total annual rainfall is 394 mm m^{-2} (DMIGM, 2009).

The vegetation in Çankırı province changes with elevation and consequently the climate (Mutlu, 2006; Ertuğrul, 2011). In the northern parts the winds coming from the Black Sea form a microclimate of eurosiberian character, therefore we come across *Pinus* L., *Abies* Miller, and *Juniperus* L. forests and woods

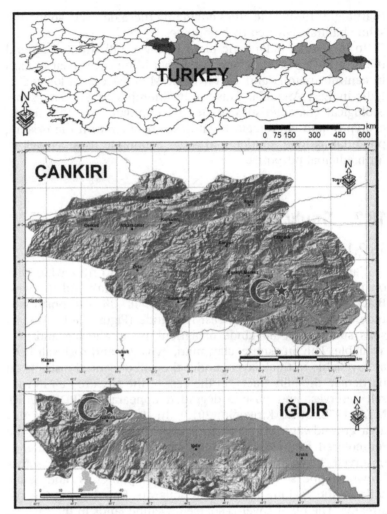

Figure 18.1 Map showing the study areas.

at higher altitudes. In the south and southwest of the province, steppes dominate large areas. Along the rivers the most common trees are *Populus* L. and *Salix* L.

Some work has been done on the floral diversity of Çankırı (Sağıroğlu, 1998; Mutlu, 2006; Ertuğrul, 2011). Approximately 1200 native plant taxa are distributed in Çankırı, and more than 180 are endemics.

Çankırı was known as Duduşna in the Hittite period, Germanikopolis and Gangra in the Byzantine era, and Kengr and Çankırı during the 1900s. The Byzantine era name "Germanikopolis" is currently found as an epithet to many of the plants such as; *Centaurea germanicopolitana* Bornm., *Gypsophila germanicopolitana* Hub.-Mor., *Onobrychis germanico-politana* Hub.-Mor. & Simon, *Tanacetum germanicopolitanum* (Bornm. & Heimerl) Grierson, *Sideritis germanicopolitana* Bornm., *Helianthemum germanicopolitanum* Bornm., *Reseda germanicopo-litana* Hub.-Mor. Inan et al. (2012) depicted the distribution of these taxa in the province. In addition to these taxa, six endemic plants show distribution in Çankırı namely: *T. germanicopolita-num*, *G. germanicopolitana*, *Gypsophila simonii* Hub.-Mor., *H. germanicopolitanum*, *Astragalus barbarae* Bornm., and *O. ger-manicopolitana*. All are regarded as the specific local endemics of Çankırı. Different investigators have defined ten new taxa apart from these six (*Acantholimon lycaonicum* Boiss. & Heldr. ssp. *Cappadocicum* Dogan & Akaydın, *Alyssum nezaketiae* Aytac & H. Duman, *Viola alba* Besser ssp. *alba*, *Genista vuralii* A. Duran & H. Dural, *Astragalus fallacinus* Podlechand *Astragalus Rausianus* Podlech & Ekici, *Centaurea cankiriense* A. Duran & H. Duman, *Erysimum jacquemoudii* Yild and *Erysimum Yaltırıkii* Yıld., *Galium cankiriensis* Yild.) taxa. This brings the total of local endemic taxa to 16 (Inan et al., 2012).

In our study site the Çan-Kaya rock salt mine area and its environs in Çankırı, we find the largest rock salt reserves of Turkey. The salt cave is located 20 km away from the central district. It has been mined all through the last 5000 years since the Hittite period. The mine was operated by the state, but was privatized after this date. The mineral reserve is 500 million tonnes and the daily pro-duction capacity is 500 tonnes.

This salt cave has several galleries and serves as the most important tourism attraction for Çankırı; it is 8 km in length. It has been turned into a treatment center for asthma and because of this it serves as a health tourism center as well.

18.2.2 IĞDIR

This province is located in the Eastern Anatolian Region around Erzurum-Kars (Figure 18.1), with an area of 3539 km^2 (Anonymous, 2008a). It is one of the highest Eastern Anatolian plateaus located near the famous glacier-covered Mount Ararat (5165 m above sea level), average altitude varies between 800–900 m. However, in the Sürmeli (Aras) pit citrus and olives, together with many fruits and vegetables, are grown in abun-dance in the area. This feature is considered a rare situation in world geography. Approximately 26% of the land in this province is plains and the remainder consists of mountainous and high regions.

Igdir is located at a large depression basin of the Aras river. The basin is divided into two by the Aras river, the part within the Turkish border is called the Sürmeli concavity and the remainder which lies within the Armenian border is called the Sahat concavity. The Sürmeli concavity includes the Tuzluca basin, İğdır Plain, and Dil Plain. Dil Plain is an extension of İğdır Plain towards the southeast and is the country's most eastern tip (44° 48′).

Aras river and the Turkey–Armenia border along the riverbed make up the northern and northeastern borders of this province, whereas Nakhchivan (Azerbaijan) is located in the east, and on the south is the Turkey–Iran border of the province. It is the only province in Turkey which has borders with three countries. Geologically, this area has Turabi and Çinçevat formations from the Early Pliocene, a Tuzluca formation from the Late Pliocene and terrace/slope debris and the alluvials from the Quaternary period (Yılmaz, 2007).

Continental climate prevails here, but it is totally different from the surrounding areas. It has typical microclimatic features, because the temperatures are higher and rainfall is less, thus constituting a "regional climate area" in the East Anatolian region. During the year, there is a short, but significant winter season (December–February), and a relatively long summer (May–September). In winter the temperatures rarely fall to −25 to −30°C. The maximum temperature in July and August, is around 35–40°C. The highest rainfall is recorded in spring, while winter is notable as the least rainy season.

Steppe vegetation is predominant in the depressions, but mountain steppes and alpine vegetation are found at higher altitudes. The plains usually have plant species of saline-alkali habitats mostly belonging to the Chenopodiaceae. In Tuzluca these are common in the lowlands, especially in the areas with high groundwater levels. We also come across reed swamps here, along the Aras river willow trees are a common sight, which reduce the effects of flooding, small birch trees are found at higher levels (İstanbulluoğlu, 2004).

Very limited number of floristic studies are available on İğdır. Nearly 1000 plant taxa are distributed naturally in this area (Altundağ, 2009). Tuzluca is described as a natural place of healing, with water sources coming from the high plateaus around the salt cave (Altundağ, 2009). The Tuzluca salt rock cave is used for the production of rock salt (Şimşek, 2005).

18.3 Halophyte Diversity

The halophytes distributed on the saline habitats around the salt mine caves and their surroundings in Tuzluca, Çan-Kaya, and Potuk are listed in Table 18.1. The list includes life forms, ecotypes, and distribution sites for each taxon in detail (Raunkier, 1934; Guvensen et al., 1996, 2006; Ozturk et al., 1995, 2006a,b, 2008a,b,c, 2011a,b, 2014). A total of 154 halophyte taxa, belonging to 34 families and 110 genera, were identified (Table 18.1). The families containing the maximum number of taxa are the Chenopodiaceae (34 taxa; 22.08%), Poaceae (21 taxa; 13.64%), Fabaceae (14 taxa; 9.09%), and Asteraceae (13 taxa; 8.44%). These four families represent 53.25% of the halophyte plants. The dominant genus is *Salsola*, which has six taxa. In terms of life forms, therophytes form the largest group (44.81%), followed by hemicryptophytes (20.78%) (Figure 18.2); however, in terms of ecotypes, the percentage distribution is as follows; xerophytes (25.97%), hygrohalophytes (21.43%), psammohalophytes (16.88%), xerohalophytes (16.88%), halophytes (12.34%), and ruderals (6.50%), respectively (Figure 18.3). Phytogeographically the highest number of taxa belong to the Irano-Turanian (33 taxa) division, others being the Euro-Siberian (11 taxa), the Mediterranean (seven taxa), and the East Mediterranean (one taxon), the phytogeographic origin of 100 taxa is unknown (Figure 18.4). The four taxa *Alyssum pateri* Nyár ssp. *pateri* (Brassicaceae), *Astragalus karamasicus* Boiss. & Bal. (Fabaceae), *Consolida glandulosa* (Boiss. & Huet) Bornm., and *Delphinium venulosum* Boiss. (Ranunculaceae) are endemics, all distributed around the salt mine area of Çankiri Salt Mines. Of 154 halophytes, 68 taxa are found only in Çankırı Province, 36 only in Iğdır Province, and 50 species flourish in both these areas (Table 18.1).

18.4 Economical Evaluations

In view of the food security problems expected to be faced in the near future, the degraded habitats around the salt mines inhabit several economically important halophytic plant species, which could be evaluated during the future food crisis programs.

The consumption of halophytes as food is not currently common practice, but 11 different halophytic taxa have been identified, of which three are consumed around the Çan-Kaya cave area, while ten taxa are consumed in the Tuzluca area, two of these taxa are used in both areas (Table 18.2). These taxa are

Table 18.1 Halophytes Distributed Around the Salt Mine Caves of Çankırı and Iğdır Provinces and Their Ecological Characteristics

Family/Taxa	Province	Life Form	Eco-Types	Salinity Tolerant	Choro-Types
Apiaceae					
1 *Eryngium campestre* L. var. *virens* Link	ÇA, IĞ	H	X	Hst	IN
2 *Daucus carota* L.	ÇA, IĞ	T	X	Hst	IN
Asteraceae					
3 *Artemisia campestris* L.	ÇA	H	PH	Lst	IN
4 *Artemisia santonicum* L.	ÇA, IĞ	CH	PH	Lst	ES
5 *Centaurea drabifolia* Sm. ssp. *detonsa* (Bornm.) Wagenitz	ÇA	T	X	Hst	IN
6 *Centaurea virgata* Lam.	ÇA, IĞ	T	XH	Hst	IN
7 *Chondrilla juncea* L. var. *juncea*	ÇA	H	X	Hst	IN
8 *Cirsium alatum* (Gmelin) Bobrov ssp. *alatum*	ÇA	H	HG	Lst	IR
9 *Crepis foetida* L. ssp. *rhoeadifolia* (Bieb.) Čelak	ÇA	T	PH	Lst	IN
10 *Crepis sancta* (L.) Babcock	ÇA, IĞ	T	XH	Hst	IN
11 *Inula aucherana* DC.	ÇA	C	XH	Hst	IR
12 *Onopordum turcicum* Danin	ÇA	H	R	Lst	IR
13 *Sonchus oleraceus* L.	ÇA	H	X	Hst	IN
14 *Xanthium strumarium* L. ssp. *strumarium*	ÇA	T	R	Lst	IN
15 *Xanthium strumarium* L. ssp. *cavanillesii* (Schouw) D. Löve & P. Dansereau	ÇA	T	R	Lst	IN
Boraginaceae					
16 *Cynoglossum creticum* Miller	ÇA	H	PH	Lst	IN
17 *Echium angustifolium* Miller	ÇA	CH	PH	Lst	EM
18 *Heliotropium dolosum* De Not.	ÇA, IĞ	T	PH	Lst	IN
19 *Heliotropium europaeum* L.	ÇA, IĞ	T	R	Lst	IN

(Continued)

Table 18.1 (Continued)

Family/Taxa	Province	Life Form	Eco-Types	Salinity Tolerant	Choro-Types
Brassicaceae					
20 *Alyssum linifolium* Steph. ex Willd. var. *linifolium*	ÇA, IĞ	T	X	Hst	IN
21 *Alyssum pateri* Nyárssp. *pateri*	ÇA	H	X	Hst	EN, IR
22 *Erysimum crassipes* Fisch. & Mey.	ÇA	H	PH	Lst	IN
23 *Malcolmia africana* (L.) R. Br.	ÇA	T	X	Hst	IN
24 *Sinapis arvensis* L.	ÇA, IĞ	T	XH	Hst	IN
25 *Sisymbrium altissimum* L.	ÇA	H	X	Hst	IN
Caryophyllaceae					
26 *Cerastium anomalum* Waldst. & Kit.	ÇA	T	HG	Lst	IN
27 *Silene supina* Bieb. var. *pruinosa* (Boiss.) Chowdh.	ÇA	H	X	Hst	IN
28 *Spergularia rubra* (L.) J. & C. Presl	ÇA	T	XH	Hst	IN
Chenopodiaceae					
29 *Aellenia glauca* (M. Bieb.) Aellen ssp. *glauca*	IĞ	CH	HA	Mst	IN
30 *Anabasis aphylla* L.	ÇA, IĞ	CH	XH	Hst	IR
31 *Arthrocnemum macrostachyum* (Moric) C. Koch	IĞ	CH	HA	Mst	IN
32 *Atriplex laevis* C.A. Meyer	ÇA	T	HA	Mst	IN
33 *Atriplex lasiantha* Boiss.	ÇA	T	R	Lst	IN
34 *Atriplex lehmanniana* Bunge	IĞ	CH	HA	Mst	IR
35 *Atriplex rosea* L.	ÇA, IĞ	T	R	Lst	IN
36 *Atriplex tatarica* L. var. *tatarica*	IĞ	T	HA	Mst	IN
37 *Bienertia cycloptera* Bunge ex Boiss.	IĞ	T	XH	Hst	IR
38 *Camphorosma monspeliaca* L. ssp. *lessingii* (Litw.) Aellen	ÇA, IĞ	CH	HG	Lst	IN
39 *Ceratocarpus arenarius* L.	IĞ	T	X	Hst	IN
40 *Chenopodium album* L. ssp. *album* var. *album*	ÇA, IĞ	T	PH	Lst	IN
41 *Chenopodium botrys* L.	ÇA, IĞ	T	PH	Lst	IN
42 *Chenopodium foliosum* (Moench) Aschers	ÇA, IĞ	T	R	Lst	IN
43 *Chenopodium murale* L.	IĞ	T	XH	Hst	IN
44 *Halanthium rarifolium* C. Koch	IĞ	T	XH	Hst	IR

(Continued)

Table 18.1 (Continued)

Family/Taxa	Province	Life Form	Eco-Types	Salinity Tolerant	Choro-Types
45 *Halogeton glomeratus* (Bieb.) Ledeb.	IĞ	T	HA	Mst	IR
46 *Halothamnus glaucus* (M. Bieb.) Botsch.	IĞ	CH	XH	Hst	IR
47 *Kalidium caspicum* (L.) Ung.-Sternb.	IĞ	CH	XH	Hst	IR
48 *Kochia prostrata* (L.) Schrad.	ÇA, IĞ	CH	X	Hst	IN
49 *Krasscheninnikovia ceratoides* (L.) Guldenst.	ÇA, IĞ	CH	X	Hst	IN
50 *Noaea mucronata* (Forssk.) Aschers & Schweinf. ssp. *mucronata*	ÇA, IĞ	CH	X	Hst	IN
51 *Panderia pilosa* Fisch. & Mey.	IĞ	T	HA	Mst	IN
52 *Salsola crassa* Bieb.	IĞ	T	HA	Mst	IR
53 *Salsola dendroides* Pall.	IĞ	CH	XH	Hst	IR
54 *Salsola kali* L.	IĞ	T	PH	Lst	IN
55 *Salsola macera* Litw.	IĞ	T	X	Hst	IR
56 *Salsola nitraria* Pallas	IĞ	T	HA	Mst	IR
57 *Salsola ruthenica* Iljin	ÇA, IĞ	T	PH	Lst	IN
58 *Seidlitzia florida* (M. Bieb.) Bunge	IĞ	T	XH	Hst	IR
59 *Suaeda altissima* (L.) Pall.	IĞ	T	HA	Mst	IN
60 *Suaeda carnossisima* Post	ÇA	T	HA	Mst	IR
61 *Suaeda confusa* Iljin	IĞ	T	HA	Mst	IN
62 *Suaeda microphylla* Pall.	IĞ	CH	HA	Mst	IR
Convolvulaceae					
63 *Convolvulus lineatus* L.	ÇA	CH	X	Hst	IN
Cuscutaceae					
64 *Cuscuta campestris* Yuncker	IĞ	T	X	Hst	IN
Cyperaceae					
65 *Bolboschoenus maritimus* (L.) Palla var. *maritimus*	ÇA	C	HG	Lst	IN
66 *Carex divisa* Hudson	ÇA, IĞ	C	HG	Lst	ES
67 *Carex flacca* Schreberssp. *serratula* (Biv.) Greuter	ÇA, IĞ	C	HG	Lst	M
68 *Cyperus longus* L.	IĞ	C	HG	Lst	IN
69 *Eleocharis palustris* (L.) Roemer & Schultes	ÇA, IĞ	C	HG	Lst	IN

(Continued)

Table 18.1 (Continued)

Family/Taxa	Province	Life Form	Eco-Types	Salinity Tolerant	Choro-Types
70 *Schoenus nigricans* L.	ÇA	H	HG	Lst	IN
71 *Schoenoplectus lacustris* (L.) Palla	IĞ	C	HG	Lst	IN
72 *Scirpoides holoschoenus* (L.) Sojak.	ÇA	C	HG	Lst	IN
Euphorbiaceae					
73 *Euphorbia falcata* L. ssp. *falcata* var. *falcata*	ÇA	T	XH	Hst	IN
74 *Euphorbia macroclada* Boiss.	ÇA, IĞ	H	HG	Lst	IR
Fabaceae					
75 *Alhagi pseudalhagi* (Bieb.) Desv.	ÇA, IĞ	CH	XH	Hst	IR
76 *Astragalus hamosus* L.	ÇA, IĞ	T	R	Lst	IN
77 *Astragalus karamasicus* Boiss. & Bal.	ÇA	H	R	Lst	EN, IR
78 *Astragalus microcephalus* Willd.	ÇA, IĞ	CH	X	Hst	IR
79 *Glycyrrhiza glabra* L. var. *glandulifera* (Waldst. & Kit) Boiss.	ÇA, IĞ	C	PH	Lst	IN
80 *Lotus corniculatus* L. var. *tenuifolius* L.	ÇA	H	HA	Mst	IN
81 *Medicago lupulina* L.	ÇA, IĞ	H	XH	Hst	IN
82 *Medicago minima* (L.) Bart var. *minima*	ÇA, IĞ	T	PH	Lst	IN
83 *Melilotus officinalis* (L.) Desr.	ÇA, IĞ	H	PH	Lst	IN
84 *Ononis spinosa* L. ssp. *leiosperma* (Boiss.) Širj.	ÇA	H	X	Hst	IN
85 *Trifolium campestre* Schreb.	IĞ	T	X	Hst	IN
86 *Trifolium repens* L. var. *repens*	ÇA, IĞ	H	X	Hst	IN
87 *Trigonella orthoceras* Kar. & Kir.	ÇA	T	X	Hst	IR
88 *Vicia sativa* L. ssp. *sativa*	ÇA	T	X	Hst	CO
Frankeniaceae					
89 *Frankenia hirsuta* L.	ÇA	CH	PH	Lst	IN
Juncaceae					
90 *Juncus articulatus* L.	ÇA	C	HG	Lst	ES
91 *Juncus gerardi* Loisel. ssp. *gerardi*	ÇA	C	HG	Lst	IN
92 *Juncus gerardi* Loisel. ssp. *libanoticus* (Thiéb.) Snog.	ÇA, IĞ	C	HG	Lst	IR
Liliaceae					
93 *Asparagus persicus* Baker	IĞ	H	XH	Hst	IR

(Continued)

Table 18.1 (Continued)

Family/Taxa	Province	Life Form	Eco-Types	Salinity Tolerant	Choro-Types
Lamiaceae					
94 *Salvia syriaca* L.	ÇA	H	XH	Hst	IR
95 *Salvia viridis* L.	ÇA	T	X	Hst	M
96 *Teucrium polium* L.	ÇA, IĞ	H	PH	Lst	IN
97 *Teucrium scordium* L. ssp. *scordioides* (Schreber) Maire & Petitmengin	ÇA	C	PH	Lst	ES
Linaceae					
98 *Linum bienne* Miller	ÇA	H	HG	Lst	M
Lythraceae					
99 *Lythrum salicaria* L.	ÇA	H	HG	Lst	ES
Orchidaceae					
100 *Orchis palustris* Jacq.	ÇA, IĞ	C	HG	Lst	IN
Orobanchaceae					
101 *Orobanche minor* Sm.	ÇA	T	X	Hst	IN
Papaveraceae					
102 *Fumaria vaillantii* Lois.	ÇA	T	R	Lst	IN
103 *Hypecoum procumbens* L.	ÇA	T	X	Hst	M
Plantaginaceae					
104 *Plantago lagopus* L.	ÇA	H	HG	Lst	M
105 *Plantago lanceolata* L.	ÇA, IĞ	H	HG	Lst	IN
106 *Plantago major* L. ssp. *intermedia* (Gilib.) Lange	ÇA, IĞ	H	HG	Lst	IN
107 *Plantago maritima* L.	ÇA	T	PH	Lst	IN
Plumbaginaceae					
108 *Limoniun gmelinii* (Willd.) O. Kuntze	ÇA, IĞ	H	HA	Mst	ES
109 *Limonium meyeri* (Boiss.) O. Kuntze	IĞ	T	HA	Mst	IR
Poaceae					
110 *Aeluropus lagopoides* (L.) Trin. ex Thwaites	IĞ	C	PH	Lst	IN
111 *Aeluropus littoralis* (Gouan) Parl.	IĞ	C	PH	Lst	IN
112 *Agrostis stolonifera* L.	ÇA	C	HG	Lst	ES

(Continued)

Table 18.1 (Continued)

Family/Taxa	Province	Life Form	Eco-Types	Salinity Tolerant	Choro-Types
113 *Alopecurus myosuroides* Hudson var. *myosuroides*	ÇA	T	HG	Lst	ES
114 *Avena sterilis* L. ssp. *sterilis*	ÇA	T	X	Hst	IN
115 *Briza maxima* L.	ÇA	T	X	Hst	IN
116 *Bromus arvensis* L.	ÇA, IĞ	T	XH	Hst	IN
117 *Bromus japonicus* Thunb. ssp. *japonicus*	ÇA, IĞ	T	X	Hst	IN
118 *Bromus madritensis* L.	IĞ	T	X	Hst	IN
119 *Bromus tectorum* L. ssp. *tectorum*	ÇA, IĞ	T	PH	Lst	IN
120 *Catabrosa aquatica* (L.) P. Beauv.	ÇA	C	HG	Lst	IN
121 *Cynodon dactylon* (L.) Pers var. *villosus* Regel	ÇA, IĞ	C	PH	Lst	IN
122 *Holcus lanatus* L.	ÇA	C	PH	Lst	ES
123 *Hordeum murinum* L. ssp. *glaucum* (Steudel) Tzvelev	ÇA	T	X	Hst	IN
124 *Lolium rigidum* Gaudin var. *rigidum*	ÇA	T	XH	Hst	IN
125 *Phleum exaratum* Hochst. ex Griseb. ssp. *exaratum*	ÇA	T	XH	Hst	IN
126 *Phragmites australis* (Cav.) Trin. ex Steudel	ÇA, IĞ	C	HG	Lst	ES
127 *Poa bulbosa* L.	ÇA, IĞ	C	X	Hst	IN
128 *Poa trivialis* L.	ÇA, IĞ	H	XH	Hst	IN
129 *Polypogon monspeliensis* (L.) Desf.	ÇA	T	XH	Hst	IN
130 *Trachynia distachya* (L.) Link	ÇA	T	PH	Lst	M
Polygonaceae					
131 *Atraphaxis spinosa* L.	IĞ	CH	XH	Hst	IR
132 *Polygonum aviculare* L.	IĞ	T	PH	Lst	CO
133 *Polygonum bellardii* All.	ÇA	T	HG	Lst	IN
Primulaceae					
134 *Anagallis arvensis* L. var. *arvensis*	ÇA	T	PH	Lst	IN
Ranunculaceae					
135 *Consolida glandulosa* (Boiss. & Huet) Bornm.	ÇA	T	X	Hst	EN, IR
136 *Delphinium venulosum* Boiss.	ÇA	T	X	Hst	EN, IR

(Continued)

Table 18.1 (Continued)

Family/Taxa	Province	Life Form	Eco-Types	Salinity Tolerant	Choro-Types
137 *Ranunculus constantinopolitanus* (DC.) d'Urv.	ÇA	C	HG	Lst	IN
138 *Thalictrum lucidum* L.	ÇA	C	HG	Lst	IN
Rosaceae					
139 *Potentilla reptans* L.	ÇA, IĞ	C	HG	Lst	IN
140 *Rubus sanctus* Schreber	ÇA	CH	X	Hst	IN
Rubiaceae					
141 *Cruciata taurica* (Willd.) Ehrend.	ÇA	CH	XH	Hst	IR
142 *Galium tricornutum* Dandy	IĞ	T	X	Hst	M
Scrophulariaceae					
143 *Verbascum cheiranthifolium* Boiss. var. ceiranthifolium	ÇA	H	X	Hst	IN
Tamaricaceae					
144 *Reaumuria alternifolia* (Lab.) Britten	IĞ	CH	XH	Hst	IR
145 *Tamarix smyrnensis* Bunge	ÇA, IĞ	P	HA	Mst	IN
146 *Tamarix tetrandra* Pall. ex Bieb.	ÇA, IĞ	P	HA	Mst	IN
Thymelaeaceae					
147 *Thymelaea passerina* (L.) Cosson & Germ.	ÇA	T	HG	Lst	IN
Typhaceae					
148 *Typha latifolia* L.	ÇA, IĞ	C	HG	Lst	IN
149 *Typha laxmannii* Lepecbin	IĞ	C	HG	Lst	ES
150 *Typha minima* Funck	IĞ	C	HG	Lst	IN
Verbenaceae					
151 *Verbena officinalis* L.	ÇA	CH	X	Hst	IN
Zygophyllaceae					
152 *Peganum harmala* L.	ÇA, IĞ	H	HA	Mst	IN
153 *Tribulus terrestris* L.	ÇA	T	X	Hst	IN
154 *Zygophyllum fabago* L.	ÇA, IĞ	H	X	Hst	IR

Province: ÇA, Çankırı; IĞ, Iğdır.
Life Forms: C, Cryptophytes; CH, Chamaephytes; H, Hemicryptophytes; P, Phanerophytes; T, Therophytes.
Ecotypes: HA, Halophyte; HG, Hygrohalophytes; PH, Psammohalophytes; X, Xerophytes; XH, Xerohalophytes; R, Ruderal.
Salinity Tolerant: Hst, High salinity tolerant; Lst, Low salinity tolerant; Mst, Medium salinity tolerant.
Chorotypes: CO, Cosmopolitan; EM: East Mediterranean; EN, Endemic; ES, Euro-Siberian; IN, Imperfectly Known; IR, Irano-Turanian; M, Mediterranean.

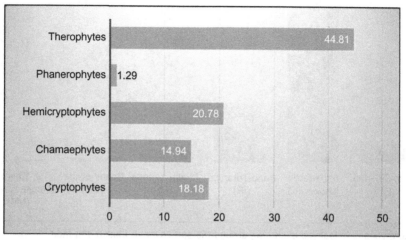

Figure 18.2 Distribution and number of halophytes around the salt caves on the basis of their life forms (%).

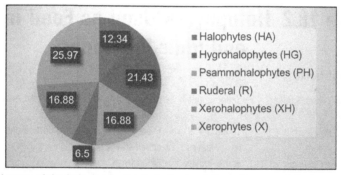

Figure 18.3 Ecological types of the halophytes distributed around the salt caves of Çankırı and Iğdır (%).

mostly eaten fresh, raw or roasted, with eggs (Table 18.2). A total of 29 halophytes have potential fodder value (Table 18.3), 11 of these plants belong to the family Poaceae and nine to the Fabaceae.

18.5 Medicinal and Aromatic Halophytes in Çankırı and Iğdır Provinces

Our findings have revealed that 17 halophytic taxa are used in folk medicine in the study areas; *Centaurea drabifolia* Sm. ssp. *detonsa* (Bornm.) Wagenitz, *Echium angustifolium* Miller,

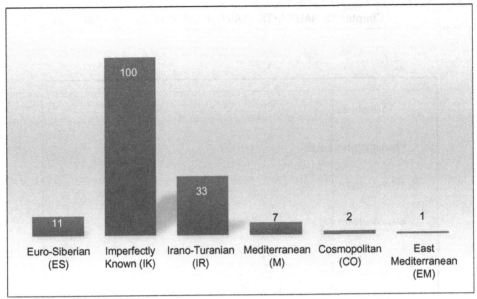

Figure 18.4 Chorotypes of halophytic plant taxa around the salt caves of Çankırı and Iğdır.

Table 18.2 Halophytes Used as Food in Çankırı and Iğdır Provinces

	Botanical Name	Part Used	Preparation	Source[a]
1	Asparagus persicus Baker	Young shoot	Boiled and fried, with eggs also added to the soup	A
2	Chenopodium album L. ssp. album var. album	Young aerial parts, leaves	Fried with eggs, filled in patties and added to the bulgur dish	A, B
3	Chenopodium foliosum (Moench) Aschers.	Fruits, leaves	Eaten fresh, cooked	A
4	Chenopodium murale L.	Young aerial parts	Boiled and fried, with eggs also added to the soup as well as bulgur dish	A
5	Crepis sancta (L.) Babcock	Leaves	Raw	B
6	Eryngium campestre L. var. virens Link	Stem	Peeled off and eaten raw	A, B
7	Glycyrrhiza glabra L.	Root	Peeled off and sucked, roots boiled to get a syrup	A
8	Plantago major L. ssp. intermedia (Gilib.) Lange	Young leaves	Consumed raw with cheese	A
9	Polygonum aviculare L.	Aerial parts	Fried and added to soups	A
10	Trifolium campestre Schreb.	Young flowers	Eaten fresh	A
11	Trifolium repens L. var. repens	Young flowers	Eaten fresh	A

[a]Source: A, Altundağ (2009); B, Karataş (2007).

Table 18.3 Halophytes with Fodder Potential in Çankırı and Iğdır Provinces

	Botanical Name	Çankırı Province	Iğdır Province	Source[a]
1	*Aeluropus littoralis* (Gouan) Parl.		x	a
2	*Agrostis stolonifera* L.	x		a, b
3	*Alhagi pseudalhagi* (Bieb.) Desv.	x	x	c, d
4	*Alopecurus myosuroides* Hudson var. *myosuroides*	x		a
5	*Atriplex lasiantha* Boiss.	x		c
6	*Bromus arvensis* L.	x	x	a
7	*Bromus japonicus* Thunb. ssp. *japonicus*	x	x	a, b
8	*Bromus tectorum* L. ssp. *tectorum*	x	x	a, b
9	*Camphorosma monspeliaca* L. ssp. *lessingii* (Litw.) Aellen	x	x	c
10	*Cerastium anomalum* Waldst. & Kit.	x		a
11	*Crepis sancta* (L.) Babcock		x	a
12	*Cynodon dactylon* (L.) Pers var. *villosus* Regel	x	x	a, b
13	*Eryngium campestre* L. var. *virens* Link		x	d
14	*Glycyrrhiza glabra* L. var. *glandulifera* (Waldst. & Kit.) Boiss.	x	x	a, d
15	*Holcus lanatus* L.	x		a
16	*Hordeum murinum* L. ssp. *glaucum* (Steudel) Tzvelev	x		b
17	*Lotus corniculatus* L. var. *tenuifolius* L.	x		a
18	*Medicago lupulina* L.	x	x	a, d
19	*Medicago minima* (L.) Bart. var. *minima*	x	x	b, d
20	*Melilotus officinalis* (L.) Desr.		x	d
21	*Ononis spinosa* L. ssp. *leiosperma* (Boiss.) Širj.	x		a
22	*Phleum exaratum* Hochst. ex Griseb. ssp. *exaratum*	x		b
23	*Plantago major* L. ssp. *intermedia* (Gilib.) Lange	x	x	a
24	*Poa bulbosa* L.	x	x	a, b
25	*Salsola dendroides* Pall.		x	d
26	*Suaeda altissima* (L.) Pall.		x	c
27	*Suaeda confusa* Iljin		x	c
28	*Trifolium campestre* Schreb.		x	d
29	*Trifolium repens* L. var. *repens*	x	x	a, b, d

[a]Source: a, Anonymous (2008b); b, Yılmaz and Göl (2012); c, Sözeri and Tıpırdamaz (2001); d, Altundağ (2009).

Plantago maritima L., and *Ranunculus constantinopolitanus* (DC.) d'Urv. are used in the Çankiri area and 11 halophytes in the Iğdır area. *Plantago lanceolata* L. and *Teucrium polium* L. are also used in folk medicine here (Table 18.4). Mostly it is the aerial parts and leaves of plants that are utilized for this purpose and the most widely used methods of preparation are decoction and infusion (Table 18.4).

Some of the halophytes are used as mixtures with other herbs in traditional folk medicine. For example, *P. lanceolata*, *Plantago major* L. ssp. *intermedia* (Gilib.) Lange, *Polygonum aviculare* L., and *Galium tricornutum* Dandy are widely used in Iğdır (Altundağ, 2009).

A *Plantago lanceolata* and *P. major* ssp. *İntermedia* infusion is prepared from the leaves, and the paste is mixed with the flour as a dough and applied to the abdominal area externally as an anti-inflammatory.

Rumex crispus L., *P. major* ssp. *intermedia*, and *P. lanceolata* leaves are ground and heated with flour and milk and used topically in women to prevent infertility. Above-ground parts of *Malva neglecta* Wallr. and *P. major* ssp. *intermedia* leaves are salted overnight, water is squeezed out and discarded, the washed and boiled plants are used as anti-inflammatory agents. *Malva neglecta* aerial parts, *Arctium platylepis* (Boiss. & Bal.) Sosn. ex Grossh. Leaves, and *P. major* ssp. *intermedia* leaves are crushed, the mixture is boiled with milk and applied topically to infected wounds. A *Plantago media* L. and *P. major* ssp. *intermedia* infusion is prepared from leaves and used internally as an anti-inflammatory agent. The aerial parts of *M. neglecta* and *P. major* ssp. *intermedia* leaves are chopped, salted overnight and in the morning cooked and eaten by women to aid attempts at pregnancy. *Rumex crispus*, *P. major* ssp. *Intermedia*, and *P. lanceolata* leaves are pounded and then boiled with milk and flour, these are applied to cure gynecological disorders. *Convolvulus arvensis* L. or *Convolvulus scammonia* L. leaves are mixed and boiled with the above-ground parts of *P. aviculare* or *Polygonum cognatum* Meissn., the juice is filtered and salted to relieve stomach discomfort. The capitula of *Achillea bieberstei-nii* Afan. and above-ground parts of *Galium humifusum* Bieb. or *G. tricornutum* are prepared as infusions and used by women with womb problems. In addition, 16 halophytic taxa are used as potential dye plants, for example, *Chenopodium album* L., *Euphorbia macroclada* Boiss., *Glycyrrhiza glabra* L., and *Peganum harmala* L., which show distribution at both sites (Table 18.5). *Halanthium rarifolium* C. Koch is used a great deal by the locals in the area for washing clothes. *Peganum harmala*

Table 18.4 Medicinal and Aromatic Halophytes in Çankırı and Iğdır Province

	Botanical Name	Part Used	Preparation	Treatment	Sources[a]
1	*Alhagi pseudalhagi* (Bieb.) Desv.	AP	DE	Tonic	a
2	*Centaurea drabifolia* Sm. ssp. *detonsa* (Bornm.) Wagenitz	FL	IN	Malaria	b
3	*Chenopodium album* L. ssp. *album* var. *album*	AP	DE	Infertility	a
4	*Chenopodium botrys* L.	AP	DE	Infertility	a
5	*Chenopodium murale* L.	AP	DE	Infertility	a
6	*Echium angustifolium* Miller	RO	OI	Injuries	b
7	*Euphorbia macroclada* Boiss.	LA		Injuries, constipation	a
8	*Galium tricornutum* Dandy	AP	IN	Gynecological diseases	a
9	*Glycyrrhiza glabra* L.	LE, RO	RW, DE, SH	Refrigerant; restorative; infertility; antitussive; kidney stones; epilepsy; cancer	a
10	*Melilotus officinalis* (L.) Desr.	YLE	RW	Anti-anemic	a
11	*Plantago lanceolata* L.	LE	DE, PO	To stop bleeding; anti-inflammatory; injuries; warts; stomach diseases; bronchitis	a,b
12	*Plantago major* L. ssp. *intermedia* (Gilib.) Lange	LE, YLE, SE	CO, DE, IN, RW	Stomach diseases; cytostatic; gastric ulcers; anti-inflammatory; hemorrhoids	a
13	*Plantago maritima* L.	LE		Wound healing	c
14	*Polygonum aviculare* L.	AP	IN	Antitussive; anti-anemic; rheumatism	a
15	*Ranunculus constantinopolitanus* (DC.) d'Urv.	FL	PN	Rheumatism	b
16	*Teucrium polium* L.	AP	DE, IN	Stomachache; facilitate digestion; appetizer; gas removed; hemorrhoids	a,b
17	*Zygophyllum fabago* L.	AP	DE	Rheumatism; infertility	a

Part used: AP, Aerial parts; FL, Flowers; LA, Latex; LE, Leaves; RO, Roots; SE, Seeds; YLE, Young leaves.
Preparation: CO, Cooked; DE, Decoction; IN, Infusion; OI, Ointment; PN, Pounded; PO, Poultice; RW, Raw; SH, Shrub.
[a]Source: a, Altundağ (2009); b, Ezer and Avcı (2004); c, Sözeri and Tıpırdamaz (2001).

Table 18.5 Other Economic Uses of Some Halophytic Plants in Çankırı and Iğdır Provinces

Botanical Name	Basket	Broom	Fuel	Mats	Ornamental	Dye Plant	Other Uses	Source[a]
Artemisia santonicum L.		x						a
Atraphaxis spinosa L.			x					a
Chenopodium album L.						x		b
Eryngium campestre L. var. virens Link			x				x	a,c
Euphorbia macroclada Boiss.						x		a,d
Glycyrrhiza glabra L.						x		e
Halanthium rarifolium							x	a
Peganum harmala L.					x	x	x	a,c,f
Phragmites australis (Cav.) Trin. ex Steudel							x	a
Poa bulbosa L.					x			c
Salsola dendroides Pall.			x					a
Tamarix smyrnensis Bunge			x				x	a
Tamarix tetrandra Pall. ex Bieb.			x				x	a
Typha latifolia L.	x			x				a
Typha laxmannii Lepecbin	x			x				a
Typha minima Funck	x			x				a

[a]Source: a, Altundağ, 2009; b, Coon, 1975; c, Karataş, 2007; d, Uysal, 1991; e, Doğan et al., 2003; f, Özgökçe and Yılmaz, 2003.

and *Poa bulbosa* L. are also use as ornamentals in Çankırı, *Eryngium campestre* L. var. *virens* Link and *Salsola dendroides* Pall. are used as fuel, and *P. harmala* seeds are burnt to evade the evil eye (Table 18.5). *Typha latifolia* L., *Typha minima* Funck, and *Typha laxmannii* Lepecbin are preferred by local people for both mat and basket making. *Artemisia santonicum* L. is used for the production of brooms, *Atraphaxis spinosa* L. and *Tamarix* species (*Tamarix. smyrnensis* Bunge and *Tamarix tetrandra* Pall. ex Bieb.) are used as fuel, and *P. harmala* is used as an ornamental (Table 18.5). The above-ground parts of *E. campestre* var. *virens* are placed on the roofs of houses, on top of manure, in order to protect from the evil eye. Similarly, the smoke from burning *P. harmala* fruity branches is thought to protect people from the evil eye; the smoke is formed and at the same time special recitations are made. *Phragmites australis* (Cav.) Trin. ex Steudel plants are collected and dried for use in roofing for houses and barns; *Tamarix* species (*T. smyrnensis* and *T. tetrandra*), especially the branches, are preferred by local people for use as skewers on grills (Altundağ, 2009).

18.6 Conclusions

To feed the world's anticipated nine billion people by 2050, and with little new productive land available, interest in halophytes has increased significantly due to the latest developments in arid and semi-arid regions of the world. Food security is becoming a theme of much interest for scientists. Studies are focusing on halophyte plant species which grow on saline soils and have great significance as indicator plants. These plants are being screened for use in the reclamation and evaluation of degraded land. Their economic importance as animal feed, vegetables, vegetable oil, ornamental plants, and chemicals has been stressed by different researchers notable among them being the following: Islam et al., 1982; Huiskes, 1995; Schmsutdinov et al., 2000; Guven, 2000; Khan and Duke, 2001; Sözeri and Tıpırdamaz, 2001; Guvensen et al., 2006; Khan and Qaiser, 2006; Weber et al., 2007; Lokhande and Suprasanna, 2012; Ozturk et al., 2014.

In recent years, studies of the mechanisms of salt resistance in plants growing on saline soils has gained great impetus. The halophytes are singled out as they are suitable candidates for evaluation in this connection. The contributions of halophytes to the national economy in light of these studies is increasing (Dolarslan and Gul, 2012).

Halophytes can also contribute to the conversion of unused and marginal land for agricultural production, which can help to improve the sustainable efficiency of available crop productivity (Hasanuzzaman et al., 2014).

Under the current global climate change scenarios, desertification and anthropogenic factors will force us in the near future to increase our research activities in the evaluation of halophytes, and especially investigations on their physiology, ecology, and biotechnology will make very important contributions towards food security. Halophytes possess the potential to use salt-affected lands for forage crops, which can increase the revenue of farmers (Sözeri and Tıpırdamaz, 2001; Temel and Şimşek, 2011; Hasanuzzaman et al., 2014). These halophytic plants are already occupying the wastelands and do not damage the natural flora. For this purpose, researchers from different disciplines should work together on joint projects, to highlight the benefits of these plants which will contribute to the world economy. Moreover the suggestions are that tree planting, deep ploughing, and the production of salt-tolerant crops is a must; it is also proposed that we need to dig drains or ditches around the affected lands to aid their recovery.

References

Adıguzel, N., Byfield, A., Duman, H., Vural, M., 2005. Tuz Gölü ve stepleri. Türkiye'nin 122 Önemli Bitki Alanı. In: Özhatay, et al., (Eds.), WWF Türkiye Doğal Hayatı Koruma Vakfı Yayını, Istanbul, pp. 287–292.

Akbaş, F., Oztürk, M., Güvensen, A., Gündüz, M., 1999. Ecology of the halophytes of Soke Plateau. VII. Kulturteknik Kongresi-Nevşehir. Ankara University Press, Turkey, pp. 246–252.

Altay, V., Ozturk, M., 2012. Land Degradation and Halophytic Plant Diversity of Milleyha Wetland Ecosystem (Samandağ-Hatay), Turkey. Pak. J. Bot. 44, 37–50 (Special issue May).

Altundağ, E., 2009. Iğdır İli'nin (Doğu Anadolu Bölgesi) doğal bitkilerinin halk tarafından kullanımı. İstanbul Üniversitesi Sağlık Bilimleri Enstitüsü, Farmasötik Botanik Programı, Doktora Tezi, İstanbul.

Anonymous, 2008a. Iğdır Valiliği İl Kültür ve Turizm Müdürlüğü. Iğdır Kültür ve Turizm Envanteri. Punto Tasarım. Ankara.

Anonymous, 2008b. Türkiye'nin çayır ve mera bitkileri. T.C. Tarım ve Köyişleri Bakanlığı, Tarımsal Üretim ve Geliştirme Genel Müdürlüğü, Çayır, mera, yem bitkileri ve Havza geliştirme Daire Başkanlığı. Nisan, 2008.

Aronson, J.A., 1984. Economic halophytes—a global review. In: Wickens, G.E. (Ed.), Plants for Arid Lands. Allen & Unwin, London, pp. 177–188.

Aronson, J.A., 1989. Halophyte, a Database of Salt Tolerant Plants of the World. Office of Arid Land Studies, University of Arizona, Tucson, AZ, p. 77.

Atamov, V., Aktoklu, E., Çetin, E., Aslan, M., Yavuz, M., 2006. Halophytication in Harran (Şanlıurfa) and Amik Plain (Hatay) in Turkey. Phytol. Balc. 12 (3), 401–412.

Aydoğdu, M., Hamzaoğlu, E., Kurt, L., 2002. New halophytic syntaxa from Central Anatolia (Turkey). Isr. J. Plant Sci. 50, 313–323.

Aydoğdu, M., Kurt, L., Hamzaoğlu, E., Ketenoğlu, O., Cansaran, A., 2004. Phytosociological studies on salt steppe communities of Central Anatolia, Turkey. Isr. J. Plant Sci. 52, 71–79.

Barutoğlu, Ö.H., 1961. Türkiye Tuz Yatakları. Madencilik. 2, 68–78.

Birand, H., 1960. Erste Ergebnisse der vegetations-Untersuchungen in der zentra-lanatolischen steppe; I- Hlophytengesellschaften des Tuz Gölü. Bot. Jahrb. Syst. 79, 255–296.

Birand, H., 1961. Tuzgölü Çorakçıl Bitkileri, vol. 103. Toprak-Su Genel Müdürlüğü Neşriyatı, Ankara, C5, pp. 1–56.

Chedlly, A., Ozturk, M., Ashraf, M., Grignon, C., 2008. In: Birkhauser, V. (Ed.), Biosaline Agriculture and High Salinity Tolerance. Springer Science, Basel, p. 367.

Choukr-Allah, R., 1991. The use of halophytes for the agricultural development of south of Morocco. Plant Salinity Research New Challenges. Hassan II-C.H. A, Agadir, Morocco, pp. 377–387.

Coon, N., 1975. The Dictionary of Useful Plants. Rodale Press, PA, USA.

DMIGM, 2009, Çankırı İstatistiki Verileri. <http://www.dmi.gov.tr/veridegerlendirme/il-ve-ilceleristatistik.aspx?m = CANKIRI> (accessed 26.11.09.).

Doğan, Y., Başlar, S., Mert, H.H., Ay, G., 2003. Plant used as natural dye sources in Turkey. Econ. Bot. 57 (4), 442–453.

Dolarslan, M., Gul, E., 2012. Toprak Bitki İlişkileri Açısından Tuzluluk. Türk Bil Der D. 5 (2), 56–59.

Ertuğrul, G., 2011. Çankırı Korubaşı Tepe ve civarındaki jipsli alanların florası. İstanbul Üniversitesi Fen Bilimleri Enstitüsü, Yüksek Lisans Tezi. Ankara.

Ezer, N., Avcı, K., 2004. Çerkeş (Çankırı) yöresinde kullanılan halk ilaçları. Hacettepe Uni, Ecz. Fac. J. 24 (2), 67–80.

Flowers, T.J., Hajibagheri, M.A., Clipson, N.J.W., 1986. Halophytes. Q. Rev. Biol. 61, 313–337.

Guven, E., 2000. İç Anadolu Bölgesi halofitik (tuzcul bitkiler) vejetasyonun ekolojisi. T.C. Başbakanlık Köy Hizmetleri Gen. Müd. Toprak ve Gübre Araş. Ens, Yayınları.

Guvensen, A., Uysal, I., Celik, A., Ozturk, M., 1996. Environmental impacts on halophytic plant cover alongside the coastal area of Karabiga and Burhaniye. In: Symposium on Environmental Problems. Çanakkale. pp. 8–13.

Guvensen, A., Gork, G., Ozturk, M., 2006. An overview of the halophytes in Turkey. In: Ajmal Khan, M., et al., (Eds.), Sabkha Ecosystems, Vol. 2. West and Central Asia, pp. 9–30.

Hamzaoğlu, E., Aksoy, A., 2006. Sultansazlığı Bataklığı halofitik toplulukları üzerine fitososyolojik bir çalışma (İç Anadolu-Kayseri). Ekoloji. 15 (60), 8–15.

Hamzaoğlu, E., Aksoy, A., 2009. Phytosociological studies on the Halophytic communities of Central Anatolia. Ekoloji. 18 (71), 1–14.

Hasanuzzaman, M., Nahar, K., Alam, Md.M., Bhowmik, P.C., Hossain, M.A., Rahman, M.M., et al., 2014. Potential use of halophytes to remediate saline affected soils in the light of phytoremediation. BioMed Res. Int. Article ID 589341, 12 p.

Huiskes, A.H.L., 1995. Saline crops, a contribution to the diversification of the production of vegetable crops by research on the cultivation methods and selection of halophytes. Annual Report. 303 p.

Inan, E., İpek, G., İpek, A., 2012. Çankırı'nın Endemik Tıbbi Bitkileri. Türk Bil Der D. 5 (2), 38–40.

Islam, M.N., Wilson, C.A., Watkins, T.R., 1982. Nutritional evaluation of seashore mallow seed, *Kosteletzkya virinica*. J. Agric. Food Chem. 30, 1197–1198.

İstanbulluoğlu, A., 2004. Investigation of the vegetation on saline-alkaline soils and marshes of Iğdır Plain in Turkey. Pak. J. Biol. Sci. 7 (5), 734–738.

Karataş, H., 2007. Ilgaz (Çankırı)'ın Etnobotaniği (M.S. thesis). Gazi University, Institute of Sciences, Ankara.

Khan, M.A., Duke, N.C., 2001. Halophytes—a resource for the future. Wetl. Ecol. Manage. 6, 455–456.

Khan, M.A., Qaiser, M., 2006. Halophytes of Pakistan: Distribution, Ecology, and Economic Importance. In: Khan, M.A., Barth, H.-J., Kust, G.C., Boer, B. (Eds.), Sabkha Ecosystems: vol. 2. The South and Central Asian Countries. Springer, Netherlands, pp. 135–160.

Kırıkoğlu, M., 2012. Çankırı'da yeraltı kaynakları. Çankırı Tuz Çalıştay Raporu, Dünden bugüne tuz çalışmaları oturumu, 11–12 Nisan 2012. T.C. Kuzey Anadolu Kalkınma Ajansı, Çankırı Yatırım Destek Ofisi, Türkiye pp. 15–20.

Lokhande, V.H., Suprasanna, P., 2012. Prospects of halophytes in understanding and managing abiotic stress tolerance. In: Ahmad, P., Prasad, M.N.V. (Eds.), Environmental Adaptations and Stress Tolerance of Plants in the Era of Climate Change. Springer, New York, NY, pp. 29–56.

Malcolm, C.V., 1993. The potential of halophytes for rehabilitation of degraded land. In: Davidson, N., Galloway, R. (Eds.), Productive Use of Saline Land, vol. 42. ACIAR, Pro, pp. 8–11.

MTA, 1978. Interim report: the neogene formations of Turkey and their economical geology. In: Gok, S. (Ed.), General Directorate of Mineral Research and Exploration. Industrial Raw Materials Division, Ankara, Turkey, pp. 40–59. (in Turkish).

MTA, 2007. <http://mta.gov.tr>.

MTA, 2010. Mineral and energy resources of Cankiri city. General Directorate of Mineral Research and Exploration. Industrial Raw Materials Division, Ankara, Turkey, pp. 1–5.

Mutlu, H., 2006. Çankırı/Yapraklı ormanlarının vasküler bitki florası. Ankara Üniversitesi Fen Bilimleri Enstitüsü, Yüksek Lisans Tezi. Ankara.

Ozturk, M., Ozcelik, H., Behcet, L., Guvensen, A., Ozdemır, F., 1995. Halophytic flora of Van Lake basin–Turkey. In: Khan, M.A., Ungar, I.A. (Eds.), Biology of Salt Tolerant Plants. Crafters, Michigan, MI, pp. 306–315.

Ozturk, M., Guvensen, A., Gork, C., Gork, G., 2006a. An overview of coastal zone plant diversity and management strategies in the Mediterranean region of Turkey. In: Ozturk, et al., (Eds.), Biosaline Agriculture & Salinity Tolerance in Plants. Birkhauser Verlag, Springer Science, Basel, pp. 89–100.

Ozturk, M., Waisel, Y., Khan, M.A., Gork, G. (Eds.), 2006b. Biosaline Agriculture and Salinity Tolerance in Plants. Birkhauser Verlag (Springer Science), Basel, p. 205.

Ozturk, M., Gucel, S., Sakcali, S., Gork, C., Yarci, C., Gork, G., 2008a. An overview of plant diversity and land degradation interactions in the eastern Mediterranean. In: Efe, et al., (Eds.), Natural Environment and Culture in the Mediterranean Region. Cambridge Scholars Publ., UK, pp. 215–239.

Ozturk, M., Guvensen, A., Gucel, S., 2008b. Ecology and Economic Potential of Halophytes—A Case Study from Turkey, Crop and Forage Production Using Saline Waters. Daya Publishing House, p. 334.

Ozturk, M., Guvensen, A., Sakcali, S., Gork, G., 2008c. Halophyte plant diversity in the Irano-Turanian phytogeographical region of Turkey. In: Abdelly, C., Öztürk, M., Ashraf, M., Grignon, C. (Eds.), Biosaline Agriculture and High Salinity Tolerance. Birkhauser Verlag, Switzerland, pp. 141–155.

Ozturk, M., Gucel, S., Guvensen, A., Kadis, C., Kounnamas, C., 2011a. Halophyte plant diversity, coastal habitat types and their conservation status in Cyprus. In: Ozturk, M., Böer, B., Barth, H.-J., Breckle, S.-W., Clüsener-Godt, M., Khan, M.A. (Eds.), Sabkha Ecosystems. vol. 3. Africa and Southern Europe, Tasks for Vegetation Science 46. Springer, Netherlands, pp. 99–111.

Ozturk, M., Boer, B., Barth, H.J., Breckle, S.W., Clüsner-Godt, M., Khan, M.A., 2011b. Sabkha Ecosystems vol. 3. Africa & Southern Europe. Tasks for Vegetation Science-46. Springer Verlag, p. 148.

Ozturk, M., Altay, V., Gücel, S., Güvensen, A., 2014. Halophytes in East Mediterranean—their medicinal and other economical values. In: Khan, M.A., et al., (Eds.), Sabkha Ecosystems: vol 4: Cash Crop Halophyte and Biodiversity Conservation, Tasks for Vegetation Science 47. Springer Science + Business Media, Dordrecht, pp. 247–272.

Özgökçe, F., Yılmaz, I., 2003. Dye plants of east anatolia region (Turkey). Econ. Bot. 57 (4), 454–460.

PDEF, 2010. Environmental status report of Cankiri city. In: Demir et al. (Eds.), Provincial Directorate of Environment and Forestry. Cankiri, Turkey (in Turkish). Available from: <http://cankiri.cevreorman.gov.tr>.

Qadir, M., Quillérou, E., Nangia, V., Murtaza, G., Singh, M., Thomas, R.J., et al., 2014. Economics of salt-induced land degradation and restoration. Nat. Resour. Forum. 38, 282–295.

Raunkier, C., 1934. Life Forms of Plants and Statistical Plant Geography. Clarendon Press, New York-London.

Sağıroğlu, M., 1998. Karlıktepe ve civarının (Çankırı) florası. Gazi Üniversitesi Fen Bilimleri Enstitüsü, Yüksek Lisans Tezi. Ankara.

Schmsutdinov, Z., Savchenko, V.I., Schmsutdinov, N.Z., 2000. Halophytes in Russia, their Ecological, Evolution and Usage. High School Press, Moscow, Russia, p. 399.

Şimşek, O., 2005. Tuzluca İlçesinin Beşeri ve Ekonomik Coğrafyası. Atatürk Üniversitesi, Sosyal Bilimler Enstitüsü, Orta Öğretim Sosyal Alanlar Eğitimi Anabilim Dalı, Erzurum.

Sözeri, S., Tıpırdamaz, R., 2001. Halophytic plants of pastures around Seyfe Lake (Kırşehir) and the possibilities of their use. Türk Herb Derg. 4 (2), 11–35.

Stuart, J.R., Tester, M., Gaxiola, R.A., Flowers, T.J., 2012. Plants of saline environments. Available from: <http://www.accessscience.com>.

Temel, S., Şimşek, U., 2011. Iğdır Ovası Toprakların Çoraklaşma Süreci ve Çözüm Önerileri. Alınteri. 21 (B), 53–59.

Tuğ, G.N., Yaprak, A.E., Vural, M., 2011. An overview of Halophyte plant diversity from Salt Lake Area, Konya, Turkey. In: Ozturk, M., et al., (Eds.), Urbanization, Land Use, Land Degradation and Environment. Daya Publishing House, Delhi.

Uysal, I., 1991. Çanakkale İli'ndeki bazı boya bitkilerinin morfolojisi, korolojisi ve boyacılıkta kullanılması. Çanakkale Edu Voc Res Jour. 2 (1), 65–104.

Weber, D.J., Ansarib, R., Gul, B., Khan, M.A., 2007. Potential of halophytes as source of edible oil. J. Arid Environ. 68, 315–321.

Yılmaz, O., 2007. Kağızman (Kars)-Tuzluca (Iğdır) tuz yataklarının jeolojisi, mineralojisi ve petrografisi. Dokuz Eylül Üniversitesi Fen Bilimleri Enstitüsü, Yüksek Lisans Tezi, Jeoloji Mühendisliği Bölümü, Ekonomik Jeoloji Anabilim Dalı, Izmir.

Yılmaz, H., Göl, C., 2012. Çankırı yöresi kurak-yarıkurak meralarında ıslah ve erozyon önleyici bitki türleri. Türk Bil Der D. 5 (2), 109–115.

HALOPHYTES AS A POSSIBLE ALTERNATIVE TO DESALINATION PLANTS: PROSPECTS OF RECYCLING SALINE WASTEWATER DURING COAL SEAM GAS OPERATIONS

Suresh Panta[1], Peter Lane[1], Richard Doyle[1], Marcus Hardie[1], Gabriel Haros[2] and Sergey Shabala[1]
[1]*School of Land and Food, University of Tasmania, Hobart, TAS, Australia*
[2]*The Punda Zoie Company Pty Ltd, Melbourne, VIC, Australia*

19.1 Introduction

Very large volumes of poor-quality water are generated by agriculture, industry, and municipal water treatment processes (Shannon et al., 1997; Glenn et al., 2009). Much of this is saline, containing from 1000 mg L^{-1} to above 7000 mg L^{-1} of total dissolved salts, with sodium being the dominant cation. These wastewaters are commonly discharged into aquifers, municipal sewage systems, or surface waters, where they represent a potential hazard to downstream water users and the environment (Gerhart et al., 2006). The result is a degradation of soil structure and detrimental effects to crops and ecosystems. Although the most "environmentally friendly" way of solving the problem of saline wastewater disposal would be the use of desalination plants, the cost of construction and operating such plants is substantial, and so is its "carbon footprint." It is estimated that the initial cost of construction of one "standard size" desalination plant to dispose of 5 ML day^{-1} of saline wastewater is about US$13M, with a further US$4.5M per annum required for operation (Siemens Water Technologies,

M.A. Khan, M. Ozturk, B. Gul, & M.Z. Ahmed (Eds): Halophytes for Food Security in Dry Lands.
DOI: http://dx.doi.org/10.1016/B978-0-12-801854-5.00019-4

personal communication). Thus, economically viable alternatives need to be found. One of these options is to use halophytes as forage or crop species.

Halophytes are naturally evolved salt-tolerant plants that are present in about half the higher plant families. Some have potential as a source of "tolerance genes" and others may be suitable as cash crops under saline conditions. Halophytes have been tested as vegetable, forage, and oilseed crops in agronomic field trials (reviewed by Panta et al., 2014). The most productive species yield 10–20 tonnes DM ha^{-1} of biomass on seawater irrigation, equivalent to the productivity of conventional crops under nonsaline conditions (Glenn et al., 1999; Masters et al., 2007; Jordan et al., 2009). Importantly, some halophytes produce high levels of biomass even under hyper-saline (i.e., above seawater level) irrigation. Halophyte forage and seed products can replace conventional ingredients in animal feeding systems, albeit with some restrictions on their use due to their high salt content and the presence of anti-nutritional compounds in some species (Glenn et al., 1999; Rogers et al., 2005). Importantly, halophytic plants are capable of accumulating high concentrations of NaCl in their tissues (e.g., up to 39% in a saltbush; Barrett-Lennard, 2002). Assuming this capacity can be matched by high biomass production, halophytic species can be a potential biological solution to rehabilitate saline-sodic or salt-affected land by extracting substantial amounts of salt from the soil, establishing plant cover, and lowering the water table (Panta et al., 2014).

The use of saline water for agricultural crop production has been explored in a number of studies over several decades (Grattan et al., 2004, 2008; Grieve et al., 2004), and methods for the use of saline agricultural drain-water have been developed for specific irrigation districts (e.g., Shannon et al., 1997). Nonetheless, the practical use of halophytes for the above purposes is still at the "testing stage," and not widely accepted by industry. Several factors may explain the current situation. First, the public perception of this practice is such that it is often labeled as "environmentally unfriendly," with concerns raised about long-lasting effects of depositing saline water on soil structure and salinization. Second, while halophytes are clearly capable of coping with excessive amounts of NaCl in their tissues (Flowers and Colmer, 2008; Shabala and Mackay, 2011), it is not clear whether they can tolerate other confounding factors and impurities associated with wastewater (e.g., coal seam gas (CSG)-produced water). The elements commonly found in oil- and gas-produced water are mainly sodium chloride (varying from 200 to more than 10,000 mg L^{-1}), sodium bicarbonate,

>2

borate, and various organic compounds (grease, oil, phenol, and benzene) (Xu and Drewes, 2006; Fakhru'l-Razi et al., 2009).

In Australia, the largest CSG production occurs in Queensland, and more than 88% of the total gas produced in that state comes from the Bowen and Surat Basins (Department of Natural Resources and Mines, 2014). The average production of low-quality CSG water is estimated at about 20,000 L per well per day (CSIRO, 2012). This may account for up to 125 GL year^{-1} when full CSG production in this region is achieved (McCormick et al., 2013). Importantly, due to logistical reasons industry often needs to dispose of saline wastewater on a regular basis and in large quantities. In the rainy seasons, that may result in either transient or prolonged soil waterlogging. If an irrigation schedule is not properly planned, plants may become flooded with saline water. Although some halophytes show a good degree of tolerance to waterlogging (Colmer and Flowers, 2008), anaerobic conditions may significantly reduce plant productivity (Bennett et al., 2009), making the use of saline wastewater commercially unviable. The aim of this study was to address some of the above issues and assess the performance of two halophyte forage species, *Atriplex halimus* and *Atriplex lentiformis*, grown under different irrigation regimens, for a range of salinity concentrations. The performance of these two species was compared to *Medicago arborea*, a traditional forage species grown in Mediterranean environments. *Medicago arborea* has a high nutritive value (comparable to lucerne, *Medicago sativa*) and is preferentially grazed by small ruminants (Lambert et al., 1989; Amato et al., 2004). Plant physiological and agronomic assessments were linked with changes in the soil chemical profile and changes in soil electrical conductivity (EC), to evaluate the environmental impact of saline irrigation.

19.2 Materials and Methods

19.2.1 Site Description and Basic Soil Properties

The study was conducted at the University of Tasmania farm located on the edge of Pitt Water, approximately 12 km northeast of Hobart in the Coal River Valley in Tasmania (42°48′ south and 147°26′ east), Australia. The soil at the study site consisted of an Aeolian-derived sandy loam (grayish yellow-brown—10YR4/2) A1 horizon (0–20 cm) and a mottled dispersive, vertic clay to sandy clay B2 subsoil with a field pH of 5.8–6.0 (Holz, 1993; Hardie et al., 2012). This soil is classified as

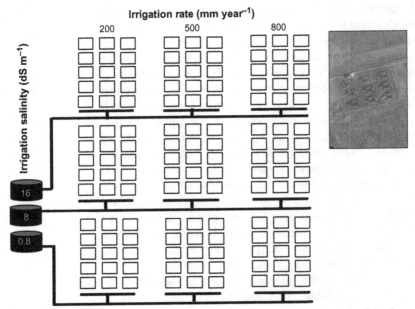

Figure 19.1 Layout of experimental plot in study site and aerial views of experimental plot (inset).

a texture-contrast soil (chromosol). The experimental design included three salinity levels (0.8, 8, and 16 dS m^{-1}), three irrigation levels (200, 500, and 800 mm year^{-1}; in addition to the natural rainfall), and three replicates (Figure 19.1). Saline irrigation was applied on a regular (weekly) basis, to match requirements of the CSG operations conducted in the Surat Basin area of the southwest gas fields of Queensland, where large amounts of saline wastewater are expected to be generated by these operations and need to be disposed of on a regular basis.

19.2.2 Climate

The Coal River Valley is classified as a "dry subhumid warm" area (Holz, 1993) and the climatic data recorded at Hobart airport (approximately 7.7 km from the experimental site) showed that the average annual rainfall and evaporation in the region were 469 and 1314 mm respectively during the period of 1981–2010 (Bureau of Meteorology, 2014). The rainfall is distributed relatively evenly throughout the year, however December had the highest rainfall (51 mm) while the lowest was in May (27.3 mm). The mean number of rain days per month is shown in Table 19.1. In 2012 and 2013, total annual

Table 19.1 The Mean Climatic Data Recorded at Hobart Airport (Station No 094008) for the Period 1981–2010

	Monthly Rainfall (mm)	Mean Daily Sunshine (h)	Mean Maximum Temperature (°C)	Mean Minimum Temperature (°C)	Mean Daily Evaporation (mm) (from 1986 to 2010)	Mean Number of Days of Rain
Jan	41	8	23	12	6	9
Feb	33	8	23	12	6	8
Mar	36	7	21	11	4	10
Apr	37	6	18	9	3	11
May	27	5	16	7	2	10
Jun	35	4	13	5	1	11
Jul	37	5	13	4	1	13
Aug	46	6	14	5	2	13
Sep	44	7	16	6	3	14
Oct	43	8	17	8	4	14
Nov	39	8	19	10	5	12
Dec	51	8	21	11	6	12
Annual	469	7	18	8	4	136

rainfall was 460 and 488 mm, respectively, which was within the range of the 30-year average.

19.2.3 Seedling Preparation and Transplantation

Seedlings of *A. lentiformis*, *A. halimus*, and *M. arborea* were grown in a potting mixture in the glasshouse for 3 months. Seedlings of approximately the same height were then selected for transplanting into the field in December 2011. Individual species plots were 2.1 m × 2.1 m with 0.6 m buffer rows between each plot. Nine plants of each species were sown at 0.7 m spacing in each plot, equivalent to a planting density of ∼20,400 plants per hectare.

19.2.4 Plant Height and Biomass Yield

Plant height and shoot canopy dimensions were measured after 14 months from time of transplanting the seedlings in the field. Forage biomass was harvested at a height of approximately 40 cm to avoid harvesting too much woody material and to encourage regrowth for further assessment. Total fresh weights were measured on site and representative samples were collected for determination of dry matter (DM) yield. Samples were dried in a forced-air oven at 60°C for 72 h, and reweighed to determine DM yield, expressed as kilograms of DM per hectare.

19.2.5 Soil Sampling and Analysis

Soil sampling was carried out before application of the treatments and during the application of treatments at 6-month intervals. Soil cores were taken to a depth of 2 m from each experimental block by hydraulic push tube (4.5 cm diameter). On one occasion soil samples were taken by hand auger up to a depth of 1.0 m. The cores were divided into 0–10, 10–20, 20–30, 30–50, 50–70, 70–100, 100–130, 130–160, and 160–200 cm fractions in the laboratory to measure the soil chemical properties. The chemical analysis was conducted on air dry soil that passed through a 2-mm sieve. The EC was measured by 1:5 soil/distilled water ratio solution stirred for 1 h at 25°C and then converted into EC values for saturated paste extract as described elsewhere (Sonmez et al., 2008).

Plant height and DM yield were analyzed by analysis of variance using Proc GLM in SAS 9.2 for the overall test and means. Pairwise differences were compared using the least significant difference method at the 5% level of probability. Salinity and water and their interaction were all fixed effect factors.

19.3 Results and Discussion

19.3.1 Plant Performance in the Field

Evaluation of plant performance under the imposed treatments showed a strong influence of salinity on plant growth and total biomass production. Plant height, measured after 14 months from time of transplanting into the field, showed, on average, a positive effect of amount (200, 500, or 800 mm year^{-1}) of irrigation water applied on the growth of *Atriplex* spp. In all treatments. The growth of *M. arborea*, however, was reduced when the amount of saline irrigation water was increased at the highest salinity level (Figure 19.2). The maximum height of *Atriplex* spp. was obtained from the high salt and high irrigation rate treatment (i.e., 16 dS m^{-1} @ 800 mm year^{-1}), while the lowest height was recorded for *M. arborea* for

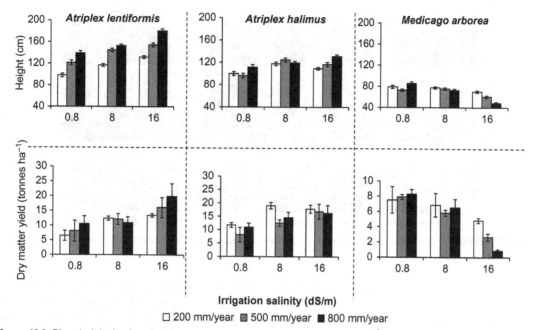

Figure 19.2 Plant height (cm) and estimated dry matter yield (tonnes DM ha ha^{-1}). Means $\pm$ SE ($n = 3$, sample measurements are comprised of nine individual plants).

the same treatment (Figure 19.3). The effect of salt on plant height of *A. halimus* was highly significant ($P = 0.0041$). There was no significant ($P < 0.05$) effect of irrigation, and there was no interaction between the two factors. For *A. lentiformis*, the effect of both salt and irrigation on plant height was highly significant ($P < 0.0001$). While there was no interaction between factors, higher rates of salt and irrigation increased plant height. For *M. arborea*, there was an overall significant ($P < 0.05$) effect of salinity level on plant height, with plants grown at 0.8 and 8 dS m^{-1} higher than the plants grown at 16 dS m^{-1}.

There was significant variation in biomass production between the treatments for all three species (Figure 19.2). The biomass yield of both *Atriplex* spp. was higher in saline conditions while the DM yield of *M. arborea* (a glycophyte species) declined with increasing salinity level in irrigation water. For *A. halimus*, the effect of salt on DM yield was highly significant ($P = 0.0017$) and positive, but there was no significant effect of irrigation, and there was no interaction between the two factors. The DM yield was not significantly different between medium- (8 dS m^{-1}) and high- (16 dS m^{-1}) salinity treatments; however, both were significantly ($P < 0.05$) higher than the mean yield of the low-salinity treatment (0.8 dS m^{-1}).

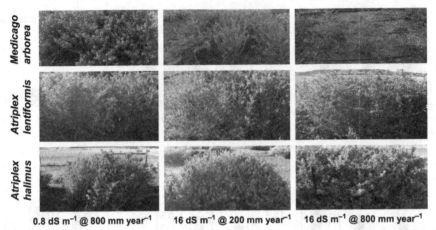

Figure 19.3 Performance of *Medicago arborea*, *Atriplex lentiformis*, and *Atriplex halimus* at different salinity and irrigation levels.

Similarly, for *A. lentiformis* the effect of salt was highly significant ($P = 0.0049$) and positive, but there was no significant effect of irrigation regimen, and there was no interaction between the two factors. For *M. arborea*, the effect of salt was very highly significant ($P < 0.0001$) but negative, and the lowest DM yield was measured in the high-salt treatment (16 dS m^{-1}).

When expressed on a per hectare basis, the highest DM yield of *A. lentiformis* was approximately 20 tonnes DM ha ha^{-1} at 16 dS m^{-1} @ 800 mm year^{-1} irrigation, while the highest yield for *A. halimus* yield (DM) was approximately 18 tonnes DM ha ha^{-1} at 8 dS m^{-1} @ 200 mm irrigation. On the other hand, the lowest DM yield (0.87 tonne DM ha ha^{-1}) of *M. arborea* was obtained for the 16 dS m^{-1} @ 800 mm year^{-1} irrigation treatment. At the same time, *M. arborea* DM production was much higher (up to 8 tonnes DM ha ha^{-1}) in the nonsaline and moderate saline treatments. It appears that at higher irrigation rates (800 mm year^{-1}) *M. arborea* plants were affected by a combination of high salinity and transient hypoxia which resulted in a large reduction in biomass production. Under the same conditions, *Atriplex* species were not affected by the excessive irrigation (and transient waterlogging) and benefited from both high-salinity/irrigation rate regimens.

19.3.2   Salinity Profiles in the Soil

The EC of the soil was measured to follow salt accumulation in the soil profile (Figure 19.4). In low-salt (0.8 dS m^{-1}) irrigation treatments, soil EC$_{se}$ values (saturated extract) were always

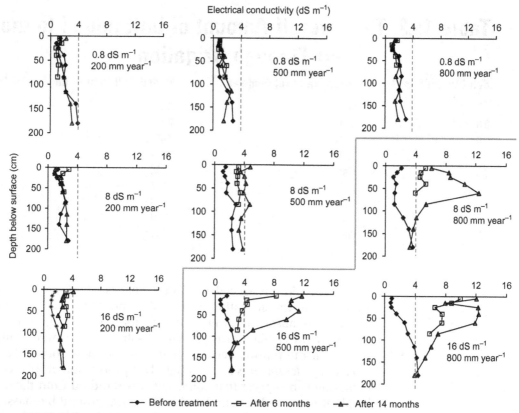

Figure 19.4 Variation of estimated saturated paste electrical conductivity of the soil before and after saline irrigation (conversion based on Sonmez et al., 2008). Dotted lines indicate salinity level threshold.

below the $4\,dS\,m^{-1}$ level considered a threshold for salinity (Shabala and Munns, 2012), irrespective of irrigation regimen (Figure 19.4). For the $8\,dS\,m^{-1}$ treatment, a significant (at $P < 0.05$) increase in soil EC above the "safe" $4\,dS\,m^{-1}$ threshold was observed only for the highest irrigation regimen (800 mm $year^{-1}$), and only in one particular horizon (20–60 cm). This increase is hardly surprising given the fact that the equivalent of ~ 35 tonnes of NaCl were added per hectare over a period of about 1 year (Table 19.2).

The highest ($16\,dS\,m^{-1}$) salinity treatment resulted in a significant increase in soil EC_{se} at both the 500 and 800 mm $year^{-1}$ irrigation regimens. Again, this increase was highest at the soil depth of 60 cm and dropped sharply afterwards. Even in this case, however, soil EC_{se} levels did not exceed those of the irrigation water treatments. Interestingly, at the highest treatment ($16\,dS\,m^{-1}$ @ 800 mm $year^{-1}$) the levels of salt measured in the

Table 19.2 The Overall Amount of Salt added to the Soil Through Irrigation

Salt Level (dS m^{-1})	Irrigation Level (mm year^{-1})	Added Salt (tonnes ha^{-1} year^{-1})
	200	0.87
0.8	500	2.18
0.8	800	3.49
8	200	8.72
8	500	21.80
8	800	34.88
16	200	17.44
16	500	43.61
16	800	69.77

soil profile did not exceed 12 dS m^{-1}, that is, about 75% of that added with the irrigation water. Two possible reasons may explain this observation. First, it appears that high amounts of irrigation water may have prevented salt build-up by leaching salts from the topsoil. Second, given the fact that both *A. halimus* and *A. lentiformis* showed high DM production at this regimen, one can suggest that some salt was removed from the soil by plant roots and accumulated in the aboveground biomass. It is widely accepted that cell turgor is maintained by storage of Na$^+$ and Cl$^-$ in vacuoles, with the solute potential of the cytosol adjusted by accumulation of K$^+$ and organic solutes (Flowers et al., 1977; Storey and Wyn Jones, 1979; Storey, 1995; Glenn et al., 1999; Shabala and Mackay, 2011). According to Glenn et al. (1999), the three major inorganic ions, Na$^+$, K$^+$, and Cl$^-$, account for 80–95% of the cell sap osmotic pressure in both halophyte grasses and dicots. As a result, halophytes accumulate substantial amounts (>10% of dry weight each) of Na$^+$ and Cl$^-$ in their shoots (Shabala and Mackay, 2011). Previous research showed that some halophyte species have the capacity to remove 1–6 tonnes NaCl ha^{-1} year^{-1} and, thus, can be used for desalination purposes (reviewed by Panta et al., 2014). Importantly, this NaCl removal capacity was reported to be strongly dependent on water availability, largely due to the difference in transpiration rate and salt delivery to the shoot (Norman et al., 2008).

Irrigation with 8 dS m^{-1} @ 500 mm year^{-1} and 16 dS m^{-1} @ 200 mm year^{-1}, both of which applied ~20 tonnes ha^{-1} year^{-1}

(Table 19.2), did not increase the soil EC level above the salinity threshold of 4 dS m^{-1}. It remains to be answered to what extent this pattern has resulted from soil chemistry/hydrology processes, or salt accumulation in the plant.

The highest three treatments (framed in Figure 19.4) brought soil EC levels above the threshold 4 dS m^{-1} of salinity and cannot therefore be considered environmentally sustainable. However, despite 14 months of irrigation at the highest (16 dS m^{-1} @ 800 mm year^{-1}) regimen applying over 80 tonnes ha^{-1} of salt to the soil (Table 19.2), the EC values did not exceed 12 dS m^{-1} (Figure 19.4, lower panel)—the levels considered to be most optimal for halophyte growth (Flowers and Colmer, 2008; Shabala and Mackay, 2011). Assuming this trend continues, and soil EC$_{se}$ values stay at 12 dS m^{-1} level and do not increase over the following years, our data suggest a possibility of long-term use of large quantities of industrial quality water for the purpose of growing halophyte species without yield penalties. The latter prediction needs to be tested in direct long-term experiments.

19.4 Conclusions

From the above results it appears that *Atriplex* spp. can be successfully grown using high-salinity irrigation water (16 dS m^{-1}) and a large quantity of water (800 mm year^{-1} in addition to the natural rainfall), without any detrimental impact on plant growth and biomass production. *Medicago arborea* plants, however, suffered from transient waterlogging when irrigated with 800 mm year^{-1} and could not be used for the purpose of disposing of high amounts of saline water. This suggests that *A. lentiformis* and *A. halimus* can be successfully grown to use large quantities of industrial-quality water. Longer-term studies (at least several more years) are needed to evaluate the environmental impact of this practice and before making recommendations to the industry. It is also important to check if these conclusions could be extrapolated to other soil types and environmental conditions that may affect both soil hydrology and plant performance.

Acknowledgments

This work was supported by the ARC Linkage grant to Sergey Shabala and Gabriel Haros. Dr. David Ratkowsky is acknowledged for his assistance with statistical analysis of data related to this work.

References

Amato, G., Stringi, L., Giambalvo, D., 2004. Productivity and canopy modification of *Medicago arborea* as affected by defoliation management and genotype in a Mediterranean environment. Grass Forage Sci. 59, 20–28.

Barrett-Lennard, E.G., 2002. Restoration of saline land through revegetation. Agric. Water Manage. 53, 213–226.

Bennett, S.J., Barrett-Lennard, E.G., Colmer, T.D., 2009. Salinity and waterlogging as constraints to saltland pasture production: a review. Agric. Ecosyst. Environ. 129, 349–360.

Bureau of Meteorology, 2014. Climate Statistics for Australian Locations—Hobart Airport. <http://bom.gov.au/climate> (accessed 07.08.14.).

Colmer, T.D., Flowers, T.J., 2008. Flooding tolerance in halophytes. New Phytol. 179, 964–974.

CSIRO, 2012. Coal Seam Gas-Produced Water and Site Management, (fact sheet). CSIRO, Melbourne (accessed 28.06.14.).

Department of Natural Resources and Mines, 2014. Queensland's Coal Seam Gas Overview. January 2014. <http://mines.industry.qld.gov.au/assets/coal-pdf/csg-update-2014.pdf>, viewed 10 September 2014.

Fakhru'l-Razi, A., Alireza, P., Luqman, C.A., Dayang, R.A.B., Sayed, S.M., Zurina, Z.A., 2009. Review of technologies for oil and gas produced water treatment. J. Hazard. Mater. 170, 530–551.

Flowers, T.J., Colmer, T.D., 2008. Salinity tolerance in halophytes. New Phytol. 179, 945–963.

Flowers, T.J., Troke, P.F., Yeo, A.R., 1977. The mechanism of salt tolerance in halophytes. Annu. Rev. Plant Physiol. 28, 89–121.

Gerhart, V., Kane, R., Glenn, E., 2006. Recycling industrial saline wastewater for landscape irrigation in a desert urban area. J. Arid Environ. 67, 473–486.

Glenn, E.P., Brown, J.J., Blumwald, E., 1999. Salt tolerance and crop potential of halophytes. Crit. Rev. Plant Sci. 18, 227–255.

Glenn, E.P., McKeon, C., Gerhart, V., Nagler, P.L., Jordan, F., Artiola, J., 2009. Deficit irrigation of a landscape halophyte for reuse of saline waste water in a desert city. Landscape Urban Plan. 89, 57–64.

Grattan, S.R., Benes, S.E., Peters, D.W., Daiz, F., 2008. Feasibility of irrigating Pickledweed (*Salicornia bigelovii*. Torr) with hyper-saline drainage water. J. Environ. Qual. 37, S-149–S-156.

Grattan, S.R., Grieve, C.M., Poss, J.A., Robinson, P.H., Suarez, D.L., Benes, S.E., 2004. Evaluation of salt-tolerant forages for sequential water reuse systems. III. Potential implications for ruminant mineral nutrition. Agric. Water Manage. 70, 137–150.

Grieve, C., Poss, J., Grattan, S., Suarez, D., Benes, S., Robinson, P., 2004. Evaluation of salt-tolerant forages for sequential water reuse systems. II. Plant-ion relations. Agric. Water Manage. 70, 121–135.

Hardie, M., Cotching, W.E., Doyle, R.B., Lisson, S., 2012. Influence of climate, water content and leaching on seasonal variations in potential water repellence. Hydrol. Process. 26, 2041–2048.

Holz, G., 1993. Principals of Soil Occurrence in the Lower Coal River Valley, S.E. Tasmania (Ph.D. dissertation). University of Tasmania, Hobart, TAS, Australia.

Jordan, F.L., Yoklic, M., Morino, K., Brown, P., Seaman, R., Glenn, E.P., 2009. Consumptive water use and stomatal conductance of *Atriplex lentiformis* irrigated with industrial brine in a desert irrigation district. Agric. For. Meteorol. 149, 899–912.

Lambert, M.G., Jung, G.A., Fletcher, R.H., Budding, P.J., Costall, D.A., 1989. Forage shrubs in North Island hill country. 2. Sheep and goat preferences. N. Z. J. Agric. Res. 32, 485–490.

Masters, D.G., Benes, S.E., Norman, H.C., 2007. Biosaline agriculture for forage and livestock production. Agric. Ecosyst. Environ. 119, 234–248.

McCormick, B., John, A., St Tomaras, J., 2013. Environment Protection and Biodiversity Conservation Amendment Bill 2013, Bills Digest, No.108 2012–13. Australian Parliamentary Library, Canberra, p. 8 <http://www.aph.gov.au/Parliamentary_Business/Bills_Legislation/bd/bd1213a/13bd108> (accessed 28.06.14.).

Norman, H.C., Masters, D.G., Wilmot, M.G., Rintoul, A.J., 2008. Effect of supplementation with grain, hay or straw on the performance of weaner Merino sheep grazing old man (*Atriplex nummularia*) or river (*Atriplex amnicola*) saltbush. Grass Forage Sci. 63, 179–192.

Panta, S., Flowers, T., Lane, P., Doyle, R., Haros, G., Shabala, S., 2014. Halophyte agriculture: success stories. Environ. Exp. Bot. 107, 71–83.

Rogers, M., Craig, A., Munns, R., Colmer, T., Nichols, P., Malcolm, C., et al., 2005. The potential for developing fodder plants for the salt-affected areas of southern and eastern Australia: an overview. Aust. J. Exp. Agric. 45, 301–329.

Shabala, S., Mackay, A., 2011. Ion transport in halophytes. Adv. Bot. Res. 57, 151–199.

Shabala, S., Munns, R., 2012. Salinity stress: physiological constraints and adaptive mechanisms. In: Shabala, S. (Ed.), Plant Stress Physiology. CABI Publishing, Wallingford, UK, pp. 59–93.

Shannon, M., Cervinka, V., Daniel, D., 1997. Drainage water reuse. In: Madramootoo, C., Johnson, W., Willardson, L. (Eds.), Management of Agricultural Drainage Water Quality. FAO, Rome (Chapter 4). http://www.fao.org/docrep/w7224e/w7224e08.htm (accessed 29.08.14.).

Sonmez, S., Buyuktas, D., Okturen, F., Citak, S., 2008. Assessment of different soil to water ratios (1:1, 1:2.5, 1:5) in soil salinity studies. Geoderma. 144, 361–369.

Storey, R., 1995. Salt tolerance, ion relations and the effect of root medium on the response of citrus to salinity. Aust. J. Plant Physiol. 22, 101–114.

Storey, R., Wyn Jones, R.G., 1979. Responses of *Atriplex spongiosa* and *Suaeda monoica* to salinity. Plant Physiol. 63, 156–162.

Xu, P., Drewes, J.E., 2006. Variability of nanofiltration and ultra-low pressure reverse osmosis membranes for multi beneficial use of methane produce water. Sep. Purif. Technol. 52, 67–76.

INDEX

Note: Page numbers followed by "*b*", "*f*", and "*t*" refer to boxes, figures, and tables, respectively.

A

ABC Transporters, 170–171
Abies Miller, 295–296
Abiotic stresses, *Chenopodium quinoa*, 262
Accessibility, in food security, 111
Achillea biebersteinii Afan., 310–313
Aeluropus lagopoides
  growth of grass, 7, 12–13
  lipid membrane peroxidation, 7
  malondialdehyde content, 5, 12–13
  Na$^+$
    flux in, 8
    measurement, 4
    secretion of, 4–5, 8–9
  plant material collection, 3
  sodium exchanger genes characterization
    cDNA isolation and sequence analysis of, 3, 12
    gene expression by qRT-PCR, 4, 10–11, 13–14
    molecular characterization, 5–7
    statistical analyses, 5
Afghanistan
  halophytes in, 52–54
  vegetation, 48–52
Agricultural land salinity, 245–246
AlaNHX, 6*f*, 10
Anti- or nonnutritional compounds, high concentration of, 250
Antioxidant responses, 208–209

Antiporters, 171–172
Aquaporin, 172
Arabidopsis, gibberellic acid (GA) genes, 170
Arctium platylepis, 310–313
Arid zone, in Morocco, 140–141
Aromatic halophytes, in Çankiri and Iğdır Provinces, 307–313, 311*t*
Artemisia santonicum L., 310–313
Arthrocnemun macrostachyum, 19–20
Ascorbate peroxidase activity, *Chenopodium quinoa*, 264
Ascorbic acid (AsA), 219–222, 224
Atraphaxis spinosa L., 310–313
Atriplex halimus, saline irrigation, 321
  climate, 322–323
  plant height and biomass yield, 323–324
  plant performance in field, 324–326
  salinity profiles in soil, 326–329
  seedlings, 323
  site description and basic soil properties, 321–322
  soil sampling and analysis, 324
Atriplex lentiformis, saline irrigation, 321
  climate, 322–323
  plant height and biomass yield, 323–324
  plant performance in field, 324–326

  salinity profiles in soil, 326–329
  seedlings, 323
  site description and basic soil properties, 321–322
  soil sampling and analysis, 324
Atriplex nummularia
  chemical composition, 188–189
  glycinebetaine in, 170
Atriplex spp., 250–251, 254–255
Availability, in food security, 111

B

Bamyan-Valley, central Afghanistan, 49*f*
Bio-climate in Morocco, 140–141
Biodiversity in Morocco, 141
Biomass yield, of *Salicornia bigelovii*
  greenhouse procedures, 68
  harvest and processing, 69
  in *Salicornia bigelovii*, 68
  seed purity and proximate analysis, 69
  survival and growth, 72
Biosphere reserves
  buffer zone, 125
  conservation function, 125
  core area, 125–126
  development function, 125
  evolution of, 126*f*
  list of, 134*t*
  logistic support function, 125
  transition zone, 126

C

Cakile maritima
  antioxidant responses,
      208–209
  cellular mechanisms,
      209–211
  dispersal and environmental
      adaptation, 201–203
  early osmotic and ionic
      effects, 206–207, 208t
  environmental adaptation,
      201–203
  latitudinal distribution,
      200–201
  as model plant, 211–212
  physiological mechanisms,
      204–206
  responses at development,
      205t
  taxonomic diversity, 200–201
Candidate genes, for salt
    tolerance, 247
Can Gio Mangrove Biosphere
    Reserve, 136
Çankiri
  chorotypes of halophytic
      plant taxa, 308f
  continental climate, 295
  distribution of halophytes,
      307f
  ecological characteristics of
      halophytes, 300t
  ecological types of
      halophytes, 307f
  economical evaluations,
      299–307
  floral diversity of, 296
  groundwater, 295
  halophyte diversity, 299
  halophytes used as food, 308t
  halophytes with fodder
      potential, 309t
  location, 295
  medicinal and aromatic
      halophytes in, 307–313,
      311t
  rock salt mine area, 297
  salt cave, 297

  vegetation in, 295–296
Cannabis sativa, 53
Cannabis sativa ssp. *indica*,
    59–60
Capacity building, 193–194
Carbon mitigation
  CO_2 rising in salt marshes,
      95–102
  electron transport rates,
      96–99, 98f
  global warming and carbon
      stocks, 91–95
  halophytes in, 86–87
  hydrological control of
      carbon stocks, 89–91
  light saturation constants,
      96–99, 98f
  out-welling carbon, 88
  photosynthetic efficiency,
      96–99, 98f
  production and losses, 87t
Catalase activity, *Chenopodium
    quinoa*, 264
Cellular mechanisms of *Cakile
    maritima*, 209–211
Chenopodiaceae, *Halocnemum
    strobilaceum*, 50f
Chenopodium album L.,
    310–313
Chenopodium quinoa, 264–265
  abiotic stresses, 262
  ascorbate peroxidase activity,
      264
  carotenoids, 268–269
  catalase activity, 264
  cation and anion content,
      269–270, 270t, 272t
  cation and anion detection,
      266
  cultivation, 261–262
  enzyme activities, 267–268
  enzyme activities
      determination, 264–265
  Folin-Ciocalteu reagent, 265
  growth chamber experiment,
      262–263
  growth parameters, 267t
  ion content, 269–270

o-guaiacol-peroxidase
    activity, 264
  photosynthetic pigments,
      268–269
  plantlet growth in pots,
      263–264
  plant material, 263
  proline detection, 266,
      268–269
  seedling growth, 266
  statistical analysis, 266
  superoxide dismutase activity,
      264–265
  total antioxidant capacity
      determination, 265,
      268–269, 270t
  total phenolic content, 265,
      270t
  water deficiency, 262–263
Chenopodium quinoa, seed
    evaluation of
  biochemical analysis of seed,
      41, 42t
  carbohydrate content, 42
  characteristics of, 38
  flavonoid content, 43–44
  irrigation protocol, 39–41
  lipid content, 44
  physical and chemical
      characteristics of soil, 40,
      40t
  polyphenol content, 42–43
  protein content, 41–42
  seed composition, 41–42
  site experiment, 39
  vitamin C, 42
  yield data, 41
Climate change, 109, 245–246
Climax community, 290
Conservation function, 125
Consumption of halophytes,
    299–307
Convention on Wetlands of
    International
    Importance, 124
Convolvulus arvensis L., 310–313
Convolvulus scammonia L.,
    310–313

Crop solutions, forage, 246–247
Cross crop species development, 116
Crude protein (CP), 249

D
Deciduous forest, 287
Deserts
  desertification, 60–61
  Ferula and *Dorema* species in, 53
  mangrove ecosystems in, 129
  nonsaline/semi-deserts, 52
Digestibility energy, 249
Distichlis spicata, 254–255
Drâa river basin
  agriculture in, 140–141
  salinity in, 145
Dryland salinity, 246
Dry matter (DM) production, 249

E
Early osmotic and ionic effects, 206–207, 208t
ECe, soil salinity, 29
Economical evaluations, 299–307
Egypt
  climate, 180
  deserts, 180
  general characteristics, 180–181
  North Sinai, 181–182
  Ras Sudr Area, 182
Electrical conductivity (ECa) of soil, 22–24, 29–30
El Tina plain, 181–182
EMI, soil salinity
  calibration and mapping, 22–24
  EMI survey, 24–33
  multi-temporal EMI measurement, 31–33
Environmental adaptation of *Cakile maritima*, 201–203
Environmentally Critical Areas Network (ECAN), 133–136

Eryngium campestre L., 310–313
Estuarine systems, 81–82
Euphorbia macroclada Boiss., 310–313
Eutrema salsugineum
  gene targets in, 173
  transposable elements, 173
Exogenous chemical treatments
  seed collection sites, 217
  seed germination, effect on, 217–219, 223–224
  on seed growth, 218–222, 224–226
  statistical analyses, 219
  test species, 216–217
  water-spray on salinity tolerance, 218–219, 222–223, 226

F
Feeding value
  of chenopod shrubs, 251–254
  herbivory and environmental stress, 248, 250
Fertilizer and livestock production, 255
Ferula assa-foetida, 53
Floating Mangroves, 132
Flood irrigation, 245–246
Fodder crops, in Afghanistan, 56
Fodder crop species
  capacity building and economic assessment, 193–194
  chemical composition of, 188–189
  evaluate the nutritive value, 189–191
  reproductive and productive performance of sheep and goat, 191–193
Food security
  accessibility, 111
  availability, 111
  climate change, 109
  cross crop species development, 116

definition, 110–111
  local cultivars development, 115
  local salt-tolerant grains, 117–118
  new crops for local consumption, 117
  in oil-rich countries, 112–113
  poorest regions, 113–114
  population growth, 109–110
  quinoa, 117
  salinity in agriculture, 114–115
  stability, 111
  usability, 111
  vegetables, 116
Forage
  and crop solutions, 246–247
  halophytes, 247

G
Galium humifusum Bieb., 310–313
Galium tricornutum, 310–313
Genes
  for cell maintenance, 167–168
  ion transporters encoding, 170–172
  LEA protein coding genes, 172–173
  mitochondrial and ROS related genes, 169
  photosynthetic genes, 169
  plant hormones encoding, 170
  proline and other amino acids, 169–170
  regulatory molecules, 172
  stress genes, 168–169
  transposable elements, 173
Genetic improvement of halophytes, 251–254
Geographical information system (GIS), 21
Germanikopolis. *See* Çankiri
Gibberellic acid (GA) genes, 170
Glycinebetaine (GB), 170
Glycophytes *vs.* halophytes, 206–209

Glycyrrhiza sp., 52
    glabra L., 59–60, 310–313
Grazing, 55–56
Green Morocco Plan, 150–151
Groundwater and seawater
    intrusion into, 246

H
Halanthium rarifolium C. Koch,
    310–313
Halimione portulacoides, 97*t*
Halocnemum strobilaceum
    (Chenopodiaceae), 50*f*
Halophytes
    biotic factors, 55
    Cannabis sativa, 53
    carbon pump, 86–87
    CO_2 rising in salt marshes,
      95–102
    Ferula assa-foetida, 53
    Glycyrrhiza species, 52
    grazing, 55–56
    medicinal plants, 53
    productivity, 47–48
    for salinity affected areas,
      146–147
    species in Afghanistan, 57*t*
    woody plants, 53–54
Haloxylon aphyllum, 51*f*
HDZip genes, 167
Heritiera fomes, 281
    germination, 282–290
    natural habitat, 284*f*
High-nonprotein nitrogen, 249
Humid zone, 140–141
Hydrogen peroxide (H_2O_2) seed
    germination, 224

I
IĞDIR
    chorotypes of halophytic
      plant taxa, 308*f*
    continental climate, 298
    distribution of halophytes,
      307*f*
    ecological characteristics of
      halophytes, 300*t*
    ecological types of
      halophytes, 307*f*

economical evaluations,
    299–307
halophyte diversity, 299
halophytes used as food, 308*t*
halophytes with fodder
    potential, 309*t*
location, 297–298
medicinal and aromatic
    halophytes in, 307–313,
    311*t*
Sahat concavity, 298
Sürmeli concavity, 298
vegetation, 298
Improved management package
    (IPM), 184–185
International Convention for
    the Protection of the
    World Cultural and
    Natural Heritage,
    123–124
Ionic effects, 206–207, 208*t*
Ion transporters encoding
    genes, 170–172

J
Juniperus L., 295–296

K
Khettaras, 145–146

L
LEA protein coding genes,
    172–173
Leucaena leucocephala,
    chemical composition,
    188
Livestock production
    anti- or nonnutritional
      compounds, 250
    dry lands, 248
    environmental manipulation
      fertilizer, 255
      salinity, 255
      water, 254–255
    genetic improvement,
      251–254
    high mineral composition,
      249–250
    limitations in, 248–251

low digestibility and
    metabolizable energy,
    249
low dry matter production,
    249
low-protein/high-nonprotein
    nitrogen, 249
in mixed farming systems,
    247
palatability, 253–254
in vitro measurements, 253
Livestock production system
    activities and achievements,
      183–194
    farmer-based seed
      production, 183, 184*t*
    fodder crop species
      capacity building and
        economic assessment,
        193–194
      chemical composition of,
        188–189
      evaluate the nutritive value,
        189–191
      reproductive and
        productive performance
        of sheep and goat,
        191–193
    fodder crop utilization and,
      187–193
    integrated management
      package, 184–187
    summer season, 186–187
    winter season, 185–186
Local cultivars development,
    115
Local salt-tolerant grains,
    117–118
Low-protein nitrogen, 249

M
Malondialdehyde (MDA)
    content, 5, 12–13
Malva neglecta Wallr., 310–313
Man and Biosphere Programme
    (MAB)
    biosphere reserves, 124–125
    buffer zone, 125
    conservation function, 125

core area, 125–126
development function, 125
evolution of, 126*f*
logistic support function, 125
transition zone, 126
Mangrove ecosystems
aquaculture, 129–130
biomass productivity, 128
Can Gio Mangrove Biosphere
Reserve, 136
desert and semi-desert
countries, 129
economic and ecological
value, 127
Floating Mangroves, 132
global distribution, 127, 127*f*
island and coastal biosphere
reserves, 132
in Latin America, 130
nutritional quality, 128
Palawan Island, 136*f*
in scenic and aesthetic
functions, 128
Securing the Future of
Mangroves, 131–132
for wave attenuation,
128–129
World Atlas of Mangroves, 131
World Network of Biosphere
Reserves, 132–133
Marginal dry areas, in Morocco
arid zone in, 140–141
bio-climate in, 140–141
biodiversity in, 141
Drâa river basin, 145
humid and sub-humid zone
in, 140–141
latitudinal extension, 140
Massa, 144
Saharan zone, 140–141
salinity in, 142–145
semiarid zone, 140–141
topography, 140
vulnerability of, 141–142
youth potential in arid areas,
147–151
Massa
agriculture in, 140–141
salinity in, 144

Medicago arborea, saline
irrigation, 321
climate, 322–323
plant height and biomass
yield, 323–324
plant performance in field,
324–326
salinity profiles in soil,
326–329
seedlings, 323
site description and basic soil
properties, 321–322
soil sampling and analysis,
324
Medicinal halophytes, 307–313,
311*t*
Medicinal plants, 53, 56–59
Mesohaline Zone, plant
communities, 285, 286*t*
Metabolizable energy, 249
Millettia pinnata gene
expression, 168–169
Mineral composition, 249–250
Mitochondrial and ROS related
genes, 169
Molinia sp., 286–287
Morocco
arid zone in, 140–141
bio-climate in, 140–141
biodiversity in, 141
crop losses, 37–38
Drâa river basin, 145
humid and sub-humid zone
in, 140–141
latitudinal extension, 140
Massa, 144
Saharan zone, 140–141
salinity in, 37–38, 142–145
semiarid zone, 140–141
topography, 140
vulnerability of, 141–142
youth potential in arid areas,
147–151

N

NaCL stress, in *Aeluropus
lagopoides*, 2–3
New crops, for local
consumption, 117

Next-generation sequencing
(NGS), 155, 163–167.
See also Transcriptomes
NHX1, 171–172
Northern Aral Sea, Kazakhstan,
51*f*
Nummularia, 251–253, 252*f*
Nutritional value
Chenopodium quinoa seed,
41–44
Nutritive value of halophytes,
255

O

O-guaiacol-peroxidase activity,
Chenopodium quinoa,
264
Oil-rich countries, food security,
112–113
Oilseed yield, of *Salicornia
bigelovii*
greenhouse procedures, 68
harvest and processing,
68–69
seed purity and proximate
analysis, 69
survival and growth, 71
Oligohaline Zone, plant
communities, 285, 286*t*
One-Way ANOVA, 7*t*
Osmotic adjustment, 225
Osmotic effects, 206–207, 208*t*
Oxidation-reduction potentials,
286–287
Oxygen diffusion rate (ODR),
286–287

P

Palatability, 253–254
Palawan Biosphere Reserve, 133
Palawan Island Biosphere
Reserve, 136*f*
Panicum turgidum, 254–255
Papaver somniferum, 59–60
Pearl millet, 186
Peganum harmala, 55*f*, 59–60,
310–313
Permanent Commission for the
South Pacific (CPPS), 132

Photosynthetic genes, 169
Phragmites australis, 310–313
Physiological mechanisms of
  Cakile maritima,
  204–206
Phytomelioration, 47
Pinus L., 295–296
Plantago lanceolata, 307–310
Plantago major ssp. *intermedia*,
  310–313
Plantago maritima L., 307–310
Plantago media L., 310–313
Plant hormones encoding
  genes, 170
PMNHX, in *Aeluropus
  lagopoides*
  cDNA isolation and sequence
    analysis of, 3, 12
  gene expression by qRT-PCR,
    4, 10–11, 13–14
  molecular characterization,
    5–7
Poa bulbosa L., 310–313
Polyhaline zone
  plant communities, 285, 286*t*
  salinity in, 286–287
Populus euphratica
  metabolic pathways, 173–174
  photosynthetic genes, 169
Populus L., 295–296
Protein phosphatase 2C PP2C,
  172

Q
qRT-PCR, *VNHX* and *PMNHX*
  gene expression by, 4,
  10–11, 13–14
Quinoa *See Chenopodium
  quinoa*, seed evaluation
  of

R
*Ranunculus
  constantinopolitanus*,
  307–310
Reactive oxygen species (ROS),
  169, 262
RNA sequencing
  alignment, 164*b*

assembly, 164*b*
De Bruijn graph approach,
  164
differential expression, 164
gene and isoforms, 163
gene annotation, 164
mapping of short reads,
  156–163
methods for, 156, 162*f*
overlap layout consensus, 164
sequence aligners, 164
transcriptome reconstruction,
  163
visualization, 164*b*
Rock salt deposits, 294–295

S
Sabkhas, 50–52
Saharan zone, 140–141
Sahl El Tina, 181–182
Salicornia, 254–255
Salicornia bigelovii
  biology of, 66
  crop observations, 68
  environmental
    measurements, 70–71
  experimental design, 66–67
  greenhouse procedures, 68
  harvest and processing,
    68–69
  oil content, 75, 78*t*
  seed purity and proximate
    analysis, 69, 78*t*
  statistical methods, 71
  survival and growth, 71–74
  temperature effects, 75, 76*f*
  wild accessions sources, 66,
    67*t*
Salicornia europaea, 54*f*
  gibberellic acid (GA) genes,
    170
  photosynthetic genes, 169
Salinas, salt lake of (Alicante,
  Spain), 17, 18*f*
Saline irrigation, 321
  materials and methods
    climate, 322–323
    plant height and biomass
      yield, 323–324

seedlings, 323
site description and basic
  soil properties, 321–322
soil sampling and analysis,
  324
results and discussion
  plant performance in field,
    324–326
  salinity profiles in soil,
    326–329
Saline wastewater disposal,
  319–321
Salinity
  in agriculture, 114–115
  large and expanding areas of,
    246–247
  and livestock production, 255
  in Polyhaline zone, 286–287
Salinization, in agricultural
  land, 245–246
Salix L., 295–296
Salsola dendroides Pall.,
  310–313
Salt marshes
  CO_2 rising in, 95–102
  hydrological control of
    carbon stocks, 89–91
  Mediterranean wetlands,
    82–83
  oxygen, 84–86
  sediment CO_2, 84–86
  sediment microbial
    communities, 83–84
  vegetated and nonvegetated
    sediments, 82–83
Salt Overly Sensitive 1 (SOS1),
  171
Salt tolerant species genotypes,
  184*t*
Salt-degraded land areas,
  293–295. *See also*
  Çankiri; IĞDIR
Salt-tolerant plants, in Morocco,
  148*t*
Saxaul *(Haloxylon aphyllum)*,
  51*f*
Schrenkiella parvula, 168
Seawater and intrusion into
  groundwater, 246

Securing the Future of Mangroves, 131–132
Sediment microbiology, in salt marshes, 83–84
Seed germination, of *Suaeda fruticosa*, 217–219, 223–224
Seed growth, of *Suaeda fruticosa*, 218–222, 224–226
Seedlings, 323
Seguias, 145–146
Semiarid zone, 140–141
Sequence aligners, 164
Shorea robusta, 287
Shrubs, halophytic, 249, 251–253
Sodium exchanger genes characterization
  growth of plant, 7, 12–13
  lipid membrane peroxidation, 7
  malondialdehyde content, 5, 12–13
  Na$^+$
    flux in, 8
    measurement, 4
    secretion of, 4–5, 8–9
  plant material, 3
  statistical analyses, 5
  VNHX and *PMNHX*
    cDNA isolation and sequence analysis of, 3, 12
    gene expression by qRT-PCR, 4, 10–11, 13–14
    molecular characterization, 5–7
Soil construction, 19–20
Soil salinity
  area of study, 19–20
  ECa, spatial variation of, 29–30
  ECe calibration and mapping, 29
  electromagnetic survey, 22–24
  EMI calibration and mapping, 22–24

EMI survey, 24–33
  multi-temporal EMI measurement, 31–33
  plants inventory, 21–22
  problems caused, 155–156
  soil properties, 26–29
  soil sampling and analysis, 22
  soil survey, 21
  vegetation inventories, 21, 26
Solonchak type saline, 144–145
Spartina
  carbon pump, 86–87
  S. maritima, 89–90, 97t
Spiculata, 251–253, 252f
Sporobolus virginicus, 254–255
Sprinkler irrigation, 245–246
Stability, in food security, 111
Stable isotope, 86f, 97f
Strategic Environmental Plan for Palawan (SEP), 133–136
Stress genes, 168–169
Stress tolerance plants, transcriptomes/genomes of, 165t
Suaeda fruticosa
  exogenous chemical treatments on
    seed collection sites, 217
    seed germination, effect on, 217–219, 223–224
    seed growth, 218–222, 224–226
    water-spray on salinity tolerance, 218–219, 222–223, 226
  halophyte transcriptomics
    f-box kelch protein, 168
    mitochondrial and ROS related genes, 169
    photosynthetic genes, 169
Suaeda maritima
  cell maintenance genes, 167–168
  mitochondrial and ROS related genes, 169
Sub-humid zone, 140–141
Sundarban mangrove forest, Bangladesh

abundance of species, 288, 289f
agro-ecological regions, 280f
climax community, 290
germination, 279–281
H. fomes, germination, 282–290
low-lying areas, 279–280
Mesohaline Zone, 285, 286t
Oligohaline Zone, 285, 286t
plant communities in, 285–286
plant density, 284f
plant diversity, values of indices, 288, 288t
Polyhaline Zone, 285, 286t
rivers and locations, 281f
situation, 279–280
species diversity, 281
vegetation of, 285–290, 286t
X. mekongensis, germination, 282–290
Superoxide dismutase activity, 264–265

T
Tamarix species, 310–313
Tamarix tree, 51f
Teucrium polium L., 307–310
Total antioxidant capacity, 265
Total phenolic content, 265
Transcriptome reconstruction, 163
Transcriptomes. *See also* Genes; RNA sequencing
  genes, 167–172
  genomic elements, 173
  LEA protein coding genes, 172–173
  pathways, 173–174
  regulatory molecules, 172
  RNA sequencing, 156, 162f
  for salt-tolerance studies, 163–167
Typha latifolia L., 310–313
Typha laxmannii Lepecbin, 310–313
Typha minima Funck, 310–313

U

UNESCO
  environmental protection,
    123–124
  Man and Biosphere
    Programme, 124–126,
    131–133
United Nations Educational,
  Scientific and Cultural
  Organization (UNESCO).
  See UNESCO
Usability, in food security, 111

V

Vegetation
  of clearing spaces, 285
  of forest proper, 285

high-saline sabkhas, 50–52
nonsaline/semi-deserts, 52
in salt lakes and saline flats,
  49, 50*t*
Vitamin C, in quinoa seeds, 42
VNHX, in *Aeluropus lagopoides*
  cDNA isolation and sequence
    analysis of, 3, 12
  gene expression by qRT-PCR,
    4, 10–11, 13–14
  molecular characterization,
    5–7

W

Water and livestock production,
  255
World Atlas of Mangroves, 131

World Network of Biosphere
  Reserves, 132–133

X

Xylocarpus mekongensis,
  281–282

Y

Young human resources, in arid
  areas, 147–151

Z

Zygophyllaceae, 55*f*